The Urban Book Series

The Urban Book Series is a resource for urban studies and geography research worldwide. It provides a unique and innovative resource for the latest developments in the field, nurturing a comprehensive and encompassing publication venue for urban studies, urban geography, planning and regional development.

The series publishes peer-reviewed volumes related to urbanization, sustainability, urban environments, sustainable urbanism, governance, globalization, urban and sustainable development, spatial and area studies, urban management, transport systems, urban infrastructure, urban dynamics, green cities and urban landscapes. It also invites research which documents urbanization processes and urban dynamics on a national, regional and local level, welcoming case studies, as well as comparative and applied research.

The series will appeal to urbanists, geographers, planners, engineers, architects, policy makers, and to all of those interested in a wide-ranging overview of contemporary urban studies and innovations in the field. It accepts monographs, edited volumes and textbooks.

Indexed by Scopus.

Kasturi Mukherjee · Tuhin Bhadra ·
Jitendra K. Pandey · Pratik Dash · Rajib Sarkar
Editors

Urban Climate

Volume 2: A Comprehensive Assessment of Climate Solutions in Cities

Springer

Editors
Kasturi Mukherjee
Department of Geography
Adamas University
Kolkata, India

Tuhin Bhadra
Department of Geography
Adamas University
Kolkata, India

Jitendra K. Pandey
University of Petroleum and Energy Studies
Dehradun, India

Pratik Dash
Department of Geography
Khejuri College
Medinipur, India

Rajib Sarkar
Department of Geography
Adamas University
Kolkata, India

ISSN 2365-757X ISSN 2365-7588 (electronic)
The Urban Book Series
ISBN 978-3-032-17894-7 ISBN 978-3-032-17895-4 (eBook)
https://doi.org/10.1007/978-3-032-17895-4

This Springer imprint is published by the registered company Springer Nature Switzerland AG
The registered company address is: Gewerbestrasse 11, 6330 Cham, Switzerland

Foreword

If knowledge of urban climate issues forms the foundation for action, then the urgent next step is to chart solutions that are sustainable, equitable, and transformative. Cities, as hubs of governance, innovation, and community life, hold immense potential to lead the global response to climate change. Harnessing this potential requires not only technical expertise but also ethical responsibility, collaborative governance, and inclusive participation.

This volume, *A Comprehensive Assessment of Climate Solutions in Cities*, offers a comprehensive and hopeful vision of how urban areas can rise to the climate challenge. Through its wide-ranging chapters, the book explores nature-based solutions, green infrastructure, disaster risk reduction strategies, renewable energy transitions, and equitable urban planning. It speaks directly to the goals of the United Nations 2030 Agenda, particularly SDG 11 (Sustainable Cities and Communities) and SDG 13 (Climate Action), while recognizing the interconnectedness with health, water, and energy systems.

What sets this book apart is its emphasis on interdisciplinarity and inclusivity. By drawing from case studies across the Global North and South, the contributors demonstrate that solutions must be context-specific and socially just. The discussions extend beyond technical innovation to highlight governance frameworks, ethical responsibilities, and the central role of community participation in shaping resilient futures.

At a time when climate crises demand urgent and collective responses, this book provides both inspiration and guidance. It is an indispensable resource for scholars, practitioners, policymakers, and urban leaders committed to transforming cities into laboratories of sustainability, resilience, and justice.

Alice J. Hovorka
Dean and Professor, Faculty
of Environmental and Urban Change
York University
Toronto, Canada

Preface

If the first volume of this series diagnosed the complex challenges of urban climate, this second volume is dedicated to the solutions. As cities worldwide grapple with the escalating impacts of climate change from heat islands and air pollution to social vulnerability and infrastructural strain—the need for actionable, integrative, and equitable strategies has never been greater.

Urban Climate—Volume 2: A Comprehensive Assessment of Climate Solutions in Cities moves from analysis to action, bringing together cutting-edge research, innovative case studies, and forward-thinking policy frameworks aimed at mitigating urban climate risks and enhancing adaptive capacity. This volume is structured around four core themes: mitigation strategies for urban heat islands, the role of policy and governance in driving sustainable transitions, the often-overlooked dimension of urban mental well-being, and integrated approaches to environmental management and disaster resilience.

Chapters in this volume explore nature-based solutions, bioclimatic design, participatory governance, climate justice, ecopsychology, AI-enabled disaster management, and education for sustainability. By weaving together perspectives from environmental science, urban planning, public health, psychology, law, and technology, this collection emphasizes that effective climate action must be interdisciplinary, inclusive, and context-sensitive.

Our aim is to provide researchers, policymakers, planners, educators, and community advocates with a robust toolkit of ideas and evidence-based practices. The solutions presented here are not merely theoretical; they are grounded in real-world applications from cities across India and beyond, illustrating that transformative change is both possible and already in motion.

This volume is an invitation to rethink, redesign, and reinvest in our urban futures—futures that are not only sustainable and resilient but also just, healthy, and restorative for all inhabitants.

Kolkata, India — Kasturi Mukherjee
Kolkata, India — Tuhin Bhadra
Dehradun, India — Jitendra K. Pandey
Medinipur, India — Pratik Dash
Kolkata, India — Rajib Sarkar

Acknowledgements

The creation of this volume has been a collective endeavour, made possible by the generosity, expertise, and encouragement of numerous individuals and institutions.

We begin by expressing our profound gratitude to the authors whose insightful contributions form the backbone of this book. Your dedication to advancing practical and transformative climate solutions spanning mitigation, governance, well-being, and resilience has provided this volume with its scholarly depth and real-world relevance. Each chapter reflects a commitment not only to understanding urban climates but to actively improving them.

We are equally indebted to the diligent peer reviewers whose critical feedback and constructive suggestions strengthened the rigour, clarity, and impact of every submission. Your thoughtful engagement ensured that this volume meets the highest standards of academic and practical scholarship.

Our heartfelt thanks go to Adamas University for its institutional support and for fostering an environment where interdisciplinary climate research can thrive. We are especially grateful to the university's leadership Prof. (Dr.) Samit Ray, Chancellor; Prof. Suranjan Das, Vice-Chancellor; Prof. Radha Tamal Goswami, Pro Vice-Chancellor; and Prof. Shouvik Roy Chowdhury, Registrar for their unwavering encouragement and for providing the resources necessary to bring this project to fruition. We also extend our appreciation to our partner institutions and research networks for their collaborative spirit.

We acknowledge with sincere thanks Springer Nature for their continued commitment to publishing impactful climate scholarship and for guiding us expertly through the editorial and production process. Their global platform ensures that these solutions reach the audiences who can turn ideas into action.

On a personal level, we are deeply thankful to our families, friends, and colleagues for their patience, understanding, and steadfast support during the intensive periods of editing and coordination that this volume required.

Finally, this book is dedicated to the visionaries and practitioners on the ground urban planners, community organizers, environmental educators, policymakers, and resilient citizens who are already building the climate-smart cities of tomorrow. Your work inspires ours, and it is to you that we offer this volume as a resource, a testament, and a catalyst for continued change.

Contents

About the Editors

Dr. Kasturi Mukherjee Associate Professor, Head, Department of Geography, Adamas University

Dr. Kasturi Mukherjee is an associate professor and head of the department of Geography, School of Basic and Applied Sciences, Adamas University, and Visiting Researcher at Stockholm Resilience Centre, Stockholm University. She received her Ph.D. in Urban Climate, from the University of Calcutta. Dr. Mukherjee has completed a post-graduation degree in Geography with a specialization in Environmental Geography and Post-graduation Diploma in Geoinformatics. She is a gold medalist for securing 1st rank in the university in post-graduation and she has received the best paper award two times in 2010 and 2018. Her research interest is focused on urban climate, sustainable development, and peri-urban development. In 2022, she has received the prestigious Core Research Grant from the Science and Engineering Research Board (SERB), DST as CoPI.

Dr. Tuhin Bhadra Assistant Professor, Department of Geography, Adamas University

Dr. Bhadra received his master's degree in Geography, an M.Phil. degree in Oceanography and Coastal Management, and Post Graduate Diploma in Remote Sensing and GIS. He did his Doctoral research on Hydrology and Water Resource Management from Jadavpur University and received a Newton-Bhabha Fellowship to attend a Ph.D. internship at the University of Edinburgh, UK. His research interest includes fluvial geomorphology, hydrology, water resource management, climate change impact assessment, remote sensing, GIS applications, etc. In 2022, he received the prestigious Core Research Grant from the Science and Engineering Research Board (SERB), DST as PI.

Prof. (Dr.) Jitendra K. Pandey Professor, University of Petroleum and Energy Studies

Prof. (Dr.) Jitendra K. Pandey completed his Ph.D. in Chemistry (Polymer Science), in 2007 from National Chemical Laboratory-Pune. His research areas are in Bio/Nanocomposites, degradation, and stabilization of polymer composites, nanomaterial for water treatment, synthesis, characterization, and analysis of reinforced composites. He has 70+ Research articles, 6 Books, and 17 Patents to his credit. He has been associated with Research and Development in Adamas University, University of petroleum and energy studies Dehradun, Manav Rachna University, Brain Pool Fellow Seoul National University, South Korea, The University of Tokushima, Japan, Max Planck Institute of Colloid and Interfaces, Glom Germany in various roles.

Dr. Pratik Dash Assistant Professor, Department of Geography, Khejuri College (Affiliated with Vidyasagar University)

Dr. Pratik Dash is currently working as an assistant professor in the Department of Geography, at Khejuri College. He had more than twelve years of research experience including more than seven years of teaching experience. He is expertise in remote sensing and GIS, disaster management and policy, hydrological modelling, flood modelling, etc. He has published several articles in Scopus-indexed journals and books.

Dr. Rajib Sarkar Assistant Professor, Department of Geography, Adamas University

Dr. Rajib Sarkar is a geographer who has completed his Graduation, Post-graduation, and Ph.D. from the University of Burdwan. He completed his Post Graduation in Geography in 2007 with a specialization in Advanced Geomorphology. He joined as Junior Research Fellow (CSIR-UGC) in 2011 and was awarded Ph.D. in 2016. He also completed his Post-graduate Diploma in Applied Remote Sensing and GIS from Jadavpur University in 2013 and his Post-graduate Diploma in Urban Management and Planning in 2017. Prior joining to as an assistant professor at Adamas University, he acted as Guest Faculty and Resource Person at Jadavpur University, Kalyani University, and Diamond Harbour Women's University. Dr. Rajib Sarkar is currently working on applications of Remote Sensing and GIS in urban planning and management, He has also an interest in Urban Growth Modelling (UGM), Geostatistics. He has more than 15 publications in reputed peer-reviewed journals like Springer Nature, Routledge publications, IOP Environmental Science, etc.

List of Figures

List of Tables

Part I
Urban Heat Island (UHI) Mitigation Strategies

Chapter 1
Issues and Challenges in the Mitigation Strategies of Urban Heat Island (UHI)

Mohammad Salman, Aditya Santhosh, Abdul Qadir, Safiullah Khan, and Zalina Samadi

Abstract Urban Heat Island (UHI) is a phenomenon where urban areas experience significantly higher temperatures than their rural surroundings due to human activities and modifications in land use. The increased heat in urban centers can exacerbate energy consumption, pose health risks, and impact environmental quality. This chapter explores the key mitigation strategies employed to address the UHI effect, including green infrastructure, cool roofs, water bodies, and urban planning. Through a systematic literature review and the analysis of four case studies, Singapore, Phoenix, Tokyo, and Nagpur, the chapter identifies the effectiveness of these strategies and the associated challenges. The findings highlight technical, economic, and socio-political constraints and the trade-offs involved in implementing UHI mitigation measures. Furthermore, the study emphasizes the need for integrated policy frameworks and innovative approaches to enhance the resilience and sustainability of cities facing rising temperatures.

Keywords Urban Heat Island (UHI) · Mitigation strategies · Urban planning

M. Salman (✉) · A. Qadir · S. Khan
Faculty of Architecture, Planning and Design, Integral University, Uttar Pradesh, Lucknow, India
e-mail: mdsalman@iul.ac.in

A. Qadir
e-mail: abdulquadir@iul.ac.in

S. Khan
e-mail: safi@iul.ac.in

A. Santhosh
Bearys Enviro Architecture Design School, Mangalore, Karnataka, India
e-mail: ar.adithyasanthosh@gmail.com

Z. Samadi
Faculty of Built Environment, Universiti Teknologi Mara, Shah Alam, Selangor, Malaysia
e-mail: zalin628@uitm.edu.my

K. Mukherjee et al. (eds.), *Urban Climate*, The Urban Book Series,
https://doi.org/10.1007/978-3-032-17895-4_1

1.1 Introduction

Urban Heat Island (UHI) refers to the phenomenon where urban areas experience significantly higher temperatures compared to their rural surroundings. This temperature disparity is primarily attributed to human activities and the modifications of land surfaces due to urbanization (Oke 1982). The increasing urban population, expanding cityscapes, and intensification of built-up areas have exacerbated the UHI effect, leading to heightened public health risks, increased energy demands, and deteriorated environmental conditions (Li et al. 2017; Oke 1982; Santamouris 2015). The urgency to understand and address the UHI effect has gained global recognition as cities grapple with rising temperatures and climate change impacts.

The UHI effect is driven by a combination of factors, including the replacement of natural landscapes with impervious materials, such as concrete and asphalt, which have low albedo and high heat-retaining properties (Li et al. 2017). The high concentration of buildings and limited vegetation cover restricts the dissipation of heat, creating localized warming effects within cities. Moreover, anthropogenic heat sources, such as vehicles, industrial activities, and air conditioning units, further amplify the urban temperatures (Mirzaei and Haghighat 2010). This intensification of UHI not only exacerbates energy consumption and increases cooling costs but also results in compromised air quality and increased thermal stress on urban populations.

Recent research underscores the importance of sustainable mitigation strategies, such as green infrastructure and reflective materials, to combat UHI effects. However, limited technological advancements and insufficient legislative frameworks have constrained their adoption in many urban settings (Irfeey et al. 2023). Furthermore, studies indicate that urban morphology, including impervious surface coverage and urban reflectivity, plays a critical role in mitigating high temperatures, especially in areas characterized by substantial anthropogenic heat emissions (Chen et al. 2022). Despite these advancements, the selection and effectiveness of mitigation strategies remain challenging due to the lack of comprehensive models and standardized evaluation metrics (Bathaei and Abdel-Raheem 2022).

Given the complexity and multidimensional nature of UHI, urban planners, architects, and policymakers are increasingly focusing on integrative approaches. Effective UHI mitigation requires a combination of policies and technological measures, such as albedo enhancement, sustainable transport, and green building envelopes (Degirmenci et al. 2021). This integrative approach has been recognized as essential for addressing the complex socio-economic and environmental challenges posed by UHI.

The purpose of this chapter is to provide a comprehensive assessment of the current mitigation strategies for UHI, while highlighting the underlying issues and challenges that limit their effectiveness. Through an examination of case studies and critical literature, the chapter aims to identify gaps in current approaches and propose directions for future research and policy development. Addressing these challenges is essential for promoting sustainable urban development and enhancing resilience to climate change impacts in cities.

1.2 Literature Review

1.2.1 Theoretical Background and UHI Overview

The Urban Heat Island (UHI) effect refers to the phenomenon where urban areas exhibit higher temperatures than their rural counterparts, primarily due to anthropogenic activities and the transformation of land use. Urban materials such as asphalt and concrete possess low albedo, absorbing solar radiation and releasing heat at night, thereby intensifying temperature gradients (Mirzaei and Haghighat 2010; Oke 1982). Urbanization contributes significantly by replacing vegetated surfaces with impervious ones, reducing evapotranspiration and increasing heat retention (Sütçüoğlu and Önaç 2023). Anthropogenic emissions from transportation, industry, and cooling systems further exacerbate this effect, particularly in dense and industrialized areas (Murayama and Wang 2023). Additionally, the compact urban form, marked by limited ventilation and high-rise buildings, traps heat, forming urban canyons that restrict thermal dissipation (Chen et al. 2022).

Impacts of UHI The consequences of UHI span environmental, health, and socioeconomic dimensions. Elevated temperatures increase the demand for cooling energy, contributing to greenhouse gas emissions and straining energy infrastructure (Santamouris 2015). Public health is compromised, especially among vulnerable groups, due to a rise in heat-related illnesses and exacerbation of respiratory and cardiovascular conditions (Rajagopal et al. 2023; Gonzalez-Trevizo et al. 2021). Deteriorating air quality from heightened ozone levels adds further stress, particularly in low-income communities with limited adaptive capacity (Aneesa and Reshma 2022; Bathaei and Abdel-Raheem 2022). Ecologically, UHI disrupts local habitats and biodiversity while also increasing urban water demand (Wang et al. 2023).

1.2.2 Causes of UHI

The root causes of UHI are intertwined with urban form, human behavior, and climate dynamics.

Urban Morphology and Land Use Urban design significantly influences thermal behavior. High-density construction, low-albedo materials, and vegetation loss collectively reduce natural cooling mechanisms. Urban expansion without adequate planning displaces green cover and amplifies thermal storage, contributing to persistent heat buildup (Lin et al. 2024; Li et al. 2017).

Anthropogenic Heat Emissions Cities are thermal hotspots due to emissions from vehicular traffic, industries, and air conditioning units. The heat released in these processes accumulates in the built environment, with limited airflow compounding the problem, especially in zones of high population density (Gonzalez-Trevizo et al. 2021; Aneesa and Reshma 2022).

Climate Change Interaction Global warming acts as a feedback mechanism, exacerbating UHI by elevating base temperatures and altering precipitation patterns. Reduced rainfall diminishes evapotranspiration capacity, undermining the efficiency of green infrastructure (Rajagopal et al. 2023; Wang et al. 2023). Therefore, UHI and climate change are interdependent phenomena that must be addressed jointly.

1.2.3 Mitigation Strategies for UHI

In response to the complex causes of UHI, mitigation strategies are developed at the intersection of design, technology, nature-based solutions, and policy.

Green Infrastructure and Urban Vegetation Nature-based interventions such as green roofs, urban forests, and parks are effective in lowering surface and ambient temperatures through shading and evapotranspiration. These solutions simultaneously enhance air quality and biodiversity, as observed in cities like Nagpur and Tokyo (Kotharkar et al. 2020; Yin et al. 2024).

Cool Roofs and Reflective Materials Technological solutions such as cool roofs and high-albedo pavements reduce heat absorption and mitigate temperature spikes, especially in high solar-radiation zones. These methods have shown measurable reductions in both surface and ambient temperatures (Irfeey et al. 2023; Khorat et al. 2024).

Water-Based Cooling Systems Water bodies and misting installations offer localized cooling via evaporative processes. While effective in arid regions, these solutions must be implemented with caution to avoid overexploitation of limited water resources (Wang et al. 2023).

Climate-Responsive Design and Urban Planning Urban morphology can be optimized to support passive cooling through ventilation corridors, street shading, and mixed-use development. Integrating these elements into masterplans ensures microclimate regulation and thermal comfort (Pigliautile et al. 2020).

Policy Instruments and Community Participation Successful UHI mitigation depends on robust policy frameworks and active citizen engagement. Incentives for green building practices and awareness campaigns foster both compliance and grassroots participation. Tokyo's green building mandates and Singapore's LUSH program exemplify policy-driven transformations (Degirmenci et al. 2021; Peng and Jim 2015).

1.2.4 Challenges in UHI Mitigation

Technical and Measurement Constraints Implementing and evaluating UHI strategies requires specialized expertise and standardized assessment protocols, which are often lacking. Long-term durability and adaptability of mitigation materials also remain ongoing concerns (Bathaei and Abdel-Raheem 2022).

Economic Barriers High upfront costs and maintenance burdens limit the scalability of UHI interventions, especially in resource-constrained cities. Additionally, the delayed realization of benefits deters investment from both the public and private sectors.

Socio-political Resistance Mitigation strategies frequently encounter institutional inertia, fragmented governance, and limited stakeholder coordination. Public resistance, fueled by low awareness and competing urban priorities, further impedes implementation (Gonzalez-Trevizo et al. 2021).

Environmental Trade-Offs Some strategies may inadvertently create new challenges, such as water-intensive green spaces in drought-prone areas or glare from reflective pavements. A balanced, site-specific approach is thus essential for sustainable outcomes (Murayama and Wang 2023).

1.3 Research Methodology

This study employs a systematic literature review and case study analysis to investigate the key challenges associated with mitigating the Urban Heat Island (UHI) effect (Fig. 1.1). The primary objective is to identify barriers hindering effective UHI mitigation and evaluate strategies through a structured approach.

Systematic Literature Review Protocol To ensure methodological rigor, the systematic review follows the Preferred Reporting Items for Systematic Reviews and Meta-Analyses (PRISMA) framework. The process consists of four key stages:

Identification A comprehensive search was conducted using multiple academic databases, including Scopus, Web of Science, and Google Scholar. The search strategy incorporated relevant keywords such as *Urban Heat Island mitigation*, *barriers to UHI mitigation*, *green infrastructure for heat reduction*, and *policy challenges in UHI adaptation.*

Screening Articles were assessed based on predefined inclusion and exclusion criteria:

Inclusion Criteria

- Peer-reviewed publications from the last 10 years (2015–2025).

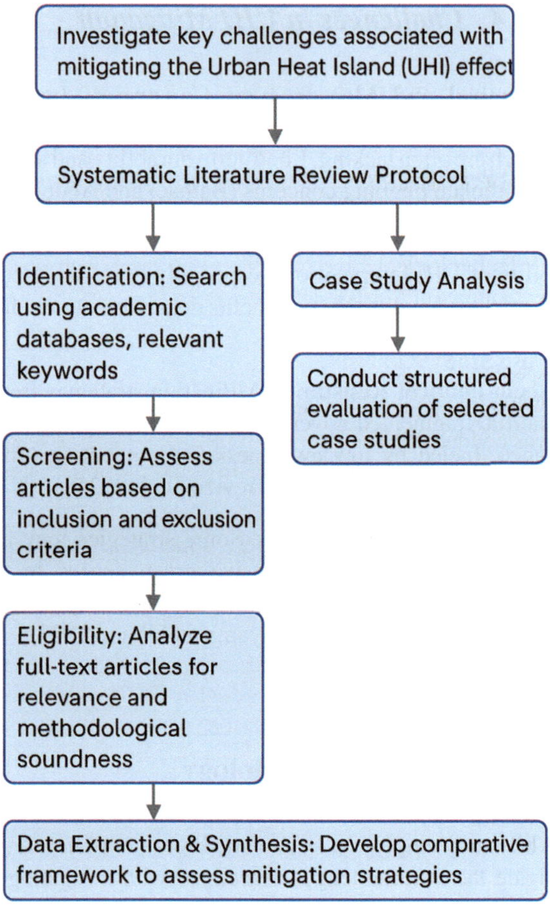

Fig. 1.1 Methodological framework for investigating UHI mitigation strategies

- Studies analyzing mitigation strategies for UHI in urban environments.
- Research employing comparative assessments of effectiveness.

Exclusion Criteria

- Studies focusing solely on theoretical models without practical implementation.
- Publications unrelated to urban climate resilience.
- Duplicate records.

Eligibility Full-text articles meeting the criteria were further analyzed for methodological soundness and relevance.

1.3.1 Data Extraction and Synthesis

A **comparative framework** was developed to assess the effectiveness of mitigation strategies, categorizing them based on:

- Policy interventions.
- Green infrastructure implementations (e.g., urban vegetation, reflective materials).
- Technological advancements in cooling solutions.
- Community-driven approaches.

Case Study Analysis

To complement the literature review, a **structured evaluation** of selected case studies was conducted. The analysis follows a comparative approach, examining real-world implementation of UHI mitigation strategies in diverse urban settings. The selected cases were assessed based on:

- Effectiveness in reducing urban temperatures.
- Adaptability to different city scales and climates.
- Policy support and implementation challenges.
- Long-term sustainability and community engagement.

This dual approach enhances the study's robustness by integrating both theoretical insights and practical applications, ensuring a comprehensive and structured evaluation of UHI mitigation strategies.

1.4 Case Study Analysis

Case studies offer valuable insights into the real-world application and effectiveness of Urban Heat Island (UHI) mitigation strategies. By examining different urban contexts, policymakers and urban planners can better understand the challenges, successes, and lessons that come from implementing these strategies.

1.4.1 Case Study 1: Singapore's Green Roof Initiative

Singapore is recognized as a global leader in integrating green infrastructure into urban planning to combat the Urban Heat Island effect. The city-state's initiative to implement green roofs began as part of its comprehensive policy to promote sustainable urban development and improve the quality of life for its citizens. The Green Roof Policy was introduced under the broader LUSH (Landscaping for Urban

Spaces and High-Rises) program, which encourages property developers to incorporate greenery into their building designs (Peng and Jim 2015). The primary objectives of this policy are to increase urban greenery, enhance the city's aesthetic appeal, and mitigate the UHI effect.

The initiative mandates that new developments, including commercial and residential buildings, dedicate a portion of their rooftop areas to greenery. Various implementation strategies were employed, including providing incentives and subsidies to developers, offering technical guidelines for roof greening, and integrating green roofs with other sustainability features like rainwater harvesting systems (Peng and Jim 2015).

Outcomes and Effectiveness The implementation of green roofs in Singapore has led to several notable outcomes. One of the primary benefits of green roofs is their ability to reduce rooftop surface temperatures by enhancing heat insulation and evapotranspiration. Studies have reported that green roofs can lower roof temperatures by up to 20 °C during peak hours, significantly reducing the need for cooling energy in buildings (Peng and Jim 2013). This temperature reduction directly contributes to lowering indoor cooling demands and decreasing energy consumption, which aligns with Singapore's sustainability goals. Additionally, the presence of green roofs has contributed to improving air quality by absorbing pollutants and dust particles, which helps alleviate the negative impacts of air pollution on residents (Wong et al. 2003). The aesthetic appeal of these green spaces also enhances mental well-being and increases recreational opportunities in a highly dense urban environment (Shafique et al. 2018).

Challenges Despite its success, the Green Roof Initiative in Singapore faced several challenges. One key barrier was the high initial cost of installing and maintaining green roofs, which deterred some developers from adopting the policy voluntarily. While subsidies and incentives provided financial relief, the perceived complexity and cost of retrofitting existing buildings posed significant challenges (Peng and Jim 2015).

Another challenge was related to technical expertise and maintenance. The selection of suitable plant species, soil conditions, and irrigation systems required technical knowledge and ongoing maintenance to ensure the sustainability of green roofs (Dauda and Alibaba 2020; Shafique et al. 2018). Additionally, ensuring the long-term durability of these green installations in Singapore's humid tropical climate was a significant concern.

Lessons learned from Singapore's initiative include the importance of providing financial incentives to encourage early adoption, offering clear technical guidelines to support developers, and integrating green roofs with other environmental initiatives such as rainwater harvesting (Peng and Jim 2013). Furthermore, public awareness campaigns helped garner support from residents and stakeholders, making green roofs a part of the city's identity.

1.4.2 Case Study 2: Cool Pavements in Phoenix, Arizona

Phoenix, Arizona, is known for its extreme heat, particularly during summer months, which exacerbates the Urban Heat Island (UHI) effect. To combat this, the city initiated a Cool Pavements Program that focuses on applying reflective coatings to road surfaces. The chosen coating, known as Cool Seal, has a higher albedo, enabling it to reflect a larger portion of solar radiation compared to traditional asphalt. By reflecting more sunlight, these cool pavements aim to lower surface and ambient temperatures, thus contributing to a more comfortable urban environment (Santamouris 2013; Sen et al. 2019). The strategy was implemented in residential neighborhoods as part of a pilot program launched in 2019, with comprehensive monitoring to evaluate its effectiveness.

Outcomes and Effectiveness The application of cool pavements in Phoenix demonstrated a significant reduction in surface temperatures. Studies reported that the reflective coatings lowered surface temperatures by up to 12 °C (54 °F) during peak hours, translating to a reduction of up to 1 °C in air temperatures within treated areas (Sen et al. 2019; Wang et al. 2021). The reduction in surface temperatures not only enhances pedestrian comfort but also has potential implications for lowering energy consumption in adjacent buildings, as less heat is radiated into the environment (Ferrari et al. 2020).

The program also contributed to improved air quality, as lower temperatures can mitigate the formation of ground-level ozone, a common urban pollutant. These benefits align with Phoenix's broader efforts to enhance sustainability and improve the quality of life for its residents.

Challenges Despite its positive outcomes, the Cool Pavements Program faced several challenges. One of the primary barriers was the high cost of implementation. Reflective coatings like CoolSeal are more expensive than traditional asphalt, and their widespread application requires substantial financial investment. Additionally, maintaining these coatings over time is necessary to sustain their cooling benefits, which leads to additional maintenance costs (Wang et al. 2021). Another challenge was related to public perception and discomfort. While residents appreciated the temperature reduction, some expressed concerns about glare from the reflective pavements, particularly during peak daylight hours, which could be visually uncomfortable for both drivers and pedestrians (Sen et al. 2019).

1.4.3 Case Study 3: Urban Greening in Tokyo, Japan

In Tokyo, Japan, urban greening policies have been a crucial part of the city's strategy to mitigate the Urban Heat Island (UHI) effect and enhance environmental sustainability. The Japanese government has adopted an integrated approach known as the Healthy Cities Initiative, which aims to create more green spaces and improve public health by promoting collaboration among local governments (Mabon and Shih 2021; Miyagawa et al. 2023). The initiative focuses on urban forestry projects, use of native plant species, increasing green spaces, and integrating greenery into urban infrastructure, such as parks, green roofs, and green walls (Shimpo 2022).

Urban planning regulations encourage developers to incorporate green elements into new constructions through incentives and subsidies. One of the notable policies is the Tokyo Green Building Program, which mandates certain green area ratios for new developments and promotes the use of native vegetation to restore ecological balance. This initiative has also led to the development of urban forests and pocket parks across densely populated areas, creating vital green corridors that connect fragmented natural spaces (Najar Ramirez 2018).

Outcomes and Effectiveness Urban greening efforts in Tokyo have shown positive results in reducing UHI intensity and improving environmental quality. Urban forests and green spaces play a significant role in lowering surface temperatures through shading and evapotranspiration processes. These measures help cool down urban areas and reduce the reliance on air conditioning, thereby lowering energy consumption and greenhouse gas emissions (Yin et al. 2024). Studies indicate that green roofs and urban forests in Tokyo contribute to lowering local temperatures by up to 2 °C during peak summer months, providing a noticeable cooling effect (Miyagawa et al. 2023).

In terms of sustainability, Tokyo's urban greening strategies have been effective in enhancing biodiversity, improving air quality, and increasing the aesthetic appeal of urban areas. The integration of green spaces into the city's urban fabric also supports mental well-being and offers recreational opportunities for residents. Community involvement through Satoyama conservation activities not only aids in the upkeep of green spaces but also promotes social cohesion and health benefits for participants (Shimpo 2022). This model of community-based greening has been cited as a successful strategy in maintaining the ecological sustainability of Tokyo's urban forests (Yang et al. 2022).

Challenges Despite its successes, Tokyo's urban greening efforts face several challenges. One significant issue is the high cost of maintenance. Urban greenery, particularly green roofs and vertical gardens, requires regular upkeep and substantial investment to maintain plant health and irrigation systems (Jain et al. 2023). Ensuring the long-term viability of these projects in a city with limited space and competing land-use demands remains a key concern.

Another challenge is the implication of policies on urban development. Expanding urban greenery often involves restricting construction in certain areas or mandating specific building requirements, which can lead to conflicts between developers and environmental advocates (Shimpo 2022). Additionally, while community participation is encouraged, achieving widespread engagement remains difficult, as not all residents have the time or resources to contribute to greening projects (Najar Ramirez 2018).

1.4.4 *Case Study 4: Case Study 4: Multi-Strategy UHI Mitigation in Nagpur, India*

Nagpur, India, has adopted an integrated approach to mitigating Urban Heat Island (UHI) effects by combining multiple strategies, including greening initiatives, cool roofs, and policy interventions. The city's efforts to address UHI encompass the enhancement of green spaces in critical local climate zones, retrofitting buildings with cool roofs, and introducing policy frameworks to encourage sustainable urban development. The integrated strategy aims to reduce surface and ambient temperatures while also improving urban livability and public health (Kotharkar et al. 2020).

The greening initiatives focus on increasing tree cover, establishing urban parks, and promoting green roofs, particularly in densely built areas. Nagpur has strategically prioritized the implementation of cool roofs, which use reflective materials to enhance the albedo of rooftops, thereby lowering heat absorption (Elnabawi et al. 2023). These measures are complemented by policy interventions that incentivize developers to incorporate green infrastructure and adopt energy-efficient building practices.

Outcomes and Effectiveness The combined impact of these multiple strategies has been significant in reducing UHI intensity in Nagpur. The greening initiatives have been particularly effective in reducing peak air temperatures by up to 5 °C in local climate zones, enhancing outdoor thermal comfort. Cool roofs have also demonstrated notable reductions in roof surface temperatures, which translates into decreased indoor cooling requirements and energy consumption (Kotharkar et al. 2020).

In addition to lowering temperatures, these strategies have contributed to improving air quality and mitigating the adverse impacts of extreme heat on public health. By incorporating green spaces and cool roofs, Nagpur has successfully created microclimatic zones that provide thermal relief to residents, particularly in high-density urban areas (Ziaul and Pal 2020).

Challenges Despite the effectiveness of these integrated strategies, Nagpur has faced several challenges in coordinating and implementing UHI mitigation measures. One of the primary challenges is the lack of coordination between different municipal departments and stakeholders, leading to delays and inconsistencies in policy implementation. Additionally, securing adequate funding for large-scale greening and cool roof projects has been an ongoing issue, particularly in economically disadvantaged neighborhoods (Fadhil et al. 2023).

Another challenge is the need for public awareness and community engagement. Although the benefits of UHI mitigation strategies are clear, promoting widespread adoption requires continuous outreach and education efforts. Moreover, policy implications indicate that stricter regulations and incentives are needed to enforce green building practices and to encourage developers to adopt cool roof technologies.

1.5 Synthesis of Literature and Case Studies

This integrates findings from the literature review and case study analyses to provide a comprehensive perspective on UHI mitigation. The synthesis highlights cross-cutting themes, emergent trends, and strategic insights that can inform future research and policy formulation.

1.5.1 Common Challenges in UHI Mitigation

One of the most consistent barriers across all case studies is economic viability. High upfront installation costs and long-term maintenance requirements were cited in both developed (Singapore and Phoenix) and developing (Nagpur) contexts. These economic burdens often deter large-scale adoption of green infrastructure or reflective technologies, especially in regions with constrained municipal budgets (Peng and Jim 2015; Wang et al. 2021).

Technical complexity is another recurring issue. The effectiveness of UHI strategies depends heavily on local factors such as climate, urban morphology, and species selection. For example, plant suitability and irrigation in Singapore, or glare management in Phoenix, necessitate highly context-specific technical planning (Dauda and Alibaba 2020; Sen et al. 2019).

Governance challenges also hinder progress. Fragmented urban policies, lack of interdepartmental coordination, and weak enforcement mechanisms prevent consistent implementation. This was evident in Nagpur, where multi-sectoral strategies struggled due to overlapping bureaucratic responsibilities (Fadhil et al. 2023).

Public awareness and perception significantly influence outcomes. Community resistance to reflective pavements in Phoenix or low engagement in Tokyo's greening projects demonstrates the importance of participatory design and sustained outreach (Yang et al. 2022; Sen et al. 2019).

Finally, environmental trade-offs must be addressed. Strategies like water-intensive urban greening or highly reflective surfaces may inadvertently generate new sustainability concerns, such as water scarcity or urban glare (Murayama and Wang 2023).

1.5.2 Interconnected Lessons and Emerging Trends

A holistic review reveals that integrated, multi-pronged approaches are more effective than isolated interventions. Nagpur's model combining greening, reflective roofs, and policy reform underscores the value of synergy across domains (Kotharkar et al. 2020).

Nature-based solutions continue to gain traction due to their ecological co-benefits. These interventions not only lower urban temperatures but also enhance biodiversity and human well-being, particularly when designed with community input (Yin et al. 2024; Kajosaari et al. 2024).

The intersection of climate change adaptation and UHI mitigation is becoming increasingly relevant. As cities experience longer and more intense heatwaves, strategies must be designed for resilience under changing climatic conditions (Rajagopal et al. 2023).

The deployment of smart technologies such as sensor networks, predictive modeling, and digital zoning tools is emerging as a powerful enabler of targeted and efficient UHI responses. However, accessibility and technical literacy remain limiting factors in some regions (Bathaei and Abdel-Raheem 2022).

1.5.3 Policy and Planning Implications

Policy frameworks need to be adaptive, scalable, and inclusive. National and municipal governments must co-develop strategies that incentivize sustainable building practices, enforce green zoning regulations, and ensure equitable distribution of UHI mitigation infrastructure. This includes subsidies for cool roofs, mandatory green cover ratios, and public–private partnerships to fund retrofits and new developments (Degirmenci et al. 2021; Shimpo 2022).

1.5.4 Toward a Contextual Framework for UHI Mitigation

The evidence supports the development of a context-sensitive UHI mitigation framework encompassing four interrelated layers: **design**, **policy**, **community**, and **technology**. The design layer focuses on green infrastructure and passive cooling strategies. The policy layer provides regulatory support and fiscal mechanisms. The community layer ensures participatory governance and local ownership. The technology layer facilitates monitoring and data-driven decision-making.

When integrated, these components enable cities to adaptively and equitably manage urban heat, ensuring that both environmental integrity and human health are safeguarded under climate stress (Table 1.1).

The comparative table provides a concise overview of four international case studies—Singapore, Phoenix, Tokyo, and Nagpur—each implementing distinct strategies to mitigate the Urban Heat Island (UHI) effect. It highlights the main interventions such as green roofs, cool pavements, urban greening, and integrated approaches, while also summarizing their implementation methods, key outcomes, temperature reduction impacts, and associated environmental and social benefits. Additionally, it outlines the challenges faced, including high costs, technical complexities, maintenance needs, and public engagement issues. Overall, the table

Table 1.1 Comparative analysis of UHI mitigation strategies across global cities

Aspect	Singapore green roof initiative	Phoenix cool pavements program	Tokyo urban greening policy	Nagpur multi-strategy approach
Location	Singapore	Phoenix, Arizona, USA	Tokyo, Japan	Nagpur, India
Main strategy	Green Roofs under LUSH Program	Reflective road coatings (CoolSeal)	Urban forests, green roofs, green walls	Greening + Cool Roofs + Policy Integration
Implementation approach	Mandates on new buildings; incentives and technical support	Pilot projects with continuous monitoring	Healthy Cities Initiative, green ratio mandates, community involvement	Local climate zone focus, building retrofits, incentivized policies
Key outcomes	Reduced rooftop temperatures, lower energy consumption, improved air quality	Surface temperature reduction, energy savings, better air quality	Temperature reduction, biodiversity, improved air quality	Lower temperatures, better air quality, energy efficiency
Cooling impact	Up to 20 °C surface temperature reduction	Up to 12 °C surface temp reduction; ~ 1 °C ambient drop	Up to 2 °C local cooling effect	Up to 5 °C air temperature reduction
Environmental benefits	Improved air quality, rainwater harvesting integration	Reduced ozone formation, lower heat radiation	Air purification, biodiversity restoration	Cleaner air, public health improvement
Social impact	Enhanced aesthetics, mental well-being	Improved pedestrian comfort	Mental well-being, community bonding	Thermal comfort in dense areas
Challenges	High cost, technical complexity, maintenance, climate adaptation	High cost, glare discomfort, maintenance needs	Maintenance cost, land-use conflict, limited public engagement	Coordination gaps, funding issues, low public awareness

offers valuable insights into how different cities address UHI through context-specific and multi-faceted strategies, contributing to sustainable and livable urban environments.

The synthesis of literature and case studies underscores the complexity of Urban Heat Island mitigation and the importance of adopting a holistic, context-specific approach. By addressing common challenges related to economics, technology, policy, and public perception, cities can create more resilient and sustainable urban environments. The findings also underscore the importance of developing integrated policy frameworks that align urban planning, environmental regulations, and public health initiatives to effectively tackle the Urban Heat Island (UHI) phenomenon. Such frameworks should prioritize multi-sectoral collaboration involving government agencies, private stakeholders, urban planners, and local communities. Innovative approaches, such as the deployment of smart technologies for real-time monitoring of urban heat and the use of data-driven models for targeted UHI interventions, are essential for enhancing urban resilience. Additionally, adopting nature-based solutions like green roofs, permeable pavements, and urban forestry must be coupled with adaptive policies that are responsive to changing climatic conditions and community needs.

1.6 Critical Synthesis for the Mitigation Strategies of Urban Heat Island (UHI)

Based on the previous discussions, the following key indicators of Urban Heat Island (UHI) mitigation can be identified. These indicators represent both the strategic levers used to address UHI and the measurable outcomes that signify mitigation success. Below is a structured summary of these indicators, followed by an analytical mapping of how they relate to the critical synthesis of UHI Mitigation Strategies (Table 1.2).

1.6.1 Relation to Critical Synthesis of UHI Mitigation Strategies

In the critical synthesis, each UHI mitigation strategy is assessed not in isolation but for its dynamic interaction with key performance indicators derived from literature and case studies. These indicators, ranging from thermal reduction and energy efficiency to public engagement and policy support, form the basis for evaluating the strategic effectiveness and sustainability of each intervention within different urban contexts.

Green infrastructure and urban vegetation, comprising green roofs, urban forests, and street trees, are closely linked with indicators such as thermal reduction, environmental performance, ecosystem integration, and social and psychological impact. These interventions directly lower surface and ambient temperatures while enhancing urban biodiversity and purifying the air (Yin et al. 2024; Kotharkar et al. 2020). They also contribute significantly to psychological well-being and improve the visual quality of urban spaces. However, such strategies often demand high water consumption and require continuous maintenance, raising concerns over implementation feasibility, particularly in water-scarce and arid urban regions. The trade-off between environmental benefits and resource sustainability becomes a crucial consideration when applying green infrastructure widely.

Table 1.2 Identified indicators of UHI mitigation

Category	Indicators
Thermal reduction	Reduction in surface temperature, ambient temperature, rooftop cooling
Energy efficiency	Lowered cooling energy demand, reduced electricity consumption
Environmental performance	Improved air quality, increased biodiversity, reduced ground-level ozone
Water management	Use of rainwater harvesting, misting systems, evapotranspiration efficiency
Infrastructure resilience	Use of reflective materials, cool roofs, adaptive building design
Policy and institutional support	Presence of mandates, subsidies, zoning regulations, green area ratios
Public engagement	Community participation, behavior change, awareness campaigns
Implementation feasibility	Cost-effectiveness, scalability, maintenance needs, retrofitting adaptability
Social and psychological impact	Mental well-being, thermal comfort, aesthetic quality of urban spaces
Ecosystem integration	Green corridors, native vegetation, multifunctional land use
Technological integration	Smart monitoring systems, GIS-based zoning, data-driven planning tools

Cool roofs and reflective surfaces demonstrate strong alignment with indicators such as thermal reduction, energy efficiency, infrastructure resilience, and implementation feasibility. These solutions provide immediate and measurable reductions in surface and indoor temperatures, contributing significantly to lowering cooling energy demands (Sen et al. 2019). Despite their technical effectiveness, the durability of reflective materials and issues such as glare discomfort for pedestrians and drivers present limitations. These challenges highlight the need to balance technical efficiency with human-centered design, emphasizing that public acceptance and experiential quality are critical to the success of technological interventions.

Water-based cooling systems, including misting devices, fountains, and urban water bodies, correspond with indicators of thermal reduction, water management, and public engagement. These systems are particularly effective in regulating microclimates through evaporative cooling (Wang et al. 2023). Nevertheless, their sustainability is heavily contingent upon local water resource availability. In regions facing water scarcity, the deployment of such systems raises critical questions of environmental equity and long-term viability. Therefore, their adoption must be carefully guided by regional hydrological assessments and integrated with broader urban water management frameworks.

Climate-sensitive urban design is integrally related to indicators such as thermal reduction, energy efficiency, infrastructure resilience, and ecosystem integration. Strategies focusing on building orientation, street geometry, ventilation corridors, and green corridor planning can significantly improve passive cooling and reduce the reliance on mechanical energy consumption (Pigliautile et al. 2020). These approaches provide durable, long-term benefits; however, retrofitting existing densely built environments presents considerable logistical and economic challenges. Land-use competition and zoning restrictions often limit the retroactive application of such design solutions, requiring careful urban governance and innovative design adaptations.

Policy instruments and governance mechanisms are fundamentally tied to indicators of policy and institutional support, implementation feasibility, and public engagement. Cases like Singapore's LUSH program and Tokyo's Green Building Initiative demonstrate how regulatory frameworks can mandate green space integration, offer subsidies, and foster large-scale UHI mitigation (Peng and Jim 2015; Shimpo 2022). However, the effectiveness of policy interventions is often undermined by fragmented governance, lack of inter-agency coordination, and weak enforcement mechanisms. Successful UHI governance requires strong political will, coherent multi-level collaboration, and adaptive regulatory structures that can respond to evolving climate risks and socio-economic needs.

Community engagement and behavioral change directly influence indicators such as public engagement, social and psychological impact, and ecosystem integration. Models like Tokyo's Satoyama conservation initiatives show that community-driven greening projects enhance the durability and socio-environmental value of UHI mitigation efforts (Yang et al. 2022). Active public participation fosters a sense of co-ownership and strengthens resilience at the neighborhood level. However, consistent and inclusive engagement remains challenging, especially among marginalized, transient, or economically disadvantaged populations. Overcoming these barriers requires deliberate capacity-building, sustained outreach programs, and mechanisms for participatory urban governance.

Finally, integration and multi-sectoral synergy touch upon all the identified indicators across categories. Integrated strategies, such as those employed in Nagpur, which combine green infrastructure, reflective materials, policy instruments, and community outreach, demonstrate superior outcomes compared to isolated interventions (Kotharkar et al. 2020). Such approaches underscore the importance of aligning technological, ecological, institutional, and social dimensions to create synergistic effects that amplify UHI mitigation benefits. Cross-sectoral alignment, data-driven planning, and flexible governance structures are essential to realizing the full potential of integrated mitigation strategies (Table 1.3).

Table 1.3 Strategic approaches to UHI mitigation and their linked indicators

Strategy	Linked indicators	Key insights
Green infrastructure and urban vegetation	Thermal reduction, environmental performance, ecosystem integration, social and psychological impact	Effective in temperature reduction and ecological enhancement, but faces water demand and maintenance challenges in arid areas
Cool roofs and reflective surfaces	Thermal reduction, energy efficiency, infrastructure resilience, implementation feasibility	Delivers immediate thermal relief and energy savings, yet glare and material longevity can affect user satisfaction
Water-based cooling systems	Thermal reduction, water management, public engagement	Localized cooling is effective but heavily depends on regional water availability, raising sustainability concerns
Climate-Sensitive urban design	Thermal reduction, energy efficiency, infrastructure resilience, ecosystem integration	Provides passive long-term cooling benefits, though difficult to retrofit in dense urban areas with zoning limitations
Policy instruments and governance mechanisms	Policy and institutional support, implementation feasibility, public engagement	Enables large-scale implementation through regulations and incentives, but may suffer from poor enforcement and fragmentation
Community engagement and behavioral change	Public engagement, social and psychological impact, ecosystem integration	Promotes co-ownership and resilience, though requires inclusive engagement strategies, especially for marginalized groups
Integration and multi-sectoral synergy	All indicators across categories	Synergistic approach across sectors maximizes impact but requires strong coordination and adaptive planning frameworks

1.7 Conclusion and Future Scope of Research

The Urban Heat Island (UHI) effect remains a critical challenge for urban areas globally, with wide-ranging implications for public health, energy consumption, and environmental sustainability. This chapter explored various aspects of UHI, including its causes, impacts, mitigation strategies, and the challenges faced in addressing it. The literature review revealed that the primary drivers of UHI include changes in urban morphology, such as the increase in impervious surfaces and high building density, along with anthropogenic heat emissions. These factors, combined with the compounding effects of climate change, intensify the UHI effect in cities.

The chapter highlighted several mitigation strategies such as green infrastructure, cool roofs, reflective pavements, water management systems, and climate-responsive urban design. These strategies have shown promising results in reducing surface and ambient temperatures, improving air quality, and enhancing urban resilience. Green infrastructure, including green roofs, urban forestry, and parks, proved to be effective in cooling urban areas and providing additional benefits like improved air quality and biodiversity. Cool roofs and reflective materials emerged as another crucial approach to enhance heat reflectance and reduce surface temperatures. Furthermore, the integration of water bodies and urban water management systems can lower temperatures through evaporative cooling, while climate-responsive design offers a holistic approach to urban planning by promoting natural ventilation and passive cooling.

However, despite the existence of these strategies, the challenges in UHI mitigation remain significant. Technical challenges include the limitations in existing technologies, difficulties in accurately measuring the success of interventions, and challenges in modeling the complex urban heat environment. Economic barriers, such as high initial costs and long-term funding requirements, restrict the adoption of comprehensive mitigation strategies. Social and political challenges, including insufficient political will, lack of public awareness, and resistance to change, further complicate the implementation of UHI mitigation measures. Lastly, environmental challenges and trade-offs, such as increased water consumption in cooling strategies, necessitate a careful balance between effective UHI mitigation and resource sustainability.

The future scope of research should prioritize the development of innovative, cost-effective materials and technologies for UHI mitigation. Additionally, comprehensive policy frameworks are needed to support the widespread adoption of sustainable urban practices. Research should also focus on understanding the socio-economic impacts of UHI mitigation strategies, particularly in low-income urban areas, to ensure equitable climate adaptation measures. Further studies could explore the integration of digital tools and smart technologies to monitor and optimize UHI interventions in real time. By adopting a holistic and inclusive approach, cities can effectively tackle UHI and enhance their resilience to climate change.

AI Discloser

ChatGPT has been used to enhance the writing quality.

References

Aneesa S, Reshma JK (2022) Various causes of urban heat islands, effects and their mitigation measures for urban ecology—A review. Ecol Environ Conservation 1489–1493

Bathaei B, Abdel-Raheem M (2022) Assessment of the relative importance of the main parameters used in the selection of the urban heat island mitigation strategies. Constr Res Congr 2022:627–636

Chen Y, Wang Y, Zhou D, Gu Z, Meng X (2022) Summer urban heat island mitigation strategy development for high-anthropogenic-heat-emission blocks. Sustain Cities Soc 87:104197

Dauda I, Alibaba HZ (2020) Green roof benefits, opportunities and challenges. Int J Civ Struct Eng Res 7:106–112

Degirmenci K, Desouza KC, Fieuw W, Watson RT, Yigitcanlar T (2021) Understanding policy and technology responses in mitigating urban heat islands: a literature review and directions for future research. Sustain Cities Soc 70:102873

Elnabawi MH, Hamza N, Raveendran R (2023) 'Super cool roofs': mitigating the UHI effect and enhancing urban thermal comfort with high albedo-coated roofs. Results Eng 19:101269

Fadhil M, Hamoodi MN, Ziboon ART (2023) Mitigating urban heat island effects in urban environments: strategies and tools. IOP Conf Ser: Earth Environ Sci 1129(1):12025

Ferrari A, Kubilay A, Derome D, Carmeliet J (2020) The use of permeable and reflective pavements as a potential strategy for urban heat island mitigation. Urban Clim 31:100534

Gonzalez-Trevizo ME, Martinez-Torres KE, Armendariz-Lopez JF, Santamouris M, Bojorquez-Morales G, Luna-Leon A (2021) Research trends on environmental, energy and vulnerability impacts of Urban Heat Islands: an overview. Energy Build 246:111051

Irfeey AMM, Chau H-W, Sumaiya MMF, Wai CY, Muttil N, Jamei E (2023) Sustainable mitigation strategies for urban heat island effects in urban areas. Sustainability 15(14):10767

Jain AM, Sharma AS., Durve AN (2023) Employing green roof infrastructure to alleviate urban heat island effect: green-scape infused urban fabrics. Int J Res Appl Sci Eng Technol 11(VII):337–344. https://doi.org/10.22214/ijraset.2023.54623

Kajosaari A, Hasanzadeh K, Fagerholm N, Nummi P, Kuusisto-Hjort P, Kyttä M (2024) Predicting context-sensitive urban green space quality to support urban green infrastructure planning. Landscape Urban Plann 242:104952. https://doi.org/10.1016/j.landurbplan.2023.104952

Khorat S, Das D, Khatun R, Aziz SM, Anand P, Khan A, Santamouris M, Niyogi D (2024) Cool roof strategies for urban thermal resilience to extreme heatwaves in tropical cities. Energy Build 302:113751. https://doi.org/10.1016/j.enbuild.2023.113751

Kotharkar R, Bagade A, Singh PR (2020) A systematic approach for urban heat island mitigation strategies in critical local climate zones of an Indian city. Urban Clim 34:100701

Li X, Zhou Y, Asrar GR, Imhoff M, Li X (2017) The surface urban heat island response to urban expansion: a panel analysis for the conterminous United States. Sci Total Environ 605:426–435

Lin A, Wu H, Luo W, Fan K, Liu H (2024) How does urban heat island differ across urban functional zones? Insights from 2D/3D urban morphology using geospatial big data. Urban Clim 53:101787. https://doi.org/10.1016/j.uclim.2023.101787

Mabon L, Shih W-Y (2021) Urban greenspace as a climate change adaptation strategy for subtropical Asian cities: a comparative study across cities in three countries. Glob Environ Chang 68:102248

Mirzaei PA, Haghighat F (2010) Approaches to study urban heat island–abilities and limitations. Build Environ 45(10):2192–2201

Miyagawa T, Otsuka N, Abe H (2023) Green space creation and utilization in coordination with policies for Healthy Cities in Japan. Cities Health 7(5):854–874. https://doi.org/10.1080/23748834.2023.2188636

Murayama Y, Wang R (2023) Special issue on geographical analysis and modeling of urban heat island formation. In: Remote sensing, vol 15, Issue 18. MDPI, p 4474

Najar Ramirez D (2018) Unsealing Tokyo suburbs a strategy to create an eco-district through co-housing and agriculture. Politecnico di Torino

Oke TR (1982) The energetic basis of the urban heat island. Q J R Meteorol Soc 108(455):1–24

Peng LLH, Jim CY (2013) Green-roof effects on neighborhood microclimate and human thermal sensation. Energies 6(2):598–618

Peng LLH, Jim CY (2015) Economic evaluation of green-roof environmental benefits in the context of climate change: the case of Hong Kong. Urban for Urban Green 14(3):554–561

Pigliautile I, Chàfer M, Pisello AL, Pérez G, Cabeza LF (2020) Inter-building assessment of urban heat island mitigation strategies: field tests and numerical modelling in a simplified-geometry experimental set-up. Renew Energy 147:1663–1675

Rajagopal P, Priya RS, Senthil R (2023) A review of recent developments in the impact of environmental measures on urban heat island. Sustain Cities Soc 88:104279

Santamouris M (2015) Analyzing the heat island magnitude and characteristics in one hundred Asian and Australian cities and regions. Sci Total Environ 512:582–598

Santamouris M (2013) Using cool pavements as a mitigation strategy to fight urban heat island—A review of the actual developments. Renew Sustain Energy Rev 26:224–240
Sen S, Roesler J, Ruddell B, Middel A (2019) Cool pavement strategies for urban heat island mitigation in suburban Phoenix, Arizona. Sustainability 11(16):4452
Shafique M, Kim R, Rafiq M (2018) Green roof benefits, opportunities and challenges—A review. Renew Sustain Energy Rev 90:757–773
Shimpo N (2022) Urban ecological life in a metropolitan area—An insight from Satoyama conservation activities in the Greater Tokyo Area. Landscape Ecol Eng 18(1):109–119
Sütçüoğlu GG, Önaç AK (2023) A comprehensive analysis of urban heat Island determination methods. In: ISSC conference proceedings, May, 342–348. https://doi.org/10.52460/issc.2023.050
Tate C, Wang R, Akaraci S, Burns C, Garcia L, Clarke M, Hunter R (2024) The contribution of urban green and blue spaces to the United Nation's Sustainable Development Goals: an evidence gap map. Cities 145:104706. https://doi.org/10.1016/j.cities.2023.104706
Wang C, Wang Z-H, Kaloush KE, Shacat J (2021) Cool pavements for urban heat island mitigation: a synthetic review. Renew Sustain Energy Rev 146:111171
Wang Z, Zhu P, Zhou Y, Li M, Lu J, Huang Y, Deng S (2023) Evidence of relieved urban heat island intensity during rapid urbanization through local climate zones. Urban Clim 49:101537
Wong NH, Tay SF, Wong R, Ong CL, Sia A (2003) Life cycle cost analysis of rooftop gardens in Singapore. Build Environ 38(3):499–509
Yang Y, Song F, Ma J, Wei Z, Song L, Cao W (2022) Spatial and temporal variation of heat islands in the main urban area of Zhengzhou under the two-way influence of urbanization and urban forestry. PLoS ONE 17(8):e0272626
Yin S, Xiao S, Ding X, Fan Y (2024). Improvement of spatial-temporal urban heat island study based on local climate zone framework: a case study of Hangzhou, China. Building Environ 248:111102. https://doi.org/10.1016/j.buildenv.2023.111102
Ziaul S, Pal S (2020) Modeling the effects of green alternative on heat island mitigation of a meso level town, West Bengal, Indai. Adv Space Res 65(7):1789–1802

Chapter 2
Nature-Based Solutions for Mitigating Combined Impacts of Heatwaves and Urban Heat Island

Shivanjali Mohite and Meenal Surawar

Abstract Tropical regions are characterized by high temperatures, humidity, and intense solar radiation and are disproportionately vulnerable to climate change. In these zones, rapid urbanization compounds environmental risks. Being a developing nation, the surge of urbanization along with climate change has intensified heat stress in Indian cities, where the interaction between Urban Heat Islands (UHIs) and heatwaves (HWs) threatens public health and ecological resilience. This chapter reviews UHI and HWs studies in India and their combined effect on microclimate and evaluates nature-based solutions (NBS) as adaptive strategies to mitigate these dual challenges through a systematic review of studies since 2000–2023, prioritizing Indian contexts. Findings reveal that urban green spaces, rooftop gardens, and wetland restoration can reduce ambient temperatures by 2–5 °C, though barriers like land scarcity and fragmented governance hinder scalability. Case studies from Delhi, Mumbai, and Chennai demonstrate the efficacy of NBS in enhancing thermal comfort and reducing energy demand. By bridging global evidence with localized insights, this chapter advocates for integrating NBS into urban policies and provides actionable pathways for heat-resilient urban planning in tropical regions.

Keywords Urban heat island · Heatwaves · Nature-based solutions · Climate · Tropical region

S. Mohite (✉) · M. Surawar
Department of Architecture and Planning, Visvesvaraya National Institute of Technology (VNIT), Nagpur, Maharashtra, India
e-mail: shivanjali.mohite@students.vnit.ac.in

M. Surawar
e-mail: meenalsurawar@arc.vnit.ac.in

K. Mukherjee et al. (eds.), *Urban Climate*, The Urban Book Series,
https://doi.org/10.1007/978-3-032-17895-4_2

2.1 Introduction

The process of urbanization is rapidly transforming our world. In just a few decades, the percentage of people living in urban areas has significantly increased from 30% in 1950 to 55% in 2018. This trend is expected to continue, with estimates suggesting that 68% of the global population will reside in urban settings by 2050 (World Cities Report 2020). The social, environmental, and developmental implications of this shift are enormous. Unfortunately, urban growth has also brought numerous challenges, including pollution, biodiversity loss, ecosystem degradation, and climate change. Developing countries are expected to experience the highest rates of urbanization, which will alter land use and land cover, affecting surface energy and water balances. Perhaps most particularly, urbanization has led to the Urban Heat Island (UHI) effect, causing higher temperatures in cities than non-urban areas. This phenomenon can significantly impact microclimates and may contribute to the increased frequency and intensity of extreme weather events in urban settings.

The Fifth Assessment Report of the Intergovernmental Panel on Climate Change 2014 projected that tropical and subtropical countries might experience more prominent seasonal and annual mean temperatures than mid-latitudes (Jarraud and Steiner 2012; Qian et al. 2022). It suggests that the tropical zones are particularly vulnerable to climate change because of their narrow environmental tolerances. Recent studies suggest that high-temperature events have increased worldwide (Horton et al. 2016). In the tropics, heatwaves are increasing in intensity and frequency, and with increasing climate change, tropical regions are projected to experience significant increases in extreme heat events (Marcotullio et al. 2021). Climate-related heat issues affect the livability of cities, exposing people to natural hazards such as heatwaves, droughts, and air quality issues. These impacts are likely to increase, putting more people at risk of extreme temperatures in the future. The Sustainable Development Goals (SDGs) highlight the need to tackle urban heat issues, particularly in goals 7, 11, and 13. Goal 7 emphasizes access to affordable and sustainable energy, which is essential for implementing effective cooling solutions in urban areas. Goal 11 focuses on making cities inclusive, safe, resilient, and sustainable, urging urban planners to incorporate strategies that mitigate heat through green infrastructure. Lastly, Goal 13 addresses the urgency of combating climate change and its impacts, encouraging cities to adopt adaptive measures to reduce heat-related risks and enhance resilience (United Nations Development Program 2021).

This chapter explores the dynamics between Urban Heat Islands (UHI) and heatwaves (HW), examining the mechanisms driving UHI effects, the projected impacts of climate change on urban heat, and the need for advanced mitigation strategies. UHI and HW synergy intensifies urban heat, as UHI sustains high nighttime temperatures, and heatwaves drive extreme daytime heat, creating prolonged and severe heat conditions. Limited vegetation, extensive impervious surfaces, and dense infrastructure exacerbate this prolonged high heat with reduced cooling breaks. This compounded heat stress degrades outdoor thermal comfort, raises greenhouse gas emissions, and worsens air quality through increased ground-level ozone. Vulnerable populations including the elderly, children, and lower-income groups are at heightened risk for heat-related illnesses due to extended exposure to these conditions, restricting outdoor activities such as walking for commuting by affecting pedestrian thermal comfort (Mohite and Surawar 2024a) and diminishing urban livability. This chapter emphasizes nature-based solutions (NBS) as a crucial strategy to address these pressing challenges, highlighting NBS's role in cooling urban areas, improving resilience, and enhancing the quality of life in cities facing heat-related threats.

2.2 Data and Method

This chapter employs a systematic methodology to review relevant literature on mitigating UHI impacts through nature-based solutions, providing a comprehensive understanding of these issues. The research uses internet search techniques to gather pertinent literature from major databases, including Google Scholar, Scopus, Web of Science, Wiley's Online Library, the Directory of Open Access Journals (DOAJ), and ScienceDirect. Keyword combinations, including *"Urban Heat Island," "Heatwaves," "Nature-Based Solutions," "microclimate,"* and *"India,"* were used to identify relevant literature published between 2000 and 2023, aligning with India's rapid urbanization and climate policy developments. Studies were selected based on their focus on UHI-HW interactions and NBS implementation in tropical urban contexts (Table 2.1). Priority was given to Indian case studies to understand the specific context. Global examples from tropical regions were included to identify transferable strategies. By synthesizing insights from this review, the chapter aims to highlight the critical importance of studying the synergy between heatwaves and UHI and nature-based solutions to minimize this impact on the urban environment and enhance the resilience of Indian cities against heat-related challenges.

Table 2.1 Research paper selection criteria

Criteria	Inclusion	Exclusion
Study type	Peer-reviewed articles, government reports, case studies, conference papers	Non-peer-reviewed blogs, opinion pieces, non-English studies
Geographic focus	Studies focused on India (60%) or global studies with tropical relevance (40%)	Studies from non-tropical regions (e.g., Europe, North America)
Intervention	NBS strategies (green/blue infrastructure) for UHI/HW mitigation	Studies on non-NBS strategies (e.g. cool pavements only)
Outcomes	Quantified temperature reduction, thermal comfort, or policy recommendations	Theoretical models without empirical validation

2.3 Urban Heat Island

Urban Heat Island (UHI) is a phenomenon where urban areas experience significantly higher temperatures than surrounding rural regions, primarily due to human activities and urban structure (Oke et al. 2017). As cities expand, dense constructions and heat-retaining materials like asphalt and concrete replace natural landscapes, trapping heat within urban cores. This effect is exacerbated by high energy use, greenhouse gas emissions, and limited vegetation, which reduce evapotranspiration and natural cooling. Additionally, tall buildings and narrow street canyons impede airflow, further increasing surface heat retention and altering local climate patterns.

UHIs are generally classified by the level or layer in which temperature is measured: surface UHI (SUHI), canopy layer UHI (CUHI), boundary layer UHI (BUHI), and subsurface UHI (UHIsub). The Surface Urban Heat Island (SUHI) and Canopy Urban Heat Island (CUHI) are defined by their assessment levels. SUHI reflects temperature differences between urban and rural surfaces, often measured using satellite imagery. CUHI pertains to temperature variations within the urban canopy layer at approximately human height (~2 m), which is crucial for evaluating outdoor thermal comfort and human health (Chew et al. 2021). While assessing these effects, SUHI and CUHI are more commonly studied than the Boundary Urban Heat Island (BUHI) due to the complexities in measuring BUHI temperatures from sensors on tall structures. Various observation methods for CUHI and SUHI include in situ measurements, modeling approaches, remote sensing, and infrared thermography (Sclunzen et al. 2023).

The effects of UHI are profound and multidimensional, affecting both urban environments and inhabitants. High temperatures within UHI decrease outdoor thermal comfort, increase energy demands for cooling, and heighten the risk of heat stress, particularly among vulnerable populations such as the elderly, low-income groups, pedestrians, and those without adequate shelter or resources for adaptation (Ren et al. 2023). Increased ground-level ozone and air pollution compound public health risks, which can trigger respiratory and cardiovascular conditions. Environmentally, UHIs alter urban climates by affecting local wind patterns, precipitation, and the frequency of extreme events like heatwaves and flash floods (Qian et al. 2022).

2.3.1 Urban Heat Island in India

UHI studies in India date back to the 1980s, initially concentrating on major metropolitan areas like Delhi and Mumbai (Padmanabhamutry and Bahl 1984). Early research predominantly highlighted the adverse effects of urbanization on local climate zones, identifying specific urban hotspots where temperatures were significantly higher than in less developed areas. This work laid the groundwork for understanding the relationship between urban development and microclimatic changes.

Over the past fifteen years, UHI studies in India have seen significant growth. The methodologies used in these studies have also evolved considerably. Initially, many researchers relied on ground-based data collection methods. However, advancements in remote sensing technology have shifted the focus toward satellite-based thermal sensors as the primary research tools. In recent years, around 80% of UHI studies have utilized satellite imagery, particularly from systems such as Landsat, the Moderate Resolution Imaging Spectroradiometer (MODIS), and the Advanced Spaceborne Thermal Emission and Reflection Radiometer (ASTER). These satellite systems offer extensive spatial coverage and provide more accessible thermal data than traditional ground-based surveys.

In Bangalore, the UHI effect was assessed by measuring land surface temperature using remote sensing techniques. An analysis of Landsat data between 1973, 1992, and 2000, as well as IRS LISS-III data between 1999 and 2006, and MODIS data between 2002 and 2007, indicated an increase in temperature ranging from 2 to 2.5 °C over the previous decade. The seasonal and annual temperature trends were studied at four stations in the Delhi NCR region. The study's trend analysis found no significant trend in the annual mean maximum temperature; however, the annual mean minimum temperature exhibited a general warming trend, highlighting significant urbanization over recent decades (Mohan et al. 2022). The thermal mapping of Delhi was performed during the day and night using MODIS satellite data collected from November to December between 2007 and 2010. The analysis revealed that central parts of Delhi formed a daytime cool island, with temperatures 4–6 °C lower than surrounding rural areas, while nighttime warmer pockets showed temperatures 4–7 °C higher than the rural areas. The UHI mapping in Ahmedabad city used thermal infrared imagery from Landsat TM and ETM+. The findings indicated UHI intensity ranging from 0.4 to 0.8 °C, demonstrating a strong linear relationship between land surface temperature (LST) and the Normalized Vegetation Index (NDVI). Areas with higher NDVI exhibited lower temperatures, indicating that built-up areas significantly contribute to UHI (Bajaj et al. 2012). The study in Kochi was done by dividing the city into local climatic zones and conducting twelve mobile surveys between 2011 and 2013. The study found that UHI intensity was greater in winter than in summer, with measurements ranging from 4.6 °C in pre-dawn to 3.7 °C in early night, particularly in the central compact midrise areas of the city, while open and sparsely built regions remained cooler (Thomas et al. 2014). In Bhatinda city, temperatures were recorded with a Mextech Digital Thermometer at five rural and three urban locations between February and April 2015. The study generated surface temperature maps using ArcGIS 10.3, showing a strong correlation between urban areas and UHI (Kaur and Pandey 2022). One study compared UHI in Delhi and Mumbai using Landsat 5 TM images from May 5, 2010, for Delhi to April 17, 2010, for Mumbai. The study revealed that built-up and fallow lands exhibited higher temperatures than vegetated areas and water bodies. The maximum temperatures in Delhi reached 35 °C in the west and a minimum of 24 °C in the east by the River Yamuna. In Mumbai, temperatures varied from 28 to 31 °C in coastal areas to 31–36 °C at Bandra Kurla Complex, with Chhatrapati Shivaji International Airport recording 38 °C (Grover and Singh 2015).

To map UHI in Ahmedabad, a temporal assessment of remote sensing data was conducted, and MODIS-derived LST data was analyzed from 2003 to 2014. Their results indicated a temperature increase ranging from 2.96 to 4.04 °C in winter to 1.27–1.93 °C in summer, justifying the inverse relationship between UHI and rural temperatures observed during the study period (Goswami et al. 2016). The negative effects of urbanization on Lucknow were examined from 2002 to 2014 using Landsat thermal data and field surveys.

The study reported an increase in temperature by 0.75 °C due to land use changes and anthropogenic activities, primarily in the city's central regions, resulting in urban ecological degradation (Singh et al. 2017). In Nagpur, the UHI formation and effect were analyzed using various methods. In the first study, mobile traverse surveys were utilized to determine how land use, land cover, and population density influence the formation of canopy Urban Heat Islands (UHI), finding that high building and population density areas exhibit the highest UHI intensity, with nocturnal mean canopy UHI intensities of 2.14 °C in summer and 2.36 °C in winter (Kotharkar and Surawar 2015). In the second study, the UHI effect was analyzed using Landsat 7 ETM+ satellite images, applying remote sensing and GIS techniques to observe how land use/land cover (LU/LC) and vegetation density impact daytime land surface temperatures. Findings indicated that increased building density and reduced vegetation cover increased UHI intensity in Nagpur from 2000 to 2013(Surawar et al. 2017). In another study, CUHI was in Nagpur, employing local climate zone classification as its methodology. The study conducted a mobile survey and fixed station measurements from December 2015 to February 2016, finding UHI temperatures ranging from 1.76 to 4.09 °C in densely populated areas (Kotharkar et al. 2018). In Mumbai, the link between informal urban settlements, such as slums, and surface UHI is made through spatial statistical analysis (Mehrotra et al. 2018). The spatial relationship between land use land cover changes and land surface temperature using remote sensing and geographic information techniques was studied across ten metropolitan areas in India. The analysis of Landsat 7 Enhanced Thematic Mapper Plus images from 2001 to 2013 revealed the emergence of multiple UHI and their increasing prevalence, particularly in Lucknow, Jaipur, and Nagpur, attributed to the growth of built-up areas. The authors suggested that further UHI studies in developing cities like Lucknow would aid urban planners in mitigating UHI effects through effective urban planning (Sultana et al. 2018). One study observed that areas with higher-density development typically exhibited elevated LST, while regions with more green cover maintained lower LST. The research investigated the efficiency of various types of green cover, such as trees and turfs. It assessed the optimal green plot ratio to reduce air temperatures using Envi-met simulation software (Chandramathy and Kitchley 2018). To understand the effectiveness of mitigation strategies, one study evaluated four strategies for mitigating UHI effects, selecting study sites across three urban morphologies: open low-rise, compact low-rise, and mid-rise residential areas.

The study simulated the impacts of various mitigation strategies, including cool pavements, cool roofs, increased urban vegetation, and a combination of the three, concluding that vegetation plays a vital role in alleviating UHI (Chatterjee et al. 2019). In Mumbai, a macro- and micro-level UHI study identified higher LST in densely built regions compared to lower temperatures in areas with vegetation and water coverage, successfully pinpointing urban heat hotspots and their primary causes (Dwivedi 2019).

A study on UHI phenomena in urban public spaces proposed various mitigation strategies to improve comfort levels. The comparative analysis of different urban open space designs involved simulations to identify the best options for enhancing thermal comfort (Kalia 2021). A literature review created a hypothetical built environment incorporating UHI mitigation design strategies, with simulations indicating that effective approaches such as vegetation, cool roofs, green roofs, and walls could significantly reduce UHI impacts (Patil and Sood 2021). Additionally, the presence of heat pockets in Chennai was validated, with recommendations for new developments to include large canopy spaces to mitigate thermal heat stress (Rajan and Amirtham 2021). One study employs remote sensing techniques, particularly using Landsat-9 data, to examine how Land Surface Temperature (LST) and Normalized Difference Vegetation Index (NDVI) impact land use dynamics and UHI effects in Delhi, India. The findings indicate a notable rise in LST and UHI values driven by rapid urbanization and extensive concrete usage in construction, highlighting the importance of strategic urban planning and climate-sensitive design to foster sustainable and resilient urban environments (Ahmad et al. 2024). A study in Bangalore and Hyderabad investigates the spatial variation of UHI effects on populations using two decades of Land Surface Temperature (LST) data from the MODIS sensor, revealing average UHI intensities of 1.9 °C and 2.44 °C, respectively. The findings indicate that approximately 1 million residents in Bangalore and 1.8 million in Hyderabad live in high-temperature zones, highlighting the urgent need for targeted UHI mitigation strategies in the most affected areas (Arunab and Mathew 2023).

2.4 Heatwaves

Heatwaves (HWs) vary according to region, there is no universal definition, but only definitions relative to the usual weather in the area and to normal temperatures for the season. The World Meteorological Organization (WMO) defines a heatwave as 5 or more consecutive days of prolonged heat. The daily maximum temperature is higher by 5 °C (9 °F) or more than the average maximum temperature. Heatwaves are classified as meteorological disasters under extreme temperatures (Heatwave 2024). The key components are that the temperature rises quickly, remains elevated for an extended period, and then decreases. Important factors in the various definitions include the proportional increase in temperature, ambient temperature measurements, the duration for which the temperature must be maintained, and the severity of the temperature changes.

A combination of natural and human-induced factors causes HW. One of the primary drivers is high atmospheric pressure systems, which can stall over a region, trapping warm air near the Earth's surface and preventing the normal circulation of cooler air, leading to prolonged periods of extreme heat. Global warming, driven by human activities such as burning fossil fuels, deforestation, and industrial emissions, has increased the frequency and intensity of heatwaves by raising average global temperatures. Prolonged drought and dry conditions further exacerbate heatwaves by drying out the soil, which heats up more quickly. Natural climate variations, like El Niño and La Niña, can shift global weather patterns, making certain regions more prone to heatwaves. El Niño events lead to warmer ocean temperatures, resulting in increased heatwaves and altered precipitation patterns, while La Niña often causes cooler ocean temperatures that shift weather patterns, potentially leading to drier conditions and increased heat in some areas. Additionally, geographic features such as valleys and mountain ranges can trap hot air, while changes in wind patterns can transport heat to areas unaccustomed to extreme temperatures. Urbanization often intensifies HW effects; in cities, the severity of HW amplifies due to the concentration of heat-absorbing materials like asphalt, concrete, and buildings, which trap and slowly release heat. This leads to higher and more persistent temperatures in urban areas, especially at night. Cities' lack of green spaces and vegetation further reduces cooling, worsening the impact. As urbanization continues to expand, cities become increasingly vulnerable to HW.

Populations in tropical and subtropical regions are frequently exposed to high temperatures, while those in mid to high latitudes experience seasonal exposure to extreme heat. Heat can impact health through direct mechanisms, such as heat stress, and indirect ones, like exacerbating pre-existing health conditions. However, research on the long-term effects of chronic exposure to high temperatures and humidity remains limited. Certain groups, such as outdoor and manual laborers, athletes, and civil protection workers, are particularly vulnerable to heat stress due to the nature of their work.

2.4.1 HW Studies in India

Heatwaves in India are a growing concern, especially due to the region's vulnerability arising from its geography, population density, rapid urbanization, and limited resilience to extreme climate events (Ullah et al. 2022). India is susceptible to HW, and their frequency and intensity have increased in recent decades. A catastrophic heatwave in 2015 claimed the lives of over 2,200 people in India. This prompted the Indian National Disaster Management Authority (NDMA) to declare heatwaves as a disaster at the national level, as nine Indian states were severely impacted. By 2020, the heatwave had spread to 23 states, highlighting the growing severity of this phenomenon due to climate change and human-induced factors (Srivastava et al. 2022). The threshold temperature for heatwave occurrence varies by region: 30 °C in hilly regions, 37 °C in coastal areas, and 40 °C in the plains. This geographic diversity results in distinct heat experiences in northern and southern India (Indian Meteorological Department 2023).

India experiences two main types of heatwaves. The first occurs in North Central India and is often linked to atmospheric blocking over the North Atlantic, preventing heat escape. The second occurs along the East Coast and is influenced by the Matsuno-Gill response, an atmospheric circulation pattern affected by Pacific Ocean temperatures (Ratnam et al. 2016). Teleconnections, such as the El Niño Southern Oscillation (ENSO) and sea surface temperature variations in the Bay of Bengal, also contribute to heatwave occurrence in India (Mandal et al. 2019). The subcontinent, located in a subtropical climate zone, receives intense solar radiation during summer. Deserts and arid regions in northwest and central India produce dry air masses that migrate across the country, intensifying heatwave conditions (Pai et al. 2013).

2.5 The Combined Effect of UHI and HW on Microclimate

The combined effects of UHI and HW create a highly intensified heat environment in urban areas, posing significant risks to human health and comfort. During HW, cities often experience a sharp increase in net all-wave radiation as incoming shortwave radiation rises due to clear skies and lower water vapor levels (Jiang et al. 2019). This additional solar radiation gets absorbed by urban surfaces, largely composed of materials like concrete, asphalt, and other impervious structures. These materials retain heat much more than the natural surfaces in rural areas, leading to a greater heat buildup in urban settings (Ngarambe et al. 2020). The geometry of cities, with densely packed buildings and narrow streets, also limits heat dissipation, meaning urban areas absorb and store a substantial amount of heat during the day. At night, this stored heat is slowly released back into the atmosphere, making urban nights warmer and extending the overall duration of heat exposure compared to rural areas. This heightened nocturnal UHI effect prolongs thermal stress, especially in dense city centers (Kotharkar and Surawar 2015).

Beyond radiative heat, human activities significantly add to this effect. In urban areas, increased use of air conditioning and other cooling technologies during HWs contributes to anthropogenic heat, releasing additional warmth into the air. This extra energy boosts the local heat flux, creating a feedback loop where more energy input leads to increased demand for cooling, further exacerbating the urban temperature. A critical aspect of the UHI-HW interaction is the difference in latent heat flux between urban and rural regions. While rural areas experience more evapotranspiration due to abundant vegetation and available soil moisture, urban environments lack this natural cooling process. Instead, most energy in cities is converted to sensible heat due to the prevalence of impervious surfaces, resulting in lower humidity and heightened temperatures. This difference becomes particularly pronounced during HWs, as rural evapotranspiration rates rise, creating a stronger cooling effect that urban areas cannot match, further widening the temperature gap between urban and rural regions.

Moreover, reduced wind speeds during HW often exacerbate the UHI effect. During these stagnant periods, lower wind velocities limit the natural cooling effect from horizontal air movement, trapping warm air over cities and reducing the ability of urban areas to cool. Studies have shown that this stagnation is more severe in high-density areas where tall buildings and narrow streets restrict airflow, blocking any chance of relief from cool breezes or ventilation. In coastal cities, for example, normal cooling breezes from the sea may be obstructed by warm air blowing in from surrounding hot land masses, increasing daytime UHI intensities. Overall, the combination of UHI and HW results in an amplified heat environment in cities, with prolonged high temperatures both day and night, escalating the risks of heat-related health issues for urban populations.

2.6 Mitigation Approaches to Minimize the Impacts of UHI and HW at the Urban Scale

Many strategies have been proposed to mitigate the negative effects that UHI and HW have on indoor environments, outdoor spaces, people in outdoor environments such as pedestrians, roadside vendors, etc., and their thermal comfort. UHI develops in areas with a high percentage of non-reflective, water-resistant surfaces and a low percentage of vegetation. Specifically, a lack of vegetation reduces heat lost due to evapotranspiration. Vegetation, particularly in the presence of high moisture levels, is vital in regulating surface temperatures, even more so than many non reflective or low-albedo surfaces. These factors eventually worsen the HW impact. There are different ways of reducing the effects of UHI and strengthening urban resilience to HW and UHI events based on short-term solutions for mitigation and adaptation and long-term strategies and policies (Keramitsoglou et al. 2017).

High albedo materials for buildings and streets reflect solar radiation, reducing heat absorption and lowering surface temperatures (Mathew et al. 2016). Cool paints can further decrease heat retention (Donthu et al. 2024). The orientation of urban canyons helps enhance natural ventilation, reducing heat buildup and supporting airflow. Optimizing the height-to-width (H/W) ratio between buildings decreases heat retention, increases ventilation, and allows more daylight to penetrate (Pothamsetty and Chundeli 2024).

Implementing urban design guidelines ensures that new developments consider climate resilience and promote the use of sustainable materials. Strengthening public health systems improves responses to heat-related health risks, while raising public awareness about these risks encourages the adoption of personal cooling strategies. Establishing early warning systems for heatwaves prepares communities for extreme temperatures, enabling timely alerts and proactive measures to minimize health impacts.

These strategies aim to create a more resilient urban environment that mitigates the effects of heat and enhances residents' well-being. This study focuses on nature-based solutions (Table 2.2), which are explained in detail in the following sections.

Table 2.2 UHI and HW mitigation strategies

Mitigation strategy	Scale	Tool	Environmental benefits
Green infrastructure	Building/urban	Green roofs	Reduces surface temperature and improves insulation
	Building	Green walls and facade	Lowers ambient air temperature, enhances aesthetic quality, increases building insulation
	Neighborhood/city	Urban parks	Provides cooling through shade and evapotranspiration
	Street/neighborhood	Street trees	Reduces air temperature, increases shade, improves air quality
Water (Blue Infrastructure)	Building/street	Effective irrigation	Cools surrounding air, promotes plant health, and reduces dry urban surfaces
	Street/pavement	Water retentive pavement	Lowers surface temperature, reduces stormwater runoff, enhances groundwater recharge
	Urban	Waterbodies	It provides cooling through evaporation and increases humidity
	Street/urban	Water mist	Reduces ambient air temperature, increases localized humidity, enhances comfort

2.6.1 Nature-Based Mitigation Measures

According to the International Union for Conservation of Nature (IUCN), nature-based solutions (NBS) are actions designed to protect, sustainably manage, and restore natural or modified ecosystems. These solutions effectively address societal challenges while benefiting both human well-being and biodiversity (Nature-Based Solutions|IUCN 2024). Due to their potential advantages, NBS has gained popularity in tackling environmental issues, particularly climate change. One significant benefit of NBS is the reduction of climate-related risks, such as the UHI effect. NBS not only alleviates heat stress but also addresses key challenges city authorities face. Studies indicate that, in the medium to long term, NBS can lead to urban compaction. This occurs as residents are attracted to areas surrounding appealing NbS features, which mitigates the impacts of urban heating and curtails urban sprawl (Magotra et al. 2021). The European Commission describes NbS as a cost-effective strategy inspired by and supported by nature. These strategies provide environmental, social, and economic benefits while helping to enhance resilience (European Commission 2019).

NBS helps the ecosystem to adapt to mitigate the impacts of climate change. Thus, academicians, scientists, and urban planners increasingly acknowledge such strategies. The Paris Agreement clearly identifies the role of nature in addressing the adverse impacts of climate change within cities. A total of 103 countries have highlighted the incorporation of at least one or more types of NBS actions for achieving adaptation under their respective Nationally Determined Contributions (NDC). The most commonly used NBS for heat adaptation in urban areas are green infrastructure (Use of greenery) (Calheiros and Stefanakis 2021), blue infrastructure (Use of water bodies) (Anupriya and Rubeena 2023), and grey infrastructure (Use of Built architecture) (Mohite and Surawar 2024b).

2.6.1.1 Green Infrastructure

Among the various NBS available, green infrastructure is one of the most cost-effective solutions to reducing the impact of HW and the UHI effect in cities. Thus, it has been increasingly promoted as a key measure to mitigate heat stress in cities caused by the UHI effect and climate.

(a) **Tree Canopy and Street Vegetation**

Street trees offer significant cooling effects through shading and evapotranspiration, reducing solar heat exposure on buildings, sidewalks, and roads. Shaded areas beneath tree canopies experience reduced surface temperatures, enhancing outdoor thermal comfort, especially in urban canyons with intense radiant temperatures. Studies indicate that under tree cover, the Physiologically Equivalent Temperature (PET) index can drop by up to 12 °C, providing notable relief from heat (Keramitsoglou et al. 2017). In urban environments, strategically planted street trees affect radiation exchange, wind flow, and convective heat loss. For example, research from Colombo reveals that street trees improve thermal comfort by modifying the Sky View Factor and lowering ground temperatures. However, maintaining street trees is essential, particularly in humid climates, where high evapotranspiration rates can raise local humidity. Some trees also emit biogenic volatile organic compounds (VOCs), which can slightly offset their air-quality benefits.

(b) **Green Open Spaces (Urban Parks)**

Urban parks and green open spaces provide extensive, uninterrupted greenery, which plays a critical role in cooling urban areas. Parks reduce temperatures, offer shaded refuge, and support carbon sequestration, aligning with climate adaptation strategies. Their large size and simplified layouts make them particularly effective for temperature mitigation, as evidenced by studies showing that parks with simpler shapes and larger sizes offer greater cooling effects. On a global scale, parks contribute to climate change adaptation by sequestering carbon and enhancing ecosystem services. As “soft engineering” climate adaptation solutions, parks offer resilience against extreme weather events such as HW and support biodiversity, aligning with ecosystem-based adaptation practices.

(c) **Green Roofs**

Green roofs, classified as intensive (deeper soil, diverse plants) or extensive (shallow soil, lightweight vegetation), provide thermal insulation by blocking solar heat and reducing rooftop ambient temperatures. They have been shown to lower ambient temperatures, contributing to reduced UHI effects in surrounding areas (Calheiros and Stefanakis 2021). In addition to cooling, green roofs manage stormwater, improve air quality, and support urban wildlife by providing green habitats in densely built areas. While green roofs have a longer lifespan than conventional roofs, they require careful installation and maintenance to avoid structural damage or water leakage, balancing their economic benefits with upkeep considerations.

(d) **Green Walls and Facades**

Vertical greening systems, such as green walls and facades, cover building exteriors with vegetation that blocks direct sunlight and prevents heat buildup in building materials. This reduces heat transmission, which is especially beneficial in tropical climates where green facades have been shown to cut heat transfer by as much as 40.3%. Green walls also improve air quality by filtering pollutants, reducing noise, and increasing urban biodiversity. In densely populated areas with limited space for horizontal greenery, green walls offer an effective alternative for cooling and environmental benefits. Proper selection and placement of climbing plants are essential to prevent structural issues, like root infiltration or damage to walls, and regular maintenance through pruning, irrigation, and inspections is necessary to sustain their effectiveness (Ratnayake et al. 2022).

2.6.1.2 Blue Infrastructure

Blue infrastructure (BI) complements green infrastructure by incorporating water-based solutions to manage heat, reduce UHI effects, and enhance urban resilience to extreme weather events. Blue infrastructure includes natural and engineered water systems such as rivers, lakes, ponds, urban wetlands, fountains, rain gardens, and permeable surfaces designed for water management. BI not only cools urban areas through evaporative cooling but also provides essential ecosystem services that improve quality of life, biodiversity, and climate adaptation (Shimazaki et al. 2021).

(a) **Urban Water Bodies (Rivers, Lakes, and Ponds)**

Natural and artificial water bodies, including rivers, lakes, and ponds, play a vital role in urban cooling through the process of evaporation, where heat from the air is absorbed by water as it evaporates. This mechanism can significantly lower surrounding air temperatures, creating "cool islands" that effectively counteract UHI effects in cities. By altering the local microclimate, these water bodies help regulate temperatures in their vicinity, extending their cooling effects into neighboring built-up areas. Well-placed urban water bodies can also reduce mean radiant temperatures, improving outdoor thermal comfort, especially in densely populated regions. However, managing water quality is crucial, as stagnant water can lead to pollution and mosquito breeding. Implementing maintenance practices, such as aeration and the establishment of native plant buffers, can help sustain water quality and ecological balance (Gao et al. 2020).

(b) **Permeable Pavements and Water-Absorbing Surfaces**

Permeable pavements and water-absorbing surfaces offer significant cooling benefits while effectively managing water in urban environments. These surfaces allow water to seep into underlying soil layers, minimizing surface runoff and promoting groundwater recharge. During hot weather, permeable pavements facilitate water retention and evaporation, cooling surrounding areas more effectively than traditional, impervious materials like asphalt. Additionally, these pavements reduce the load on stormwater infrastructure, decreasing flood risks during heavy rains. Water infiltrates and supports nearby vegetation, further contributing to cooling through evapotranspiration. Although permeable pavements may have higher initial installation costs than conventional materials, their long-term advantages in heat reduction, flood mitigation, and enhanced urban aesthetics make them cost-effective for urban heat management (Erell et al. 2014).

(c) **Constructed Water Features (Fountains, Reflecting Pools)**

Constructed water features, such as fountains and reflecting pools, provide essential cooling and ecological benefits in urban settings. These features enhance air moisture content and reduce localized temperatures, creating a more comfortable outdoor environment. Fountains and reflecting pools not only cool the air through evaporative processes but also add aesthetic appeal and improve overall air humidity, especially in arid or warm climates. Beyond their cooling capabilities, urban wetlands support biodiversity by providing habitats for various species while improving air quality, reducing noise pollution, and offering recreational spaces that enhance urban livability. However, regular maintenance is essential to ensure water quality, prevent mosquito breeding, and address issues related to seasonal drying. Sustainable design approaches, such as recirculating water in fountains, can improve efficiency and reduce resource use, further supporting the functionality of these water features (Anupriya and Rubeena 2023).

2.7 Summary

The rate of urbanization has increased in the past couple of decades, resulting in a rise in average annual temperatures and severe heat impacts. This review presents an understanding of Urban Heat Islands (UHIs) and heatwave (HW) related impacts in cities within tropical regions, highlighting the potential of nature-based solutions (NBS) for heat mitigation. In India, due to socio-economic and climatic conditions, the population is vulnerable to escalating heat-related issues due to a heavy reliance on natural systems, unplanned urbanization, and inadequate resilient infrastructure.

The interaction between UHI and HW exacerbates thermal discomfort in urban settings. During heatwaves, cities experience intensified UHI effects, leading to significantly elevated temperatures that can exceed surrounding rural areas. This intensification results in increased heat-related health risks and energy demands for cooling. Furthermore, the vulnerable populations in these urban areas face increased exposure to extreme heat events, accentuating the serious need for effective mitigation strategies. Integrating NBS can enhance urban resilience against the dual impacts of UHI and HW, promoting better thermal comfort and reducing heat exposure for residents.

NBSs are essential for protecting and restoring ecosystems while addressing climate change challenges, especially the urban heat-related issues. Integrating green and blue infrastructure is crucial; emphasizing street trees, urban parks, green roofs, and vertical gardens can mitigate heat stress through shade, enhanced thermal comfort, and carbon sequestration. Planners should focus on the extent and distribution of green cover and select native plant species, as increasing greenery can significantly lower urban temperatures and improve air quality. Urban planning should prioritize reflective and lighter-colored materials for roofs and pavements to minimize heat absorption and passive cooling strategies like natural ventilation and shading to reduce reliance on air conditioning. Additionally, blue infrastructure, including rivers, lakes, and urban wetlands, enhances cooling and urban quality of life through evaporative processes, creating "cool islands" that alleviate UHI effects. Permeable pavements and constructed water features improve aesthetics, support groundwater recharge, and increase humidity.

NBS can effectively mitigate UHI and HW effects and enhance urban resilience by improving thermal comfort, supporting biodiversity, and regulating local climates through ecosystem restoration and management. It provides essential guidelines for urban planners to incorporate sustainable practices into city designs.

Authors Contribution All authors have contributed equally to this work.

Funding The research was not supported by any funding.

Potential conflicts of interest On behalf of all authors, the corresponding author states that there is no conflict of interest.

Data Availability Statement The datasets used and/or analyzed during the current study are available from the corresponding author on reasonable request.

References

Ahmad B, Najar MB, Ahmad S (2024) Analysis of LST, NDVI, and UHI patterns for urban climate using Landsat-9 satellite data in Delhi. J Atmos Solar Terr Phys 265:106359. https://doi.org/10.1016/J.JASTP.2024.106359

Anupriya RS, Rubeena TA (2023) Spatio-temporal urban land surface temperature variations and heat stress vulnerability index in Thiruvananthapuram city of Kerala, India. Geol Ecol Landsc. https://doi.org/10.1080/24749508.2023.2182088

Arunab KS, Mathew A (2023) Geospatial and statistical analysis of urban heat islands and thermally vulnerable zones in Bangalore and Hyderabad cities in India. Remote Sens Appl Soc Environ 32:101049. https://doi.org/10.1016/J.RSASE.2023.101049

Bajaj DN, Inamdar AB, Vaibhav V (2012) Temporal variation of urban heat island using landsat data: a case study of Ahmedabad, India. Asian Association on Remote Sensing, pp 797–804. https://researchers.mq.edu.au/en/publications/temporal-variation-of-urban-heat-island-using-landsat-data-a-case

Calheiros CSC, Stefanakis AI (2021) Green roofs towards circular and resilient cities. Circ Econ Sustain 1(1):395–411. https://doi.org/10.1007/S43615-021-00033-0

Chandramathy I, Kitchley JL (2018) Study and analysis of efficient green cover types for mitigating the air temperature and urban heat island effect. Int J Glob Warm 14(2):238–259. https://doi.org/10.1504/IJGW.2018.090182

Chatterjee S, Khan A, Dinda A, Mithun S, Khatun R, Akbari H, Kusaka H, Mitra C, Bhatti SS, Doan QV, Wang Y (2019) Simulating micro-scale thermal interactions in different building environments for mitigating urban heat islands. Sci Total Environ 663:610–631. https://doi.org/10.1016/J.SCITOTENV.2019.01.299

Chew LW, Liu X, Li XX, Norford LK (2021) Interaction between heat wave and urban heat island: a case study in a tropical coastal city Singapore. Atmos Res 247:105134. https://doi.org/10.1016/J.ATMOSRES.2020.105134

Donthu EVSKK, Long YP, Wan MP, Zhou M, Ng BF (2024) Dynamics of cool surface performance on urban microclimate: a full-scale experimental study in Singapore. Sustain Cities Soc 102:105218. https://doi.org/10.1016/J.SCS.2024.105218

Dwivedi A (2019) Macro-and micro-level studies using Urban Heat Islands to simulate effects of greening, building materials and other mitigating factors in Mumbai city. Archit Sci Rev 62(2):126–144. https://doi.org/10.1080/00038628.2019.1578193

Erell E, Pearlmutter D, Boneh D, Kutiel PB (2014) Effect of high-albedo materials on pedestrian heat stress in urban street canyons. Urban Climate 10(P2):367–386. https://doi.org/10.1016/j.uclim.2013.10.005

European Commission (2019) Nature-based solutions|European commission. In European commission. https://rea.ec.europa.eu/publications/nature-based-solutions-1_en

Gao K, Santamouris M, Feng J (2020) On the cooling potential of irrigation to mitigate urban heat island. Sci Total Environ 740:139754. https://doi.org/10.1016/J.SCITOTENV.2020.139754

Goswami A, Mohammad P, Sattar A (2016) A temporal study of urban heat island (UHI)-a evaluation of Ahmedabad city, Gujarat. Int Conf Clim Chang Mitig Technol Adapt 63–68. https://www.researchgate.net/publication/323486794_A_temporal_study_of_Urban_Heat_Island_UHI-A_Evaluation_of_Ahmedabad_city_Gujarat

Grover A, Singh RB (2015) Analysis of urban heat island (UHI) in relation to normalized difference vegetation index (NDVI): a comparative study of Delhi and Mumbai. Environments 2(2):125–138. https://doi.org/10.3390/ENVIRONMENTS2020125

Heatwave (2024). https://wmo.int/topics/heatwave

Horton RM, Mankin JS, Lesk C, Coffel E, Raymond C (2016) A review of recent advances in research on extreme heat events. Curr Clim Chang Rep 2(4):242–259. https://doi.org/10.1007/S40641-016-0042-X

Indian Meteorological Department IM (2023) Indian Meteorological Department. https://internal.imd.gov.in/pages/heatwave_mausam.php

Jarraud M, Steiner A (2012) Summary for policymakers. In Managing the risks of extreme events and disasters to advance climate change adaptation: special report of the intergovernmental panel on climate change, 9781107025. https://doi.org/10.1017/CBO9781139177245.003

Jiang S, Lee X, Wang J, Wang K (2019) Amplified urban heat islands during heat wave periods. J Geophys Res Atmos 124(14):7797–7812. https://doi.org/10.1029/2018JD030230

Kalia A (2021) Effect of vegetation cover and shading on thermal and visual comfort in institutional campus outdoor spaces in India. Int J Recent Technol Eng (IJRTE) 2277–3878. https://doi.org/10.35940/ijrte.B6078.0710221

Kaur R, Pandey P (2022) Spatial trends of surface urban heat island in Bathinda: a semiarid city of northwestern India. Int J Environ Sci Technol 19(11):10911–10932. https://doi.org/10.1007/S13762-021-03742-Z/FIGURES/16

Keramitsoglou I, Sismanidis P, Analitis A, Butler T, Founda D, Giannakopoulos C, Giannatou E, Karali A, Katsouyanni K, Kendrovski V, Lemesios G, Myrivili E, Ordoñez D, Varotsos KV, Vlastou G, Kiranoudis CT (2017) Urban thermal risk reduction: developing and implementing spatially explicit services for resilient cities. Sustain Cities Soc 34:56–68. https://doi.org/10.1016/J.SCS.2017.06.006

Kotharkar R, Ramesh A, Bagade A (2018) Urban heat island studies in South Asia: a critical review. Urban Climate 24:1011–1026. https://doi.org/10.1016/J.UCLIM.2017.12.006

Kotharkar R, Surawar M (2015) Land use, land cover, and population density impact on the formation of canopy urban heat islands through traverse survey in the Nagpur urban area, India. J Urban Plan Dev 142(1):04015003. https://doi.org/10.1061/(asce)up.1943-5444.0000277

Magotra R, Sharma Y, Raj V (2021) Nature based solutions for heat stress management in cities

Mandal R, Joseph S, Sahai AK, Phani R, Dey A, Chattopadhyay R, Pattanaik DR (2019) Real time extended range prediction of heat waves over India. Sci Rep 9(1):1–11. https://doi.org/10.1038/s41598-019-45430-6

Marcotullio PJ, Keßler C, Quintero Gonzalez R, Schmeltz M (2021) Urban growth and heat in tropical climates. Front Ecol Evol 9:616626. https://doi.org/10.3389/FEVO.2021.616626/BIBTEX

Mathew A, Sreekumar S, Khandelwal S, Kaul N, Kumar R (2016) Prediction of surface temperatures for the assessment of urban heat island effect over Ahmedabad city using linear time series model. Energy Build 128:605–616. https://doi.org/10.1016/J.ENBUILD.2016.07.004

Mehrotra S, Bardhan R, Ramamritham K (2018) Urban informal housing and surface urban heat island intensity 9(2):158–177. https://doi.org/10.1177/0975425318783548

Mohan M, Bhati S, Sati AP (2022) Urban heat island effect in India: assessment, impacts, and mitigation. Glob Urban Heat Isl Mitig 199–250. https://doi.org/10.1016/B978-0-323-85539-6.00007-X

Mohite S, Surawar M (2024a) Assessing pedestrian thermal comfort to improve walkability in the urban tropical environment of Nagpur City. Geogr Pannonica 28(1):71–84. https://doi.org/10.5937/gp28-48166

Mohite S, Surawar M (2024b) Impact of urban street geometry on outdoor pedestrian thermal comfort during heatwave in Nagpur city. Sustain Cities Soc 108(April):105450. https://doi.org/10.1016/j.scs.2024.105450

Nature-based Solutions|IUCN (2024). https://iucn.org/our-work/nature-based-solutions

Ngarambe J, Nganyiyimana J, Kim I, Santamouris M, Young Yun G (2020) Synergies between urban heat island and heat waves in Seoul: the role of wind speed and land use characteristics. PLoS ONE 15(12):e0243571. https://doi.org/10.1371/JOURNAL.PONE.0243571

Oke TR, Mills GM, Christen A, Voogt JA (2017) Urban climates 525. https://books.google.com/books/about/Urban_Climates.html?id=7h0xDwAAQBAJ

Padmanabhamutry B, Bahl HD (1984) Isothermal Isohyetal patterns at Delhi as a sequel of urbanization. Mausam 35(4):539–541. https://doi.org/10.54302/MAUSAM.V35I4.2271

Pai DS, Nair SA, Ramanathan AN (2013) Long term climatology and trends of heat waves over India during the recent 50 years (1961–2010). Mausam 64(4):585–604. https://doi.org/10.54302/MAUSAM.V64I4.742

Patil AS, Sood R (2021) Mitigation of urban heat island effect through design. https://doi.org/10.46529/darch.202125

Pothamsetty A, Chundeli FA (2024) Optimizing urban morphology to mitigate urban heat islands: a case of Hyderabad. 198–206. https://doi.org/10.62576/NAQY4939

Qian Y, Chakraborty TC, Li J, Li D, He C, Sarangi C, Chen F, Yang X, Leung LR (2022) Urbanization impact on regional climate and extreme weather: current understanding, uncertainties, and future research directions. Adv Atmos Sci 39(6):819–860. https://doi.org/10.1007/s00376-021-1371-9
Rajan EHS, Amirtham LR (2021) Urban heat island intensity and evaluation of outdoor thermal comfort in Chennai, India. Environ Dev Sustain: Multidiscip Approach Theory Pract Sustain Dev 23(11):16304–16324. https://doi.org/10.1007/S10668-021-01344-W
Ratnam JV, Behera SK, Ratna SB, Rajeevan M, Yamagata T (2016) Anatomy of Indian heatwaves. Sci Rep 6(1):1–11. https://doi.org/10.1038/srep24395
Ratnayake CW, Perera NGR, Emmanuel R (2022) Street tree planting patterns to modify the sky view factor for outdoor thermal comfort enhancement. FARU J 9(2):1–1. https://doi.org/10.4038/FARUJ.V9I2.167
Ren J, Shi K, Li Z, Kong X, Zhou H (2023) A review on the impacts of urban heat islands on outdoor thermal comfort. Buildings 13(6):1368. https://doi.org/10.3390/BUILDINGS13061368
Sclunzen KH, Grimmond S, Bakkanov A (2023) Guidance on measuring, modelling and monitoring the canopy layer urban heat island (CL-UHI). https://www.preventionweb.net/publication/guidance-measuring-modelling-and-monitoring-canopy-layer-urban-heat-island-cl-uhi
Shimazaki Y, Aoki M, Nitta J, Okajima H, Yoshida A (2021) Experimental determination of pedestrian thermal comfort on water-retaining pavement for UHI adaptation strategy. Atmosphere 12(2):127. https://doi.org/10.3390/ATMOS12020127
Singh P, Kikon N, Verma P (2017) Impact of land use change and urbanization on urban heat island in Lucknow city, central India. A remote sensing based estimate. Sustain Cities Soc 32:100–114. https://doi.org/10.1016/J.SCS.2017.02.018
Srivastava A, Mohapatra M, Kumar N (2022) Hot weather hazard analysis over India. Sci Rep 12(1):1–15. https://doi.org/10.1038/s41598-022-24065-0
Sultana S, Achanta NVS (2018) Assessment of urban heat island intensity over metropolitan cities of india for summer and winter seasons: change in climate. EGUGA 20:12455. https://ui.adsabs.harvard.edu/abs/2018EGUGA..2012455S/abstract
Surawar M, Kotharkar R (2017) Assessment of urban heat island through remote sensing in Nagpur urban area using landsat 7 ETM+ satellite images. Int J Urban Civ Enginering 11(7):868–874. https://doi.org/10.5281/ZENODO.1131073
Thomas G, Sherin AP, Ansar S, Zachariah EJ (2014) Analysis of urban heat island in Kochi, India, using a modified local climate zone classification. Procedia Environ Sci 21:3–13. https://doi.org/10.1016/J.PROENV.2014.09.002
Ullah I, Saleem F, Iyakaremye V, Yin J, Ma X, Syed S, Hina S, Asfaw TG, Omer A (2022) Projected changes in socioeconomic exposure to heatwaves in south Asia under changing climate. Earth's Future 10(2):e2021EF002240. https://doi.org/10.1029/2021EF002240
United Nations Development Programme (2021) Sustainable Development Goals|United Nations Development Programme. https://www.undp.org/sustainable-development-goals
World Cities Report (2020) Unpacking the value of sustainable urbanization. https://doi.org/10.18356/c41ab67e-en

Chapter 3
Bioclimatic Design as a Strategy to Mitigate the Impacts of Urban Heat Island: An Overview

Mahadev Bera and Pranab Kumar Nag

Abstract The urban heat island (UHI) effect is a climatic phenomenon in which urban areas experience localized increases in temperature due to factors such as urbanization, dense built environments, and extensive impervious surfaces. The impacts of UHI are particularly significant for vulnerable populations, including older adults, children, and individuals with pre-existing health conditions. Addressing these effects necessitates the integration of bioclimatic design strategies, which enhance urban resilience through improved ventilation, passive design approaches, site analysis, effective open space utilization, passive solar techniques, and the use of sustainable building materials. This chapter examines the adoption and quantification of these bioclimatic strategies, highlighting their importance in optimizing energy efficiency, human thermal comfort, and indoor environmental quality across various local climatic conditions. A deeper understanding of these strategies can assist policymakers, researchers, and building professionals in crafting sustainable, energy-efficient, and climate-responsive urban environments, ultimately alleviating the adverse effects of UHI on human health and well-being.

Keywords Urban heat island · Climate change · Bioclimatic design · Built environment · Energy efficiency · Thermal comfort

3.1 Introduction

Urban heat islands (UHIs) are characterized by a noticeable rise in ambient air temperatures within specific city regions, often surpassing temperatures in nearby semi-urban and rural areas (Hayes et al. 2022; Mirabi and Davies 2022). The temperature differences are pronounced at nighttime and during light wind conditions. The UHI effect disproportionately impacts individuals residing in urban settings during

M. Bera (✉) · P. K. Nag
School of Environment and Disaster Management, Ramakrishna Mission Vivekananda Educational and Research Institute, Narendrapur, Kolkata, West Bengal, India
e-mail: mahadev.bera@gm.rkmvu.ac.in

K. Mukherjee et al. (eds.), *Urban Climate*, The Urban Book Series,
https://doi.org/10.1007/978-3-032-17895-4_3

winter and summer (Abrar et al. 2022). The UHI threatens urban life (Piracha and Chaudhary 2022) with the decline in human health due to bioclimatic discomfort and disorders (Yin et al. 2023; Wang et al. 2022). With the evidence of a direct association between the UHI phenomenon and the population size and density of metropolitan areas, the increasing trend of population concentration in cities, and rapid urbanization, the UHI is anticipated to intensify in the coming years (Hidalgo-García and Arco-Díaz 2022). By 2050, over 2/3rd of the projected world population will inhabit urban areas (Bolay and Bolay 2020).

Whereas metropolitan areas face many social, health, environmental, and economic challenges, ample opportunities for the proximity of the urban locale's people, businesses, and services enhance resource efficiency. Conversely, poor urban environments may exacerbate the climate effects, making the sustainability of the populace more vulnerable (Chowdhury et al. 2023; Zhong et al. 2022). Underscoring the creation of environmentally friendly, human-centric, and energy-efficient building environments (Nag 2019a, b; Bera and Nag 2022; Bera et al. 2023; Bera et al. 2024a, b), bioclimatic strategies are pivotal to achieving an interplay between architectural design and the psychological and physiological aspects of comfort and well-being (Zhong et al. 2022; Bera and Nag 2022; Bera et al. 2024a, b). The overarching objective of bioclimatic design is to enhance building comfort levels by leveraging insights into the microclimate. That is, minimizing energy consumption for heating, cooling, air conditioning, and lighting, fostering a harmonious relationship with the natural environment, such as daylighting, natural ventilation, and passive heating and cooling (Tamaskani Esfehankalateh et al. 2022). Structural elements such as building floors, walls, windows, and roofs are crucial to capture and distribute solar thermal energy, ensuring comfortable conditions (Bugenings and Kumari 2022).

Despite extensive research on urban heat island (UHI) mitigation, there are still significant gaps in our understanding of how effective bioclimatic strategies are in different climatic conditions and urban designs. Previous studies have highlighted the benefits of passive design and urban greening; however, there is a lack of research that systematically measures their impacts across various urban densities and geographical contexts (Bressane et al. 2024). Moreover, an integrated approach is essential—one that not only focuses on reducing temperatures but also considers long-term sustainability, energy efficiency, and the well-being of residents. This chapter aims to fill these gaps by examining how bioclimatic strategies can be tailored for specific urban environments. It will address important questions regarding their adaptability, effectiveness, and the challenges involved in their implementation. Gaining a deeper understanding of these factors will support the development of comprehensive, data-driven strategies for mitigating UHI in future urban planning efforts. The main objective of this chapter is to identify the different bioclimatic design strategies that mitigate the adverse impacts of the UHI effect. Overall, the present contribution brings an overview of understanding bioclimatic strategies for UHI mitigation, encompassing urban form features, building porosity, ventilation, greenery, and waterfronts. A comprehensive understanding would help policymakers, scientists, and building engineers mitigate the adverse effects of UHI on both people and the environment.

3.2 Causes and Effects of Urban Heat Island

The formation of urban heat islands (UHIs) is influenced by various factors (Table 3.1). These factors include the trapping of both shortwave and longwave radiation between buildings, reduced radiative heat losses due to limited exposure to the sky, and the increased storage of sensible heat in building materials. Additionally, human activities such as combustion and vehicular traffic contribute to anthropogenic heat emissions, while lower wind speeds lead to reduced convective heat dissipation and evapotranspiration. The properties of urban construction materials, such as their solar reflectance, thermal emissivity, and heat capacity, also play a critical role in the development of UHIs. The impermeable and dry surfaces, like asphalt and concrete, exacerbate the absorption of solar radiation and heat storage and elevate surface and air temperatures, contributing to the UHI phenomenon in urban regions (Rahaman et al. 2022). These surfaces retain excess heat and emit significant longwave radiation during nighttime (Hong and Heo 2023). Moreover, the quantity, orientation, and spacing of buildings create microclimates around the infrastructures that profoundly affect the absorption of solar radiation by urban surfaces and temperature fluctuations across a city (Ulpiani 2021; He et al. 2022). The scarcity of green spaces, including trees and parks, further exacerbates the problem by diminishing shade and the cooling effects of evapotranspiration, which otherwise would help maintain an equilibrium in energy levels. Also, the escalating waste heat generation from the sources of energy consumption (e.g., heating and cooling systems, power plants, transportation) directly contributes to the rise in surface and ambient air temperature and increased incidences of heatwaves (Rahaman et al. 2022).

The UHI significantly affects the environment and human health (Fig. 3.1). The elevated temperatures escalate energy consumption for cooling purposes, leading to increased electricity demands and potential strain on power infrastructure. Moreover, fluctuations in local temperatures may influence regional weather patterns, changes in wind direction, and the distribution of precipitation. Such microclimate patterns affect the localized urban areas, including aquatic ecosystems and water quality, by influencing water temperatures in nearby rivers and lakes (Briciu et al. 2020). The quality of life and human diversity within metropolitan areas become susceptible to potential climate impacts (Abrar et al. 2022). The thermal effect of UHI can worsen urban air pollution by promoting the formation of ground-level ozone and other pollutants, thereby affecting air quality. Poor air quality can manifest in city dwellers' respiratory and cardiovascular health problems (Ulpiani 2021). A widespread prevalence of CNS disorders such as depression, irritability, memory impairment, insomnia, and dyspepsia, and a range of other health effects have been reported due to the impact of UHI (Meade and Garvey 2022).

The manifestation of UHI effects varies across countries (Table 3.2). A combination of local climate conditions, urbanization patterns, and mitigation approaches influences it. The bioclimatic approach to mitigate the impacts of UHI involves adopting sustainable urban planning and design, thereby enhancing natural cooling

Table 3.1 Factors influencing the development of the urban heat island

Factors	Description	References
Impervious surfaces	Construction materials like concrete and asphalt create impervious surfaces that absorb and retain heat, resulting in higher daytime temperatures and slower cooling at nighttime	Morabito et al. (2021); Hua et al. (2020)
Buildings and infrastructure	Tall buildings and dense urban infrastructure block natural ventilation, trap heat, and create microclimates within the city, leading to elevated temperatures	Abdulateef and Al-Alwan (2022)
Lack of vegetation	Urban areas often have fewer green spaces. Lack of vegetation reduces shade and the cooling effect of evapotranspiration (plants release water vapour through leaves)	Al-Saadi et al. (2020)
Waste heat	Human activities, industrial processes, energy consumption, and transportation release substantial heat into the environment, resulting in the UHI effect	Hidalgo-García and Arco-Díaz (2022); Abrar et al. (2022); Feng et al. (2021)
Modification of natural landscapes	Urbanization alters the natural landscape through deforestation and land development, which disrupts the balance of the thermal properties of the natural ecosystems	Feng et al. (2021)
Urban design and layout	Urban area layout and design have an impact on how much solar radiation is absorbed, retained, and reflected. Certain setups have the potential to increase the UHI impact	Mansouri and Zarghami (2023)

mechanisms, improving air quality, and creating more livable and resilient urban environments.

3.3 Quantifying Urban Heat Island Effects

Measuring the UHI effect entails evaluating temperature discrepancies between urban and non-urban regions. Methodologies include deploying temperature sensors across urban and rural locales to record air temperatures at different positions and elevations. Weather station and satellite remote sensing data assess surface temperature patterns between urban and non-urban sites over time. Thermal infrared imagery enables the visualization of surface temperature variations and aids in identifying and mapping UHI hotspots. Calculating urban heat flux involves considering incoming and outgoing longwave radiation, solar radiation, latent heat flux, and sensible heat flux. Computational models simulate urban environments to forecast temperature fluctuations under different scenarios. Satellite-derived land surface temperature (LST) data can be leveraged to produce maps highlighting temperature differentials between rural and urban areas. The disparity in temperature between these

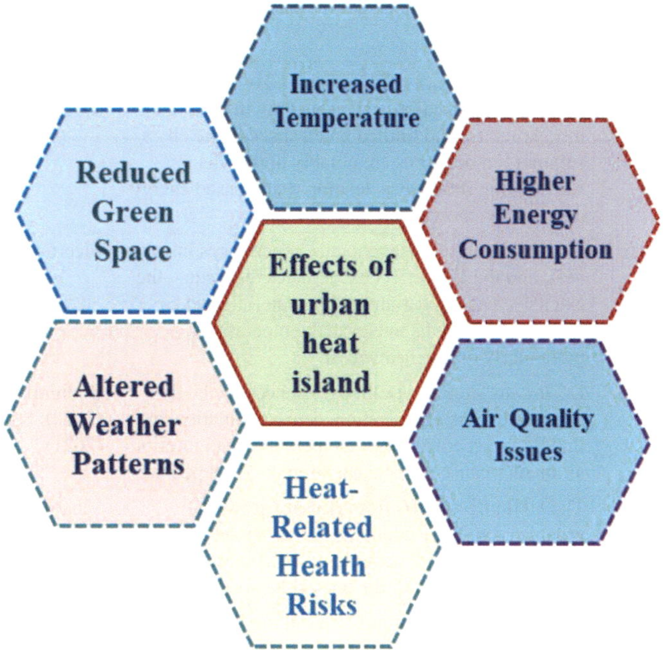

Fig. 3.1 Various effects of urban heat island

regions quantifies the magnitude of the UHI effect—long-term temperature trends discern seasonal and temporal fluctuations in UHI intensity, and diurnal temperature variations determine UHI dynamics.

Quantifying UHI effects is crucial to formulating mitigation strategies and urban planning policies for potential ramifications on public health. Evaluating how UHI impacts human comfort and well-being entails assessing heat stress and disorders using indices of heat stress and strain and perspectives of the health-related indicators between urban and non-urban areas (Table 3.3). Based on the comprehension of UHI's spatial and temporal dynamics, cities can implement measures to mitigate heat island intensity and enhance urban areas' overall sustainability and livability. Geographic Information Systems (GIS) and remote sensing (RS) are practical tools for measuring and examining the UHI effect. Some examples of how remote sensing and GIS have been applied to determine the UHI intensity, monitor land surface temperatures, and collect spatial data are given in Table 3.4.

Moreover, thermo-physiological indices (as depicted in Table 3.5) evaluate the combined influence of climatic elements on human health, comfort, and overall well-being. These indices typically consider various weather-related parameters, clothing worn, and physical activity levels to evaluate human reactions to hot environmental conditions. They have found applications in environmental science, architecture, occupational health, and urban planning.

Table 3.2 Urban heat island (UHI) effects in different countries

City/country	Description	References
Tokyo, Japan	The city experiences a significant UHI effect due to its dense urbanization, extensive growth in urban infrastructure, and limited green spaces. The city's summer temperatures are notably higher than surrounding rural areas, leading to increased energy demand for cooling	Irie (2022); Kim et al. (2022)
Phoenix, United States	In a hot desert climate, phoenix experiences intense heat, and the UHI effect exacerbates the issue—the extensive use of heat-absorbing materials and air conditioning results in elevated temperatures, particularly during heatwaves	Meerow et al. (2021)
Delhi, India	During the summer, Delhi experiences the UHI effect due to its urbanization, dense population and limited green cover. High temperatures and inferior air quality cause adverse health impacts on people	Budhiraja et al. (2020); Siddiqui et al. (2021)
Mumbai, India	The UHI effect arises from densely growing urbanization, large population, heat-absorbing materials, industrial activities, and temperature patterns influenced by the coastal location	Vinayak et al. (2022)
São Paulo, Brazil	One of the most populous cities in the Southern Hemisphere faces challenges from the UHI effect. The city's expansive urban landscape and industrial activities have increased temperatures and air pollution, negatively impacting residents and the environment	Moreira et al. (2023)
Singapore	The city-state faces UHI effects due to high urbanization, densely populated city centres, and limited green spaces. To mitigate its impact and enhance the quality of urban life, the government has introduced initiatives such as using green building designs and extensive plantations in the city	Jung (2024)
Melbourne, Australia	The city's central business district experiences the UHI effects due to growth and development, resulting in increased temperatures that impact energy consumption, air quality, and the overall urban environment	Jamei et al. (2019)

3.4 Bioclimatic Design Strategies: Adaptation and Mitigation of Urban Heat Island

The term "Bioclimate" was introduced by Victor Olgyay (1963) to describe the interaction between climate and living organisms. Bioclimate encompasses how climate influences living organisms with particular consideration for enhancing human comfort, energy efficiency, and indoor environmental quality (Cristiano et al. 2021), as depicted in Fig. 3.2. The bioclimatic design involves local climate factors

Table 3.3 Indices to quantify the UHI effect

Name of index	Description	References
Urban heat island intensity (UHII)	The UHII is a simple index of the temperature difference between urban and surrounding rural areas, widely used in research on urban heat islands and climatology	Kim and Brown (2021); Li et al. (2020)
Normalized difference vegetation index (NDVI) *NDVI = (NIR−Red)/(NIR + Red), where NIR = near-infrared reflectance and Red = red reflectance*	Satellite imagery is utilized for evaluating the health and quantity of vegetation in a particular area using the normalized difference vegetation index (NDVI). Furthermore, NDVI acts as an indirect indicator of impervious surfaces, which contribute to the urban heat island (UHI) effect	Huang et al. (2021); Pei et al. (2021)
Surface urban heat island intensity (SUHII) *SUHII = LSTurban−LSTrural*	SUHII, like UHII, explicitly focuses on the temperature difference between rural and urban areas. Unlike UHII, SUHII concentrates solely on land surface temperature, with a specific emphasis on the thermal characteristics of surfaces	Siddiqui et al. (2021); Dewan et al. (2021)
Apparent temperature (AT)	The AT index combines wind speed, ambient temperature, and vapor pressure in both radiant and non-radiant forms to determine overall thermal resistance. It is utilized to predict physiological heat stress and assess human comfort levels	Fan et al. (2022); Niu et al. (2020)
Sky view factor (SVF) *SVF = (Visible Sky Area/Total Sky Area) * 100*	The SVF measures the portion of visible sky from a specific location, which evaluates how urban geography and structures affect the amount of shading and solar exposure	Chiang et al. (2023); Chen et al. (2023)
Composite urban heat island index (CUHI)	The CUHI index considers multiple factors, such as land cover, vegetation, and surface temperature, that are weighted to reflect the UHI intensity	Kumar et al. (2023); Dewan et al. (2021)
Greenness Index*Greenness Index = (NIR−Red)/(NIR + Red + Blue)*	The Greenness Index assesses vegetation cover in a particular area by analyzing the spectral reflectance in satellite imagery. Based on remote sensing and vegetation monitoring, the index accurately measures the presence and health of vegetation cover. It provides information on the potential cooling effect of green spaces on UHI	Lamchin et al. (2018)
Modified local climate zones (LCZ) index	The LCZ system, developed by the world meteorological organization and the international association for urban climate (IAUC), classifies land cover types into zones. The modified index combines land cover characteristics and differences in surface temperatures to understand UHI variations in different local climate zones	Eldesoky et al. (2021); Yang et al. (2020a, b)

Table 3.4 GIS and remote sensing (RS) tools for determining the UHI intensity

Tools	Methods and application	References
Land surface temperature (LST) mapping	LST data, collected by remote sensing satellites fitted with thermal infrared sensors, are processed and analyzed using GIS to produce spatial maps showing temperature variations within cities and locating UHI hotspots	Hussain et al. (2023); Li et al. (2023a, b)
Temporal analysis of LST	Time-series satellite imagery is used to track variations in LST over time. By analyzing temporal trends with GIS tools, one can comprehend how the intensity of UHI varies throughout the day, seasons, and over time	Kimothi et al. (2023)
Spatial analysis of land cover changes	High-resolution imagery is provided by remote sensing to track changes in land cover. Using GIS and spatial analysis, one can locate and quantify land cover changes from natural vegetation to impermeable surfaces over time, contributing to UHI	Ren et al. (2023); Mansour et al. (2020)
Normalized difference vegetation index (NDVI)	Remote sensing devices using visible and near-infrared light calculate the NDVI, which determines the vegetation health. Higher vegetation cover areas might have less impact from UHI, which can be visualized using GIS	Pei et al. (2021)
Urban morphology and layout analysis	Combined with GIS, high-resolution satellite or aerial imagery examines the form and structure of urban areas and how building density and arrangement affect the UHI intensity. Accordingly, based on urban morphology, UHI mitigation plans may be targeted	Musiaka and Nalej (2021); Wang et al. (2020a, b)
Development of urban heat island indices	Combining remote sensing-derived parameters and GIS, indices about LST, land cover, and impermeable surfaces assess the intensity of UHI	Halder et al. (2021)
Local climate zones (LCZ) classification	Remote sensing data and GIS help analyze the relationship between the categorized LCZs and the intensity of UHI for focused urban planning interventions	Quan and Bansal (2021)
3D Urban modeling	3D models of cities produced using LiDAR technology and remote sensing data simulate urban building morphology and visualize interactions to UHI using GIS	Hu et al. (2020); Wang et al. (2020a, b)
Spatial statistical analysis	The spatial autocorrelation and clustering of temperature data using GIS-based spatial statistics provide an understanding of localized UHI effects and identify hotspots and cold spots of UHI intensity	Sun et al. (2020)

such as sunlight, wind, and precipitation to mitigate the environmental impact of buildings. Given its significance in mitigating climate change, energy efficiency is paramount in this approach. Adopting sustainable building design practices becomes imperative, with the construction industry accounting for 36% of the world's energy consumed and 39% of CO_2 emissions produced globally (IEA 2020). Bioclimatic architecture aims to create healthy, comfortable buildings while upholding environmental sustainability (Nag 2019a, b). In other words, it entails crafting buildings and infrastructure surroundings that harmonize with the local climate, natural

Table 3.5 Thermo-physiological indices for quantification of UHI effects

Indices	Description	References
Predicted mean vote (PMV)	The PMV equation uses volunteers' thermal sensation votes and analyzes environmental, physiological, and behavioral parameters to provide a detailed score representing comfort levels in an environment. This score is displayed on a psychometric chart and is rated on a seven-point scale from hot to cold, reflecting individuals' well-being in a particular climate. The PMV and the predicted percentage of dissatisfied (PPD) individuals at each PMV level are incorporated into indoor thermal comfort standards such as ISO 7730:2005 and ASHRAE 55:2010. Additionally, other comfort standards, like exceedance metrics, indicate the duration of time that conditions remain within or outside the comfort range in a building	Fanger (1970)
Physiological equivalent temperature (PET)	The Munich energy balance model for individuals forms the basis for the physiologically equivalent temperature (PET), which provides the equivalent temperature of an indoor isothermal reference environment relative to an individual's personal experience and behavioral adaptation. PET assesses how hot a person feels and the level of strain they experience, taking into account their clothing, metabolic rate, and surroundings. This index is recommended in German guidelines for use by regional and urban planners	Sen and Nag (2019); Höppe (1999)
Universal thermal climate index (UTCI)	The UTCI combines meteorological parameters, clothing insulation, and human metabolic rate to estimate people's thermal perception in various environments	Blazejczyk et al. (2012)
Discomfort index (DI)	The DI index is more sophisticated and straightforward, called a temperature–humidity index. Based on the range, it can alert the public of possible heat-related discomfort in weather forecasting and public health	Nurmaya et al. (2022)
Heat stress index (HSI)	Heat stress index (HSI) evaluates wind speed, air temperature, solar radiation, humidity, and other factors to determine the risk of heat-related illnesses. The index has wide applications in heat stress prevention measures at workplaces	Zhang et al. (2023)
Wet-bulb globe temperature (WBGT)	The WBGT, as a composite index, evaluates heat stress in outdoor and work environments. The index has been incorporated as an ISO 7243 (2003) standard to optimize heat-related risks to different exposures	Brimicombe et al. (2023)

surroundings and resources, fostering energy efficiency, thermal comfort, and overall well-being (Nag 2019a, b; Watson 2020), avoiding environmentally harmful materials, promoting local biodiversity, and optimizing energy, water, and material usage (Bugenings and Kamari 2022; Chiesa 2021).

The bioclimatic design stands out as a sustainable and effective strategy to address the negative ramifications of the UHI phenomenon, underscoring the importance of implementing efficient mitigation measures. Such principles involve integrating green spaces, permeable surfaces, and energy-saving technologies into urban planning and architecture. For example,

Green Roofs and Walls: Implementing greenery on rooftops and walls can decrease surface temperatures by providing shading and evapotranspiration. Vegetation is a natural insulator, absorbing and obstructing solar radiation and lowering building temperatures. Green roofs serve as passive cooling systems by utilizing heat transfer principles like radiation, conduction, convection, and evapotranspiration. The vegetation layer intercepts solar radiation, reflecting some and providing shade, which reduces heat absorption by the building. The soil and substrate further insulate by minimizing heat conduction. Additionally, evapotranspiration cools the air as plants release water vapor, drawing energy from the surroundings and mitigating the urban heat island effect. Overall, green roofs enhance energy efficiency, lower surface temperatures, and improve thermal comfort in urban areas (Harbiankova and Manso 2025).

Urban Green Spaces and Landscaping: Introducing parks, green belts, and open areas within urban settings provides shade and facilitates natural cooling of the surrounding air through transpiration and shading. These spaces promote biodiversity and contribute to overall environmental health. Urban green spaces and landscaping serve as natural cooling systems by managing heat transfer through processes like radiation, convection, and evapotranspiration. Vegetation in parks and open areas intercepts solar radiation, reducing heat absorption by concrete and asphalt. Trees provide shade, lowering ground and air temperatures. Evapotranspiration, the release of water vapor from leaves combined with soil evaporation, absorbs heat and cools

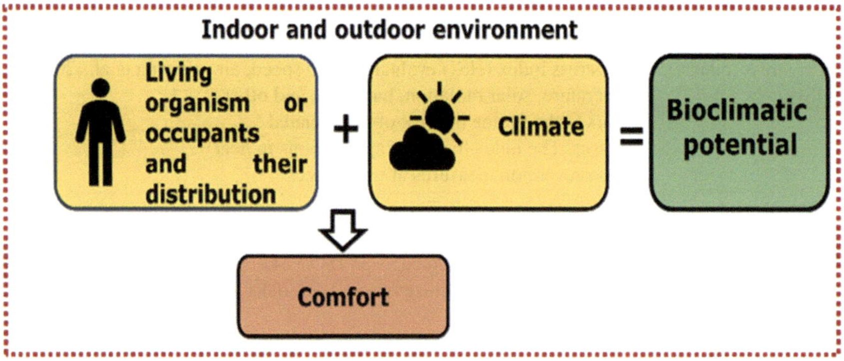

Fig. 3.2 Schematics of bioclimatic potential

the surrounding air. Additionally, greenery affects air movement, enhancing airflow and promoting heat dissipation in urban areas. By leveraging these processes, urban green spaces mitigate the urban heat island effect and improve overall environmental sustainability (Wu et al. 2025).

Permeable Surfaces: Utilizing permeable materials for pavements and surfaces enables water infiltration, which reduces surface temperatures and heat retention. Conversely, buildings with high thermal mass materials, such as stone, absorb and store heat during the day and release it at nighttime gradually. Permeable surfaces help regulate urban temperatures by utilizing heat transfer, water infiltration, and evaporative cooling. Unlike traditional impermeable materials, these pavements allow rainwater to seep through, reducing runoff and increasing ground moisture. This moisture aids in evaporative cooling, lowering surface temperatures by drawing heat from the surroundings. Additionally, permeable surfaces retain less heat than asphalt or concrete, which contributes to the urban heat island effect. Buildings with high thermal mass materials, like stone or concrete, further stabilize temperatures by absorbing heat during the day and releasing it at night. Combining permeable surfaces with thermal mass strategies can enhance urban thermal regulation and resilience against excessive heat (Zhao et al. 2025).

Natural Ventilation: Designing buildings to encourage natural ventilation reduces dependence on air conditioning, lowering energy consumption and mitigating heat generation. Passive design promotes natural ventilation by strategically placing windows, vents, and openings. Orienting buildings to capture prevailing winds enhances natural airflow (Bera and Nag 2022).

Shading Strategies: Employing passive shading methods, such as incorporating appropriate overhangs in building design and proper orientation in the placement of windows and shading devices, can prevent direct sunlight from entering indoor spaces, thus reducing the need for artificial cooling.

Cool and Reflective Materials: Choosing cool roofing materials with high solar reflectance, infrared emittance, and reflective exterior surfaces, such as light-colored pavements and walls, helps minimize heat absorption and sunlight reflection, thereby mitigating UHI effects.

Energy-Efficient Design: Integrating energy-efficient technologies like solar panels and efficient HVAC systems reduces heat generation from energy usage.

Urban Planning: Strategically planning and designing urban areas to optimize building layout and arrangement enhances airflow and reduces the concentration of heat-absorbing surfaces.

Incorporating the stated bioclimatic design strategies allows cities to alleviate the adverse effects of the UHI phenomenon, advance sustainability objectives, and establish more habitable and resilient environments. Moreover, these strategies align with broader aims of environmental preservation and adaptation to climate change. Furthermore, passive design techniques employing natural means to cool and regulate temperatures within buildings and urban areas substantially curtail energy consumption and reduce dependence on mechanical systems, thus curbing heat generation and, fostering more sustainable urban landscapes and counteracting UHI's adverse effects (Aghimien et al. 2022). Various tools and techniques (Table 3.6) have been

Table 3.6 Tools and techniques to identify appropriate bioclimatic design strategies

Toots/description	References
Climate Analysis Software, such as climate consultant, designbuilder, and EnergyPlus allow designers to analyze local climate data—such as temperature variations, solar radiation, wind patterns, and humidity levels	Keibach and Shayesteh (2022); Bera and Nag (2022)
Sun path diagrams illustrate the sun's path across the sky at different times of the day and throughout the year, thus helping designers understand solar exposure and shading requirements and optimize building orientation and placement	Yousef et al. (2023)
Wind Tunnel Testing, physical or computational wind tunnel testing, helps assess local wind patterns around buildings and landscapes. This information helps design natural ventilation strategies and optimize building forms for wind comfort	Li et al. (2023a, b)
Thermal Simulation Software, such as Autodesk Ecotect, IES VE, and DesignBuilder, have thermal simulation capabilities. Designers can simulate building performance under different conditions to evaluate passive design strategies such as natural ventilation, shading, and thermal mass	Xue et al. (2024); Fernandez-Antolin et al. (2022)
Daylighting Analysis Tools like DIALux, radiance, and DIVA provide information about building daylighting conditions and assist in optimizing fenestration design for natural light penetration and minimizing solar heat gain	Aguilar-Carrasco et al. (2023)
Green Building Certification, like BREEAM and LEED, incorporate bioclimatic design principles and define guidelines and criteria for sustainable building design, encouraging passive strategies	Mazur et al. (2023); Nag (2019a, b)
Climate-based design Guidelines and standards, such as ASHRAE Standard 169 and ASHRAE Climate Zones, provide climate-specific recommendations for building design	Wang et al. (2023a, b)

employed to ascertain the local conditions, building necessities and suitability of bioclimatic design strategies. Applying bioclimatic design strategies to mitigate the UHI effect can be tailored to different countries' and regions' specific climatic conditions, building practices, and urban contexts, as given in Table 3.7.

3.5 Conclusion and Future Perspectives to Mitigate the Urban Heat Island

This contribution aimed to bring a comprehensive understanding of the pressing issue of the causes and consequences of UHI, emphasizing urban infrastructures and advocating bioclimatic design strategies to create eco-friendly and people-friendly urban environments combating UHI effects. The contribution encompasses several areas of importance, such as (a) factors that influence the development of the UHI, (b) indices

Table 3.7 Bioclimatic design strategies applied to address UHI in different countries

Countries	Design strategies	References
Australia	The climates of Australia range from temperate in the south to tropical in the north. Cool roofing, passive solar design, and native landscaping are examples of addressing climate challenges and applying energy-efficient solutions	Ozarisoy and Altan (2021); McNeilly Smith et al. (2023)
Brazil	Using natural ventilation, green building envelopes, and adding water bodies are some urban design solutions in tropical areas. Vegetation reduces heat absorption and improves thermal comfort	Gobo et al. (2022)
China	A diverse climate demands different bioclimatic design strategies. An emphasis on natural ventilation, green spaces, and reflective surfaces aids heat management in hot and humid regions (southern China). For energy-efficient heating in colder climates (northern China), thermal mass and passive solar design are essential	Daemei et al. (2019); Yang et al. (2020a, b)
Germany	Passive solar design for winter heating, effective insulation, green roofs, and the like help promote energy efficiency in temperate climates where winters can be very harsh	Schmid and Säumel (2021)
India	In India's tropical and subtropical climates, cool roofing, passive solar design, and optimizing natural ventilation are effective solutions. Water features and urban green areas also help lessen UHI effects	Bera and Nag (2022); Bera and Nag (2021)
Japan	Natural ventilation, building orientation, green roofs, and conventional architectural elements like sliding doors for flexible control of indoor environments are effective designs in Japan with diverse climates, from humid subtropical to snowy winters	Bienvenido-Huertas et al. (2020)
Singapore	High temperatures and humidity are commonplace in Singapore, thus essentially requiring solutions to reduce heat absorption and improve evaporative cooling. Designing buildings with elevated thermal mass, using shading devices, and adding greenery throughout urban areas are helpful	Yang et al. (2018)

(continued)

Table 3.7 (continued)

Countries	Design strategies	References
United Arab Emirates (UAE)	Cool roofing, shading techniques, and adding water features are practical solutions for evaporative cooling in desert climates, where intense heat is a significant concern. Initiatives for green building and sustainable urban planning help reduce UHI	Bande et al. (2021); Rahmani et al. (2023)
United States	Passive cooling techniques (cool roofs, shading devices, and low-water landscaping) help mitigate heat absorption and lower energy demand in hot and arid regions of the southwestern United States (e.g., Arizona, Nevada). Natural ventilation, reflective surfaces, and green infrastructure (green walls and roofs) help reduce UHI effects in hot-humid cities (Miami and New Orleans)	Aparicio-Ruiz et al. (2018)

to quantify the UHI intensity and its effects, including GIS and remote sensing tools and thermo-physiological indices, and (c) tools and techniques to identify bioclimatic design strategies applied in different countries to mitigate UHI effects.

One would recognize that the UHI effect intersects with several UN Sustainable Development Goals (UN 2023). UHI denotes the situation where urban areas exhibit significantly higher temperatures with adverse impacts on peoples' living conditions, such as increased energy consumption, heightened emission levels, compromised human health, and disturbance of urban ecosystems (SDG 11, Sustainable Cities and Communities). Elevated temperatures can exacerbate pre-existing health conditions and lead to heat-related illnesses (SDG 3, Good Health and Well-Being). Given that urban warming exacerbates global warming and necessitates mitigation efforts, SDG 13 (Climate Action) is pertinent to the UHI effect toward promoting cooler, greener urban designs and supporting biodiversity (SDG 15, Life on Land).

Future research should focus on developing adaptive bioclimatic strategies that consider local climate variations, urban morphology, and socio-economic factors. Utilizing data-driven approaches, like AI-powered climate modeling and real-time urban heat monitoring, can enhance Urban Heat Island (UHI) assessments and guide targeted mitigation efforts. Collaboration among urban planners, policymakers, and climate scientists is vital for translating research into practical solutions. Additionally, implementing pilot projects and long-term monitoring in diverse urban environments will help validate these strategies and support evidence-based decision-making for sustainable urban development.

On the one hand, exploring the temporal aspects of UHI, including its evolution over time and under different climate change scenarios, would provide a more holistic understanding of its dynamics. On the other hand, practical considerations, economic constraints, and potential conflicts with existing urban infrastructure need addressing for a realistic view of strategy feasibility for the acceptance and efficacy of bioclimatic design. Effectively addressing the challenges through concerted efforts from public and private stakeholders would strengthen a more comprehensive approach aiming to mitigate UHI impacts.

AI Discloser: The authors declare that no artificial intelligence (AI) tools were used in the writing, editing, data analysis, or content generation of this book chapter. All work presented is entirely the result of the authors' own effort and scholarly contribution.

Acknowledgements The authors would like to express sincere gratitude to Revered Swami Atmapriyanandaji, Pro-Chancellor of Ramakrishna Mission Vivekananda Educational and Research Institute (RKMVERI), West Bengal, India, for his unwavering encouragement and inspiration. We are also profoundly grateful to Revered Swami Shivapurnanandaji, Assistant Administrative Head, RKMVERI, Narendrapur campus, Kolkata, West Bengal, India, for his inspiration and support throughout this work. Our heartfelt appreciation goes to all faculty members and staff at RKMVERI, Narendrapur campus, for their support, collaboration, and the academic environment that significantly aided the completion of this study. We also acknowledge with gratitude the contributions of the scientific community, whose remarkable research and dedication have provided the foundation upon which this chapter stands.

Authors' Contribution: *Mahadev Bera:* Conceptualisation, Literature Review, Methodology, Validation, Visualisation, Resources, Project Administration, Writing-Original Draft, Writing-Review & Editing; *Pranab Kumar Nag:* Conceptualisation, Methodology, Supervision, Writing-Review & Editing.

References

Abdulateef MF, Al-Alwan HA (2022) The effectiveness of urban green infrastructure in reducing surface urban heat island. Ain Shams Eng J 13(1):101526

Abrar R, Sarkar SK, Nishtha KT, Talukdar S, Shahfahad Rahman A, Mosavi A (2022) Assessing the spatial mapping of heat vulnerability under Urban Heat Island (UHI) Effect in the Dhaka metropolitan area. Sustainability 14(9):4945

Aghimien EI, Li DHW, Tsang EKW (2022) Bioclimatic architecture and its energy-saving potentials: a review and future directions. Eng Constr Archit Manag 29(2):961–988

Aguilar-Carrasco MT, Díaz-Borrego J, Acosta I, Campano MÁ, Domínguez-Amarillo S (2023) Validation of lighting parametric workflow tools of Ladybug and Solemma using CIE test cases. J Build Eng 64:105608

Al-Saadi LM, Jaber SH, Al-Jiboori MH (2020) Variation of urban vegetation cover and its impact on minimum and maximum heat islands. Urban Clim 34:100707

Aparicio-Ruiz P, Schiano-Phan R, Salmerón-Lissén JM (2018) Climatic applicability of downdraught evaporative cooling in the United States of America. Build Environ 136:162–176

Bande L, Adan R, Young K, Ghazal R, Jha M, Aldarmaki A, Khalifa M (2021) Outdoor thermal comfort study on a district level as part of the housing programs in Abu Dhabi and Al Ain United Arab Emirates. Land 10(3):264

Bera M, Nag PK (2021) Bioclimate in built environment. Ergon Int J 5(5):000277

Bera M, Nag PK (2022) Bioclimatic design of low-cost rural dwellings. Front Built Environ 8:773108

Bera M, Nag PK, Das S (2023) Quantification of bioclimatic performance of rural coastal low cost dwellings in the sundarbans. Agric Rural Stud 1(3):0015. https://doi.org/10.59978/ar01030015

Bera M, Das S, Nag PK (2024a) Quantification of the interactions between indoor environmental conditions and human thermal comfort and susceptibility in tropical coastal environments. Indoor Built Environ 33(9):1624–1644

Bera M, Das S, Dutta S, Nag PK, Roy Choudhury M (2024b) Progress and challenges of bioclimatic design strategies for enhancing building environmental performance: a systematic review. Smart Sustain Built Environ. https://doi.org/10.1108/SASBE-07-2024-0250

Bienvenido-Huertas D, Pulido-Arcas JA, Rubio-Bellido C, Perez-Fargallo A (2020) Influence of future climate changes scenarios on the feasibility of the adaptive comfort model in Japan. Sustain Cities Soc 61:102303

Blazejczyk K, Epstein Y, Jendritzky G, Staiger H, Tinz B (2012) Comparison of UTCI to selected thermal indices. Int J Biometeorol 56:515–535

Bolay JC, Bolay JC (2020) Urban facts. Urban planning against poverty: how to think and do better cities in the global south, pp 7–55

Bressane A, dos Santos Galvão AL, Loureiro AIS, Ferreira MEG, Monstans MC, de Castro Medeiros LC (2024) Valuing urban green spaces for enhanced public health and sustainability: a study on public willingness-to-pay in an emerging economy. Urban for Urban Green 98:128386

Briciu AE, Mihăilă D, Graur A, Oprea DI, Prisăcariu A, Bistricean PI (2020) Changes in the water temperature of rivers impacted by the urban heat island: case study of Suceava city. Water 12(5):1343

Brimicombe C, Lo CHB, Pappenberger F, Di Napoli C, Maciel P, Quintino T, Cloke HL (2023) Wet bulb globe temperature: indicating extreme heat risk on a global grid. GeoHealth 7(2):e2022GH000701

Budhiraja B, Agrawal G, Pathak P (2020) Urban heat island effect of a polynuclear megacity Delhi-compactness and thermal evaluation of four sub-cities. Urban Climate 32:100634

Bugenings LA, Kamari A (2022) Bioclimatic architecture strategies in Denmark: a review of current and future directions. Buildings 12(2):224

Chen Q, Liu R, Cheng Q, Chen Y, Cao S, Du M, Li K (2023) Evaluating the impact of sky view factor and building shadow ratio on air temperature in different residential and commercial building scenarios: a case study of Beijing China. Urban Climate 49:101509

Chiang YC, Liu HH, Li D, Ho LC (2023) Quantification through deep learning of sky view factor and greenery on urban streets during hot and cool seasons. Landsc Urban Plan 232:104679

Chiesa G (ed) (2021) Bioclimatic approaches in urban and building design. Springer, Berlin/Heidelberg, Germany

Chowdhury S, Noguchi M, Doloi H (2023) Methodological approach of environmental experience design to enhancing occupants' well-being Bangladesh. Buildings 13(2):542

Cristiano S, Ulgiati S, Gonella F (2021) Systemic sustainability and resilience assessment of health systems, addressing global societal priorities: learnings from a top nonprofit hospital in a bioclimatic building in Africa. Renew Sustain Energy Rev 141:110765

Daemei AB, Eghbali SR, Khotbehsara EM (2019) Bioclimatic design strategies: a guideline to enhance human thermal comfort in Cfa climate zones. J Build Eng 25:100758

Dewan A, Kiselev G, Botje D, Mahmud GI, Bhuian MH, Hassan QK (2021) Surface urban heat island intensity in five major cities of Bangladesh: patterns, drivers and trends. Sustain Cities Soc 71:102926

Eldesoky AH, Gil J, Pont MB (2021) The suitability of the urban local climate zone classification scheme for surface temperature studies in distinct macroclimate regions. Urban Climate 37:100823

Fan N, Reichstein M, Koirala S, Ahrens B, Mahecha MD, Carvalhais N (2022) Global apparent temperature sensitivity of terrestrial carbon turnover modulated by hydrometeorological factors. Nat Geosci 15(12):989–994

Fanger PO (1970) Thermal comfort. Analysis and applications in environmental engineering. Danish Technical Press, Copenhagen

Feng R, Wang F, Wang K, Wang H, Li L (2021) Urban ecological land and natural-anthropogenic environment interactively drive surface urban heat island: an urban agglomeration-level study in China. Environ Int 157:106857

Fernandez-Antolin MM, del Río JM, Gonzalez-Lezcano RA (2022) Building performance simulation tools as part of architectural design: breaking the gap through software simulation. Int J Technol des Educ 32(2):1227–1245

Gobo JPA, Wollmann CA, Celuppi MC, Galvani E, Faria MR, Mendes D, Gonçalves FLT (2022) The bioclimate present and future in the state of São Paulo/Brazil: space-time analysis of human thermal comfort. Sustain Cities Soc 78:103611

Halder B, Bandyopadhyay J, Banik P (2021) Monitoring the effect of urban development on urban heat island based on remote sensing and geo-spatial approach in Kolkata and adjacent areas India. Sustain Cities Soc 74:103186

Hayes AT, Jandaghian Z, Lacasse MA, Gaur A, Lu H, Laouadi A, Wang L (2022) Nature-based solutions (nbss) to mitigate urban heat island (UHI) effects in Canadian cities. Buildings 12(7):925

Harbiankova A, Manso M (2025) Integrating green roofs and green walls to enhance buildings thermal performance: a literature review. Build Environ 112524

He BJ, Wang J, Zhu J, Qi J (2022) Beating the urban heat: situation, background, impacts and the way forward in China. Renew Sustain Energy Rev 161:112350

Hidalgo-García D, Arco-Díaz J (2022) Modeling the surface urban heat island (SUHI) to study of its relationship with variations in the thermal field and with the indices of land use in the metropolitan area of Granada (Spain). Sustain Cities Soc 87:104166

Hong T, Heo Y (2023) Exploring the impact of urban factors on land surface temperature and outdoor air temperature: a case study in Seoul Korea. Build Environ 243:110645

Höppe P (1999) The physiological equivalent temperature–a universal index for the biometeorological assessment of the thermal environment. Int J Biometeorol 43(2):71–75

Hu Y, Dai Z, Guldmann JM (2020) Modeling the impact of 2D/3D urban indicators on the urban heat island over different seasons: a boosted regression tree approach. J Environ Manage 266:110424

Hua L, Zhang X, Nie Q, Sun F, Tang L (2020) The impacts of the expansion of urban impervious surfaces on urban heat islands in a coastal city in China. Sustainability 12(2):475

Huang S, Tang L, Hupy JP, Wang Y, Shao G (2021) A commentary review on the use of normalized difference vegetation index (NDVI) in the era of popular remote sensing. J for Res 32(1):1–6

Hussain S, Mubeen M, Ahmad A, Majeed H, Qaisrani SA, Hammad HM, Nasim W (2023) Assessment of land use/land cover changes and its effect on land surface temperature using remote sensing techniques in Southern Punjab Pakistan. Environ Sci Pollut Res 30(44):99202–99218

IEA (2020) Tracking Buildings 2020. IEA, Paris https://www.iea.org/reports/tracking-buildings-2020

Irie T (2022) The cooling effect of green infrastructure in mitigating nocturnal urban heat islands: a case study of Yoyogi Park and Meiji Jingu Shrine in Tokyo. Landsc Res 47(5):559–583

Jamei Y, Rajagopalan P, Sun QC (2019) Spatial structure of surface urban heat island and its relationship with vegetation and built-up areas in Melbourne, Australia. Sci Total Environ 659:1335–1351

Jung Y (2024) Urban heat islands and the transformation of Singapore. Urban Stud 00420980231217391

Keibach E, Shayesteh H (2022) BIM for landscape design improving climate adaptation planning: the evaluation of software tools based on the ISO 25010 standard. Appl Sci 12(2):739

Kim SW, Brown RD (2021) Urban heat island (UHI) intensity and magnitude estimations: a systematic literature review. Sci Total Environ 779:146389

Kim Y, Yu S, Li D, Gatson SN, Brown RD (2022) Linking landscape spatial heterogeneity to urban heat island and outdoor human thermal comfort in Tokyo: application of the outdoor thermal comfort index. Sustain Cities Soc 87:104262

Kimothi S, Thapliyal A, Gehlot A, Aledaily AN, Bilandi N, Singh R, Akram SV (2023) Spatio-temporal fluctuations analysis of land surface temperature (LST) using remote sensing data

(LANDSAT TM5/8) and multifractal technique to characterize the urban heat Islands (UHIs). Sustain Energy Technol Assess 55:102956
Kumar A, Mukherjee M, Goswami A (2023) Inter-seasonal characterization and correlation of surface urban heat island (SUHI) and canopy urban heat island (CUHI) in the urbanized environment of Delhi. Remote Sens Appl Soc Environ 30:100970
Lamchin M, Lee WK, Jeon SW, Wang SW, Lim CH, Song C, Sung M (2018) Long-term trend and correlation between vegetation greenness and climate variables in Asia based on satellite data. Sci Total Environ 618:1089–1095
Li Y, Schubert S, Kropp JP, Rybski D (2020) On the influence of density and morphology on the urban heat island intensity. Nat Commun 11(1):2647
Li Y, Zhu Y, Chen FB, Li QS (2023) Aerodynamic loads of tapered tall buildings: insights from wind tunnel test and CFD. In: Structures vol 56. Elsevier, p 104975
Li ZL, Wu H, Duan SB, Zhao W, Ren H, Liu X, Zhou C (2023) Satellite remote sensing of global land surface temperature: definition, methods, products, and applications. Rev Geophys 61(1):e2022RG000777
Mansour S, Al-Belushi M, Al-Awadhi T (2020) Monitoring land use and land cover changes in the mountainous cities of Oman using GIS and CA-Markov modelling techniques. Land Use Policy 91:104414
Mansouri ST, Zarghami E (2023) Investigating the effect of the physical layout of the architecture of high-rise buildings, residential complexes, and urban heat islands. Energy Built Environ
Mazur Ł, Resler M, Koda E, Walasek D, Vaverková MD (2023) Energy saving and green building certification: case study of commercial buildings in Warsaw Poland. Sustain Energy Technol Assess 60:103520
McNeilly Smith R, Tavares S, Stevens N (2023) Urban design and planning for extreme heat: an empirical study of built environment professionals' perceptions in South East Queensland, Australia. Cities Health 1–13
Meade E, Garvey M (2022) The role of neuro-immune interaction in chronic pain conditions; functional somatic syndrome, neurogenic inflammation, and peripheral neuropathy. Int J Mol Sci 23(15):8574
Meerow S, Helmrich AM, Andrade R, Larson KL (2021) How do heat and flood risk drive residential green infrastructure implementation in Phoenix, Arizona? Urban Ecosystems 1–12
Mirabi E, Davies PJ (2022) A systematic review investigating linear infrastructure effects on Urban Heat Island (UHIULI) and its interaction with UHI typologies. Urban Climate 45:101261
Morabito M, Crisci A, Guerri G, Messeri A, Congedo L, Munafò M (2021) Surface urban heat islands in Italian metropolitan cities: tree cover and impervious surface influences. Sci Total Environ 751:142334
Moreira AB, Duarte CC, Matzarakis A (2023) A review of urban heat studies in recife—PE (Brazil): urban heat scales landscape inequalities and future perspectives. Atmosphere 14(11):1596
Musiaka Ł, Nalej M (2021) Application of GIS tools in the measurement analysis of urban spatial layouts using the square grid method. ISPRS Int J Geo Inf 10(8):558
Nag PK (2019) Bioclimatic approach: thermal environment. In: Office buildings. design science and innovation. Springer, Singapore. https://doi.org/10.1007/978-981-13-2577-9_9
Nag PK (2019) Green building and assessment systems. In: Office buildings. Design science and innovation. Springer, Singapore. https://doi.org/10.1007/978-981-13-2577-9_15
Niu Y, Gao Y, Yang J, Qi L, Xue T, Guo M, Liu Q (2020) Short-term effect of apparent temperature on daily emergency visits for mental and behavioral disorders in Beijing, China: a time-series study. Sci Total Environ 733:139040
Nurmaya EM, Abidin AU, Hasanah NAI, Asmara AA (2022) Heat stress analysis using the discomfort index method: impact on macro environmental in yogyakarta. J Ecol Eng 23(1)
Olgyay V (1963) Design with climate. Princeton Univ Press, New Jersey, USA
Ozarisoy B, Altan H (2021) Systematic literature review of bioclimatic design elements: theories, methodologies and cases in the south-eastern mediterranean climate. Energy Build 250:111281

Pei F, Zhou Y, Xia Y (2021) Application of normalized difference vegetation index (NDVI) for the detection of extreme precipitation change. Forests 12(5):594

Piracha A, Chaudhary MT (2022) Urban air pollution, urban heat island and human health: a review of the literature. Sustainability 14(15):9234

Quan SJ, Bansal P (2021) A systematic review of GIS-based local climate zone mapping studies. Build Environ 196:107791

Rahaman ZA, Kafy AA, Saha M, Rahim AA, Almulhim AI, Rahaman SN, Al Rakib A (2022) Assessing the impacts of vegetation cover loss on surface temperature, urban heat island and carbon emission in Penang city Malaysia. Build Environ 222:109335

Rahmani M, Al-Sallal KA, Alkhatib O, Khoukhi M (2023) Potential of form-pocketed design in mitigating heat in a university educational glazed-curtained building in the desert climate of UAE. Energy Build 297:113424

Ren J, Yang J, Wu F, Sun W, Xiao X, Xia JC (2023) Regional thermal environment changes: integration of satellite data and land use/land cover. Iscience 26(2)

Schmid HL, Säumel I (2021) Outlook and insights: perception of residential greenery in multistorey housing estates in Berlin Germany. Urban for Urban Green 63:127231

Sen J, Nag PK (2019) Human susceptibility to outdoor hot environment. Sci Total Environ 649:866–875

Siddiqui A, Kushwaha G, Nikam B, Srivastav SK, Shelar A, Kumar P (2021) Analyzing the day/night seasonal and annual changes and trends in land surface temperature and surface urban heat island intensity (SUHII) for Indian cities. Sustain Cities Soc 75:103374

Sun S, Zhu J, Zhou X (2020) Statistical analysis of spatial expression patterns for spatially resolved transcriptomic studies. Nat Methods 17(2):193–200

Tamaskani Esfehankalateh A, Farrokhzad M, Tamaskani Esfehankalateh F, Soflaei F (2022) Bioclimatic passive design strategies of traditional houses in cold climate regions. Environ Dev Sustain 1–42

Ulpiani G (2021) On the linkage between urban heat island and urban pollution island: three-decade literature review towards a conceptual framework. Sci Total Environ 751:141727

UN (2023) The sustainable development goals report (Special edition). https://desapublications.un.org/publications/sustainable-development-goals-report-2023-special-edition

Vinayak B, Lee HS, Gedam S, Latha R (2022) Impacts of future urbanization on urban microclimate and thermal comfort over the Mumbai metropolitan region India. Sustain Cities Soc 79:103703

Wang C, Ren Z, Dong Y, Zhang P, Guo Y, Wang W, Bao G (2022) Efficient cooling of cities at global scale using urban green space to mitigate urban heat island effects in different climatic regions. Urban for Urban Green 74:127635

Wang C, Wen C, Dai Y, Yu S, Liu M (2020a) Urban 3D modeling with mobile laser scanning: a review. Virtual Rity Intell Hardw 2(3):175–212

Wang M, Wang Y, Geng M, Shang S, Zhang B, Liu L, Wen G (2023a) Effect of cold storage backfill on thermal comfort of mine stope based on standard effective temperature. Energy Build 300:113632

Wang S, Yang L, Meng X, Lyu K, Cao Q, Liu Y (2023) Can the ASHRAE standard 169 zoning method be applied to country-level energy-efficient building design in China?. In: Building simulation. Tsinghua University Press, Beijing, pp 1–18

Wang X, Song Y, Tang P (2020b) Generative urban design using shape grammar and block morphological analysis. Front Arch Res 9(4):914–924

Watson D (2020) Bioclimatic design. In: Sustainable built environments. Springer US, New York, NY, pp 19–41

Wu Z, Wang Y, Ren Y (2025) Optimizing green space-building landscape characteristics of key urban functional zones for comprehensive thermal environment mitigation. Landsc Urban Plan 257:105314

Xue W, Wang Y, Liang Y, Wang T, Ren B (2024) Efficient hydraulic and thermal simulation model of the multi-phase natural gas production system with variable speed compressors. Appl Therm Eng 242:122411

Yang J, Pyrgou A, Chong A, Santamouris M, Kolokotsa D, Lee SE (2018) Green and cool roofs' urban heat island mitigation potential in tropical climate. Sol Energy 173:597–609

Yang J, Zhan Y, Xiao X, Xia JC, Sun W, Li X (2020a) Investigating the diversity of land surface temperature characteristics in different scale cities based on local climate zones. Urban Climate 34:100700

Yang L, Fu R, He W, He Q, Liu Y (2020b) Adaptive thermal comfort and climate responsive building design strategies in dry–hot and dry–cold areas: case study in Turpan China. Energy Build 209:109678

Yin Z, Liu Z, Liu X, Zheng W, Yin L (2023) Urban heat islands and their effects on thermal comfort in the US: New York and New Jersey. Ecol Ind 154:110765

Yousef BA, Radwan A, Olabi AG, Abdelkareem MA (2023) Sun composition, solar angles, and estimation of solar radiation. In: Renewable energy-volume 1: solar, wind, and hydropower. Academic Press, pp 3–22

Zhang G, Han L, Yao J, Yang J, Xu Z, Cai X,... & Pei, L. (2023). Assessing future heat stress across China: combined effects of heat and relative humidity on mortality. *Frontiers in Public Health, 11.*

Zhao J, Lu J, Ge J, Fan Y, Wang H, Gu M, Luo X (2025) Influences of permeable pavements with different hydraulic properties on evaporative cooling and outdoor thermal environment: Field experiments. Build Environ 270:112525

Zhong W, Schröder T, Bekkering J (2022) Biophilic design in architecture and its contributions to health, well-being, and sustainability: A critical review. Frontiers of Architectural Research 11(1):114–141

Chapter 4
Investigating Urban Heat Island Effects and Mitigation Strategies in Ahmedabad, India

Pradeep Kumar Rajput and Durgesh Singh

Abstract Urban heat islands (UHIs) are urban areas that experience significantly higher temperatures compared to their rural surroundings, primarily due to anthropogenic activities and unplanned urban development. This temperature disparity is often more pronounced at night and during the summer and winter months. The key drivers of UHIs include the disruption of natural settings and modifications to land surface areas. This study investigates the relationship between land surface temperature (LST) and biophysical parameters in a selected urban area. Satellite data from 1990, 2000, and 2019 were utilized to retrieve LST and biophysical parameters such as the Normalized Difference Vegetation Index (NDVI), Normalized Difference Water Index (NDWI), Normalized Difference Built-up Index (NDBI), and Normalized Difference Built-up and Bareness Index (NDBaI). The analysis focused on understanding the temporal changes in these indices and their influence on LST. The results indicate a strong negative correlation between LST and NDVI across all time periods, signifying that higher LST levels are associated with lower vegetation cover. In 1990, the correlation coefficient between LST and NDVI was − 0.89362, while positive correlations were observed between LST and other indices such as NDBI (0.996885), NDBaI (0.9991), and NDWI (0.9825). Similar trends were noted for 2000, with an even stronger negative correlation between LST and NDVI (− 0.97869) and positive correlations with NDBI (0.952897), NDBaI (0.993283), and NDWI (0.960556). By 2019, the correlation between LST and NDVI remained significantly negative (− 0.9411), and the positive correlations with NDBI (0.94735), NDBaI (0.991199), and NDWI (0.933233) persisted.

Keywords Urban planning · LST · Vegetation · Environmental parameters · Regional planning

P. K. Rajput (✉) · D. Singh
Department of Geography, School of Basic Science, Chhatrapati Shahu Ji Maharaj University, Kanpur, UP, India
e-mail: pkrseven@gmail.com

K. Mukherjee et al. (eds.), *Urban Climate*, The Urban Book Series,
https://doi.org/10.1007/978-3-032-17895-4_4

4.1 Introduction

An urban heat island (UHI) is an urban or metropolitan area that experiences significantly higher temperatures than its surrounding rural areas due to human activities. The temperature difference is usually more pronounced at night than during the day and is most noticeable when winds are weak. UHI effects are most significant during summer and winter (Carlson 1986). The alteration and use of land surfaces is the primary cause of the urban heat island effect; waste heat from energy use is a secondary factor. As urban populations grow, the area of UHIs expands and their average temperature increases (Mustafa et al. 2020). Essentially, a heat island refers to any area, whether populated or not, that is consistently hotter than its surroundings. According to a United Nations report, rapid urbanization and globalization have increased the world's urban population from around 33% fifty years ago to over 54% today, with projections of up to 66% by 2050 (Bokaie et al. 2016). The report also suggests that Asian and African countries will experience faster rates of urbanization compared to other continents. While urbanization brings prosperity and development, it also negatively impacts the global ecological environment. Among the environmental effects of rapid urbanization, the urban thermal environment, exemplified by the UHI phenomenon, has become a significant urban environmental issue (Ozelkan et al. 2014). Research in the United States shows a relationship between extreme temperatures and mortality, with heat posing a greater risk in northern cities than in southern regions. Concerns have been raised about the potential contribution of UHIs to global warming. Studies from China and India indicate that the UHI effect contributes to climate warming by about 30%. However, a 1999 study comparing urban and rural areas found that UHI effects have little impact on global mean temperature trends (Khandelwal et al. 2018). Many studies indicate that the severity of UHIs increases with the progression of climate change. Research has shown a relationship between UHIs and patterns of land cover changes, with vegetation and water presence reducing UHI intensity, while increased urbanization intensifies it. Various indices such as the Normalized Difference Vegetation Index (NDVI), Soil-Adjusted Vegetation Index (SAVI), Normalized Difference Water Index (NDWI), and Normalized Difference Built-up Index (NDBI) have been used to represent land cover changes over analyzed time periods (Chen et al. 2022). The Normalized Difference Vegetation Index (NDVI) is a simple graphical indicator used to analyze remote sensing measurements, typically from space platforms, to assess the presence of live green vegetation in a target location. To improve upon NDVI, Huete developed the Soil-Adjusted Vegetation Index (SAVI), which accounts for the differential extinction of red and near-infrared light through vegetation canopies. SAVI minimizes soil brightness influences in spectral vegetation indices involving red and near-infrared (NIR) wavelengths. The Normalized Difference Water Index (NDWI) is used to monitor changes related to water content in water bodies and to assess whether a target location is experiencing floods or scarcity. The Difference Normalized Urban areas tend to show higher reflectance in the shortwave-infrared (SWIR) region than in the near-infrared (NIR) region; this is why the Built-up Index (NDBI) emphasizes these areas

(Engel et al. 2023). The surface and atmosphere's thermal, radiative, moisture, and aerodynamic qualities change as green spaces are replaced by structures and roadways (Santamouris et al. 2025). Urban construction materials have different thermal and radiative properties compared to rural or vegetated areas, resulting in greater absorption and storage of the sun's energy in urban surfaces. Additionally, the height and arrangement of buildings affect the rate at which the absorbed energy escapes at night. As a result, metropolitan areas continue to have relatively higher nighttime air temperatures because they cool more slowly than rural locations (Aflaki et al. 2017). The Normalized Difference Vegetation Index (NDVI) is a simple graphical indicator used to analyze remote sensing measurements, typically from space platforms, to assess vegetation health. Additional factors contributing to the urban heat island (UHI) effect include scattered and emitted radiation from atmospheric pollutants in urban areas, waste heat production from air conditioning and refrigeration systems, industrial processes, motorized vehicular traffic (i.e., anthropogenic heat), and the obstruction of rural air flows by the windward faces of built-up surfaces. The impacts of urbanization, such as pollution production, waste heat from human activities (notably from air conditioners and internal combustion engines), modifications to the physical and chemical properties of the atmosphere, and soil surface covering, have become more apparent, leading to the UHI phenomenon (Harlan et al. 2006). There is a direct relationship between UHI intensity peaks and heat-related illnesses and fatalities, as thermal discomfort affects the human cardiovascular and respiratory systems.

4.2 Study Area

Ahmedabad City is located in the western part of India, in the state of Gujarat. It lies on the banks of the Sabarmati River and serves as a prominent economic and industrial hub. Geographically, Ahmedabad is situated at 23.0225° N latitude and 72.5714° E longitude, and the area of Ahmedabad city is 480.88 km^2 (Fig. 4.1). The city experiences a semi-arid climate and has seen rapid urbanization, making it a significant area for studying urban heat island effects and related environmental changes. Ahmedabad stands as Gujarat's economic powerhouse, with industries such as textiles, pharmaceuticals, and information technology driving its economy. The city's entrepreneurial spirit has attracted significant investments, both domestic and international, while its thriving informal sector provides essential livelihoods to a large portion of the population. Ahmedabad's rapid urbanization has led to considerable infrastructural development, including notable projects like the Sabarmati Riverfront and the Bus Rapid Transit System (BRTS), improving public spaces and transportation connectivity. However, urban sprawl and inadequate planning raise concerns about sustainability and equitable growth. The city's cultural heritage, recognized by its UNESCO World Heritage City designation in 2017, includes remarkable architectural landmarks like the Jama Masjid and Sidi Saiyyed Mosque, alongside traditional pols (housing clusters). Festivals such as Uttarayan and Navratri

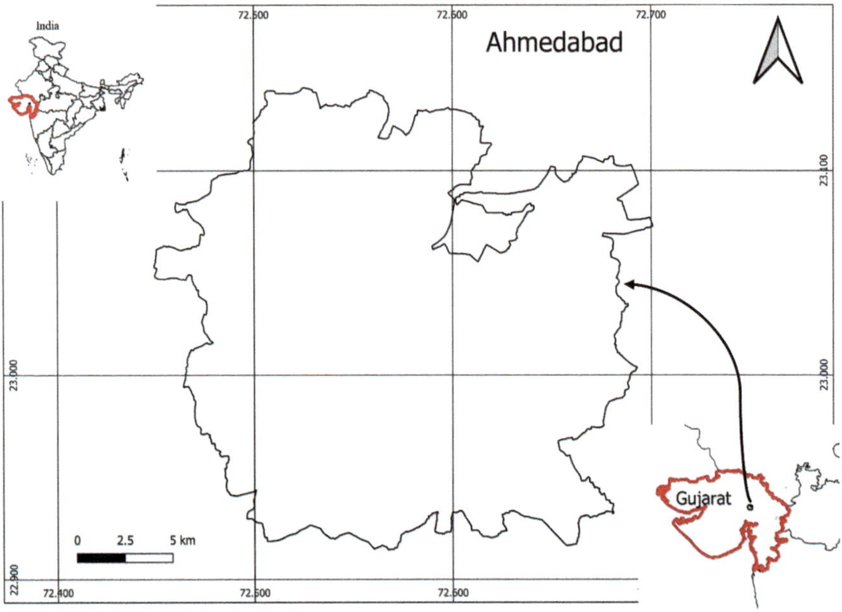

Fig. 4.1 Study area location map

reflect Ahmedabad's vibrant cultural fabric, attracting tourists worldwide. Despite its many advancements, the city faces environmental challenges, particularly air and water pollution due to industrial activity and population growth. Climate change further aggravates these issues, highlighting the need for climate adaptation and mitigation strategies. Ahmedabad exemplifies the fusion of tradition and modernity, where its rich historical past coexists with contemporary aspirations. As the city navigates the complexities of urbanization and globalization, efforts toward sustainable development, inclusive growth, and the preservation of its unique identity are crucial for its future progress.

4.3 Data and Method

4.3.1 Data Used

The data description covers imagery captured by three generations of Landsat satellites' data such as Landsat 5 Thematic Mapper (TM), Landsat 7 Enhanced Thematic Mapper Plus (ETM+), and Landsat 8. Each satellite features advanced sensors capable of capturing multispectral imagery across various wavelengths, enabling comprehensive earth observation. Landsat 5 TM, launched in 1984 and operational until 2013, provided seven spectral bands with spatial resolutions band 1–5, band 7 30 m and band 6 120 m. Landsat 7 ETM+, launched in 1999, introduced enhancements such as a new panchromatic band and improved thermal sensitivity. Although an issue with its scan line corrector caused systematic gaps in the imagery, Landsat 7 continued to supply valuable data until the launch of Landsat 8. Landsat 8, launched in 2013, offers significant advancements, including improved radiometric and spatial resolution compared to its predecessors. Equipped with the Operational Land Imager (OLI) and Thermal Infrared Sensor (TIRS), Landsat 8 provides a total of eleven spectral bands, including two thermal bands, allowing for more detailed and accurate observation of the Earth's surface. These datasets from Landsat 5 TM, Landsat 7 ETM+, and Landsat 8 are invaluable resources for various applications, including land cover mapping, environmental monitoring, urban planning, agricultural management, and scientific research.

4.3.2 Estimation of Parameters

Figure 4.2 shows the outline extraction of parameter use satellite data. Here use of thermal infrared bands from different Landsat image types, including band 6 of Landsat 5 TM, Landsat 7 ETM+, and band 10 of Landsat 8, were leveraged for land surface temperature (LST) estimation in the inter-municipal grouping of Guelma. Due to striping issues, only band 10 of Landsat 8 was utilized from the Landsat OLI-TIRS thermal bands. A single window algorithm, as proposed by Geng et al. (2022) and based on NDVI, was employed to derive Land Surface Emissivity (LSE). The following steps outline the process utilized to retrieve LST from thermal and NDVI images

4.3.2.1 Radiance Image Calculation

Radiance is a measure of the energy emitted by a surface or object in a given direction per unit area and solid angle. It is commonly used in remote sensing to quantify the amount of electromagnetic radiation (light) detected by a sensor. Calculating radiance from digital numbers (DN) in remote sensing imagery requires specific calibration parameters. Here's a general process to calculate radiance from satellite image data [Eqs. (4.1) and (4.2)].

$$L_{\lambda}=\frac{(\mathrm{LMAX}_{\lambda}-\mathrm{LMIN}_{\lambda})}{(\mathrm{QCAL}_{max}-\mathrm{QCAL}_{min})}*(\mathrm{QCAL}-\mathrm{QCAL}_{min})+\mathrm{LMIN}_{\lambda} \quad (4.1)$$

L_{λ}: Spectral radiance at the sensor's aperture (measured in W/(m^2·sr·μm)), LMAX_{λ}: Maximum spectral radiance, LMIN_{λ}: Minimum spectral radiance, QCAL: Quantized calibrated pixel value, QCAL_{max}: Maximum quantized calibrated pixel

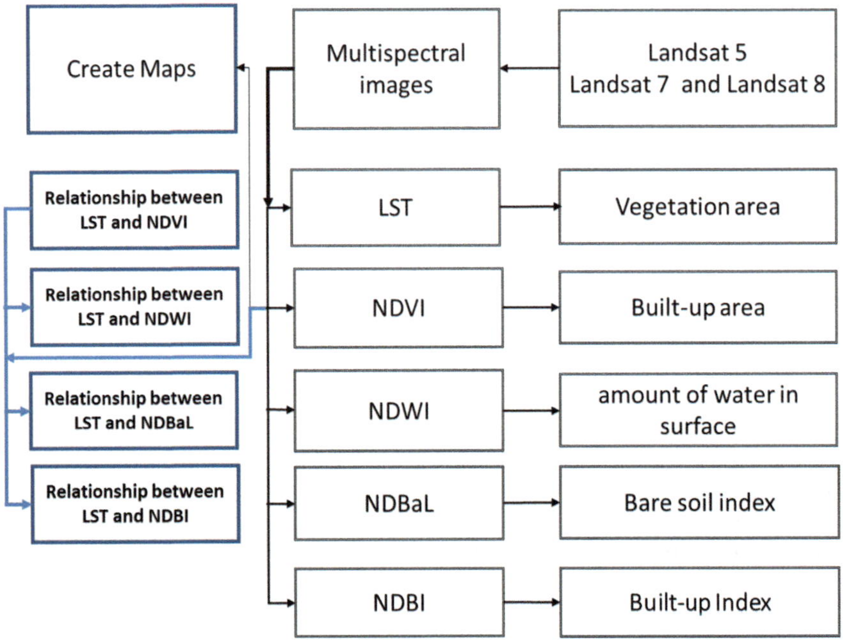

Fig. 4.2 Methodology flow chart

value, $QCAL_{min}$: Minimum quantized calibrated pixel value.

$$L_\lambda = M_L * Q_{CAL} + A_{L(2)} \tag{4.2}$$

L_λ: Spectral Radiance, M_L: Radiance Multiplicative Scaling Factor, Q_{CAL}: Quantized and Calibrated Standard Product Pixel Values A_L: Radiance Additive Scaling Factor

4.3.2.2 Radiance Temperature Calculation

In remote sensing research, radiance temperature is critical for monitoring surface energy balances, climate studies, and land surface temperature (LST) estimations. For example, in a study, researchers used thermal infrared satellite data to estimate surface temperatures across urban areas, providing insights into the urban heat island effect. Similarly, Li et al. (2023), focused on how radiance temperature can be employed to measure evapotranspiration in agricultural fields, improving water resource management. In many research papers, radiance temperature is calculated by correcting for atmospheric effects and sensor-specific factors. Weng et al. (2004), explored the relationship between radiance temperature, land surface temperature, and emissivity, showing how corrections for atmospheric interference can improve the accuracy of temperature measurements in satellite imagery. These methods help derive accurate surface temperature data from thermal sensors such as those on Landsat or MODIS satellites [Eq. (4.3)].

$$T = \frac{k_2}{\text{In}(K_1 + 1)} \tag{4.3}$$
$$L\lambda$$

T: Output variable, k_2: A constant or empirical coefficient (depends on calibration or physical laws), K_1: A coefficient or constant, often related to calibration or a scaling factor, In: Natural logarithm function (logarithm to the base *e*), $L\lambda$ (*L-lambda*): Spectral radiance at a given wavelength (λ).

4.3.2.3 Emissivity Calculation

There are several methods to calculate emissivity, ranging from laboratory-based measurements to satellite-based retrievals. Laboratory measurements typically involve spectrometers that measure emitted radiation at different wavelengths, which are then compared to a blackbody reference. However, for large-scale environmental studies, satellite-based methods are more practical. Zhang et al. (2025), proposed using thermal infrared satellite data to calculate surface emissivity, where they employed the Temperature Emissivity Separation (TES) algorithm. This method involves using multiple spectral bands to estimate both temperature and emissivity simultaneously, making it possible to derive these parameters from remote sensing data in various environmental conditions [Eqs. (4.4) and (4.5)].

$$p_v = \left(\frac{\text{NVDI} - \text{NVDI}_{\text{min}}}{\text{NDVI}_{\text{max}} - \text{NVDI}_{\text{min}}}\right)^2 \tag{4.4}$$

P_V: Vegetation Fraction or Proportion of Vegetation, NDVI: Normalized Difference Vegetation Index, NDVI_{min}: Minimum NDVI, NDVI_{max}: Maximum NDVI.

$$\varepsilon = 0.004 p_{\nu + 0.986} \tag{4.5}$$

ε is emissivity, *p* is probability or pressure, and *v* is a parameter or degrees of freedom.

4.3.2.4 LST Calculation

LST is widely used to study the urban heat island effect, where urban areas tend to be significantly warmer than their rural surroundings due to human activities and infrastructure. Voogt and Oke (2003), provided one of the foundational frameworks for analyzing UHI using satellite-derived LST. Recent studies like Peng et al.

(2022), have further explored how LST data can inform urban planning and green infrastructure development to mitigate UHI effects [Eq. (4.6)].

$$\mathrm{LST} = \frac{T_K}{1 + \left(\frac{\lambda T_K}{p}\right) \mathrm{In}\, \varepsilon} \tag{4.6}$$

LST: Land Surface Temperature, $T\kappa$: Brightness Temperature, λ: Wavelength, ρ (or p): hc/σ (Planck's constant $\times$ speed of light $\div$ Boltzmann constant), ε: Surface Emissivity.

4.3.2.5 Biophysical Indices

Biophysical indices related to radiation play an essential role in understanding the energy balance of ecosystems (Figs. 4.4, 4.5, 4.6, 4.7). Indices like the Photochemical Reflectance Index (PRI) are used to assess photosynthetic efficiency and light-use efficiency in vegetation. PRI is sensitive to changes in the xanthophyll cycle pigments, which are directly related to plant stress responses to excess light or water limitations (Gamon et al. 1997). This index is useful in studying how ecosystems respond to environmental stresses such as drought, heat waves, or nutrient deficiencies, making it a powerful tool for assessing the impact of climate change on plant productivity [Eqs. (4.7)–(4.10)].

$$\mathrm{NVDI} = \frac{(\mathrm{NIR} - \mathrm{RED})}{(\mathrm{NIR} + \mathrm{RED})} \tag{4.7}$$

Normalized Difference Vegetation Index (NDVI) is calculated using Near-Infrared Reflectance (NIR) and Red Reflectance (RED).

$$\mathrm{NDBI} = \frac{(\mathrm{MIR} - \mathrm{NIR})}{(\mathrm{MIR} + \mathrm{NIR})} \tag{4.8}$$

Normalized Difference Built-up Index (NDBI) is calculated using Mid-Infrared (MIR) and Near-Infrared (NIR) reflectance values.

$$\mathrm{NDBal} = \frac{\mathrm{SWIR} - \mathrm{TIRS1}}{\mathrm{SWIR} + \mathrm{TIRS1}} \tag{4.9}$$

Normalized Difference Built-up Index (NDBal) uses Shortwave Infrared (SWIR) and Thermal Infrared Sensor Band 1 (TIRS1) to identify built-up areas.

$$\mathrm{NDWI} = \frac{(\mathrm{GREEN} - \mathrm{NIR})}{(\mathrm{GREEN} + \mathrm{NIR})} \tag{4.10}$$

Normalized Difference Water Index (NDWI) uses Green Band Reflectance (GREEN) and Near-Infrared Band Reflectance (NIR) to detect water features.

4.4 Result and Discussion

4.4.1 Consequence of Urban Heat Island

The urban heat island (UHI) phenomenon describes the tendency for urban areas to experience higher temperatures than surrounding rural regions due to human activities and continuous urbanization. A primary factor in the formation and intensification of UHI is the replacement of natural land covers, such as vegetation and water bodies, with impervious surfaces like asphalt and concrete (Booth 1997). These surfaces absorb and retain heat, leading to increased surface temperatures. Additionally, the dense concentration of buildings and infrastructure in urban areas reduces vegetation cover, which in turn decreases evapotranspiration and cooling effects. Other contributors to UHI include heat emissions from vehicles, industries, and air conditioning systems, as well as changes in local wind patterns and atmospheric circulation (Cervero and Murakami 2009). The impacts of UHI are significant for the environment, society, and public health related problems occur. Higher temperatures in urban areas can increase the incidence of heat-related illnesses and mortality, particularly during heatwaves, posing greater risks to vulnerable populations such as the elderly, children, and individuals with pre-existing health conditions. UHI effects can also disrupt ecosystems, alter biodiversity patterns, and negatively impact agricultural productivity. Additionally, UHI can worsen air pollution levels by promoting the formation of ground-level ozone and other pollutants, which can have harmful effects on respiratory health and exacerbate conditions like asthma to humans. Studying the causes and consequences of UHI requires comprehensive future urban planning and mitigation strategies. These may include increasing green spaces and vegetation cover through urban forestry initiatives, promoting sustainable building design and materials to reduce heat absorption and improve energy efficiency, implementing cool roof and cool pavement technologies to mitigate surface heat retention, and integrating green infrastructure such as green roofs and permeable pavements to enhance urban cooling and stormwater management. Additionally, community-based adaptation measures such as public awareness campaigns, heat emergency response plans, and the provision of cooling centers can help mitigate the health impacts of UHI on vulnerable populations.

4.4.2 City Planning and Development

City planning and development is a multifaceted process that involves shaping the physical, social, economic, and environmental aspects of urban areas. This process includes the formulation, implementation, and evaluation of strategies and policies aimed at guiding urban growth and transformation in a sustainable and equitable manner. Effective city planning and development are crucial for creating livable, resilient, and inclusive urban environments that cater to the needs and aspirations of both current and future generations needs planning. City planning and development encompass various components such as land use planning, infrastructure development, transportation systems, housing provision, environmental management, economic development, and social equity (Cervero and Murakami 2009). These components are interconnected and interdependent, requiring integrated and holistic approaches. Land use planning is a fundamental aspect of city planning, involving the allocation and regulation of land for residential, commercial, industrial, recreational, and green spaces (Ahern 2012). It aims to optimize land utilization, promote compact and mixed-use development, prevent urban sprawl, and protect valuable natural and cultural assets. To reduce the effects of heat island, land use planning fosters sustainable development, promotes social inclusion, and enhances the quality of life for urban residents. Infrastructure development, another key component, includes the provision of essential services and facilities such as water supply, sanitation, energy, transportation, telecommunications, and waste management. Investments in infrastructure are vital for supporting urban growth, enhancing connectivity, improving accessibility, and ensuring the well-being of urban populations. Sustainable and resilient infrastructure is essential for mitigating climate change impacts, reducing vulnerability to natural hazards, and promoting environmental sustainability (Nelson 2021). Hence, city planning and development play a crucial role in shaping the future of urban areas, influencing the quality of life, sustainability, and resilience of cities. Effective city planning and development require integrated and participatory approaches that involve collaboration between government agencies, private sector stakeholders, civil society organizations, and local communities. By adopting sustainable, inclusive, and equitable practices, cities can create vibrant, resilient, and livable urban environments that promote the well-being and prosperity of all residents.

4.4.3 Mitigation Planning

Mitigation planning for land surface temperature (LST) involves strategic interventions aimed at reducing the impacts of urban heat islands (UHIs), heatwaves, and climate change on human health, ecosystems, and urban infrastructure. LST, a key indicator of land surface thermal properties, is influenced by factors such as land cover, land use, vegetation, building materials, and urban morphology (Fig. 4.3). As urbanization intensifies globally, mitigating the adverse effects of rising LST has become crucial in urban planning and climate resilience strategies (Vergara Perucich 2019) Urban areas typically experience higher LST than surrounding rural areas due to the urban heat island effect, characterized by the absorption and retention of solar radiation by built-up surfaces, reduced vegetation cover, and anthropogenic heat emissions. Elevated LST contributes to various environmental and societal challenges, including increased energy consumption, heat-related illnesses, reduced air quality, and biodiversity loss. Therefore, effective mitigation planning strategies are essential to alleviate these impacts and create healthier, more sustainable urban environments.

One effective approach to mitigating the impacts of elevated land surface temperature (LST) is through urban greening initiatives that increase vegetation cover and enhance natural cooling mechanisms. Green spaces such as parks, urban forests, and green roofs help mitigate urban heat island (UHI) effects by providing shade, evaporative cooling, and reducing surface temperatures through transpiration. Research indicates that the strategic placement of green infrastructure can significantly lower LST and improve thermal comfort in urban areas (Anguelovski et al. 2018). Additionally, cool roofing and cool pavement technologies are emerging as effective strategies to reduce LST. These materials, characterized by higher solar reflectance and lower thermal conductivity compared to traditional surfaces, reduce the absorption of solar radiation and lower surface temperatures. Implementing these technologies at a city scale can lead to substantial reductions in LST, energy consumption, and greenhouse gas emissions. Moreover, land use planning and urban design strategies are crucial in mitigating LST by promoting compact, mixed-use development patterns, preserving green spaces, and incorporating passive cooling measures into building design. Techniques such as increased building setbacks, optimizing street orientation, and using high-albedo materials can minimize heat retention and enhance natural ventilation in urban environments (Ouyang et al. 2020).

In conclusion, mitigation planning in the context of LST is essential for addressing the challenges posed by urban heat islands, heatwaves, and climate change in cities worldwide. By implementing a combination of nature-based solutions, cool technologies, urban design strategies, community engagement initiatives, and advanced technologies, cities can effectively reduce LST, enhance urban livability, and build climate-resilient communities. However, successful mitigation planning requires holistic, interdisciplinary approaches that integrate scientific knowledge, stakeholder perspectives, and policy frameworks to create sustainable, equitable, and resilient urban environments for current and future generations.

4.4.4 Vegetation Structure

The relationship between vegetation structure and land surface temperature (LST) is a crucial aspect of understanding the dynamics of terrestrial ecosystems and their

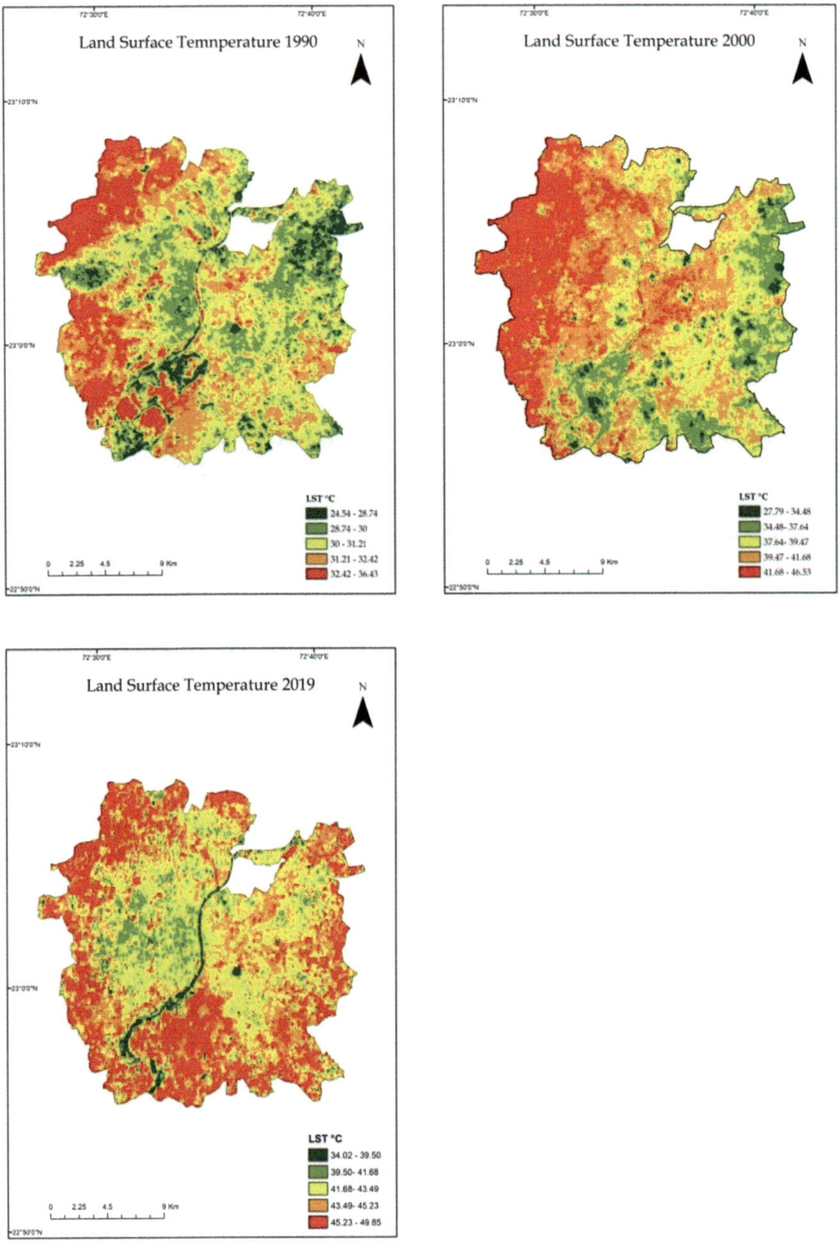

Fig. 4.3 Analysis examines land surface temperature variations over the city of Ahmedabad from the period spanning 1990, 2000, and 2019

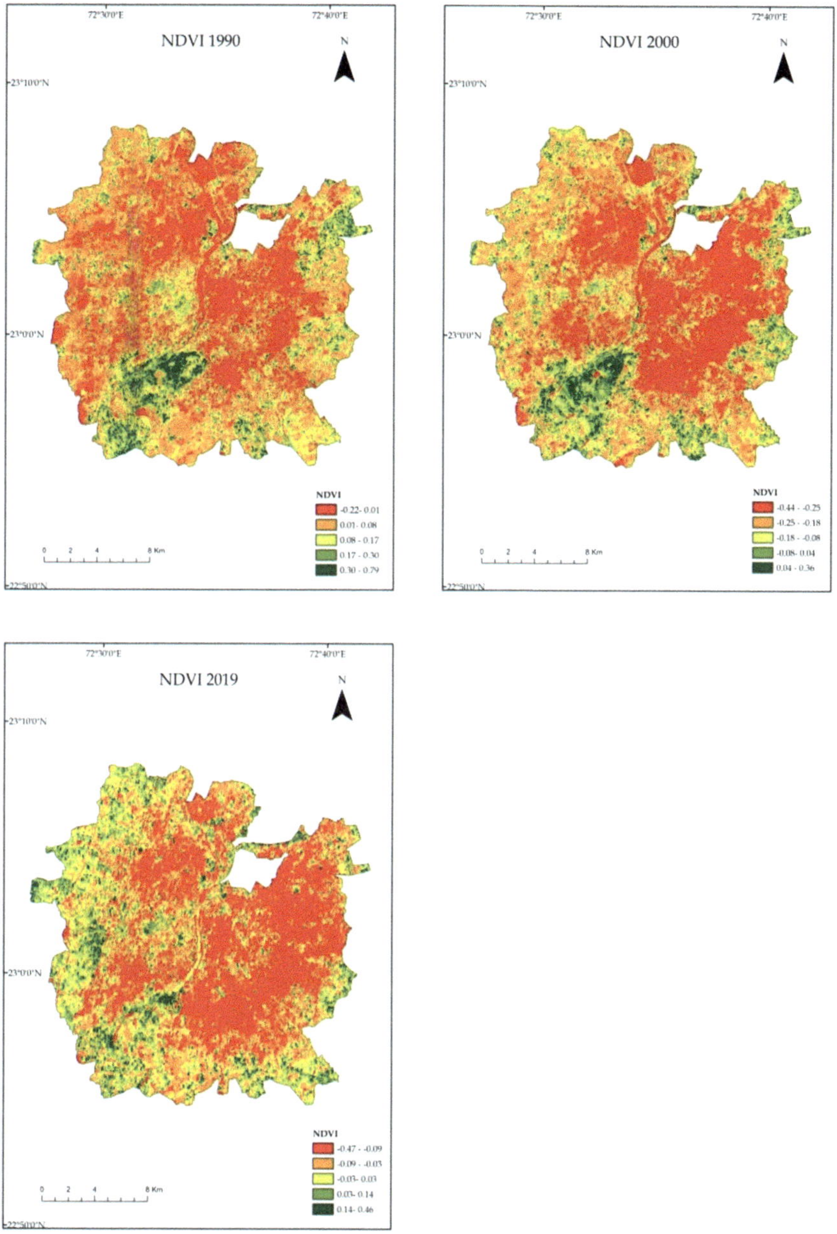

Fig. 4.4 Normalized Difference Vegetation Index (NDVI) for Ahmedabad city was analyzed over the period between 1990, 2000, and 2019

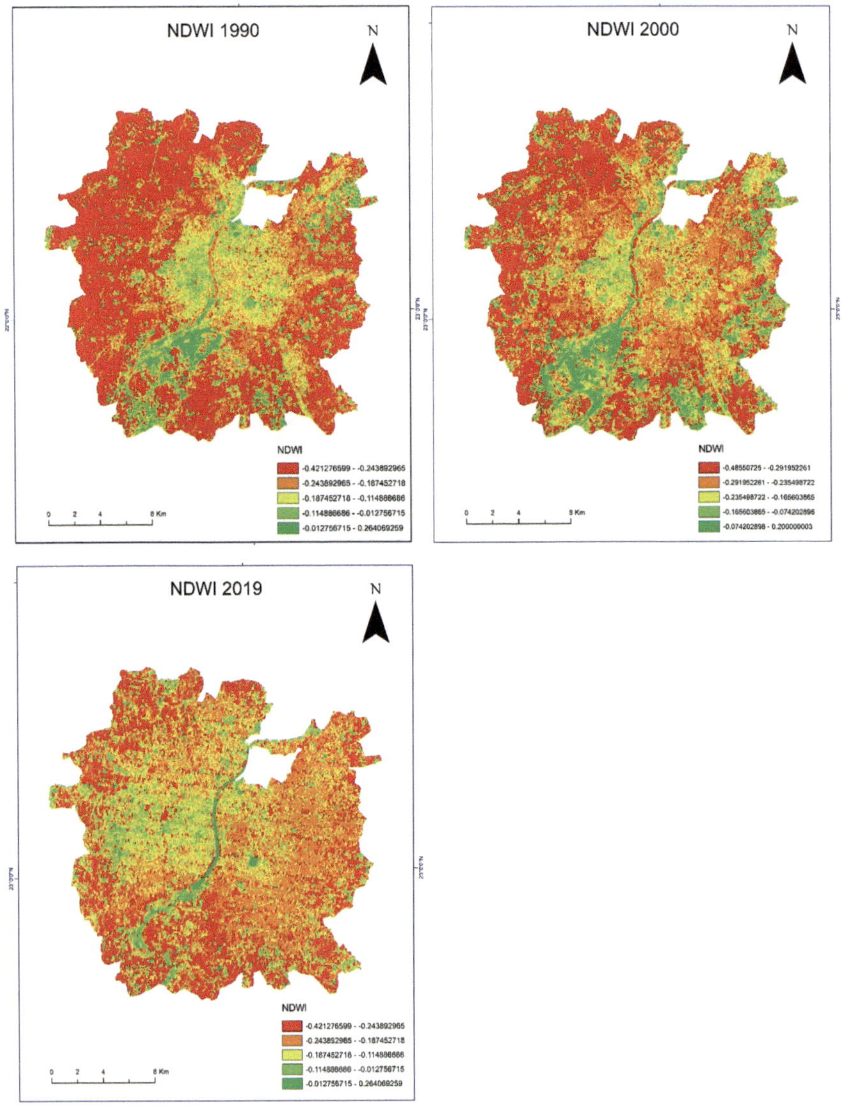

Fig. 4.5 Normalized Difference Water Index (NDWI) was calculated for the city of Ahmedabad spanning the period between 1990, 2000, and 2019

response to environmental changes. Vegetation structure encompasses various physical characteristics of plant communities, including canopy cover, leaf area index (LAI), canopy height, and biomass distribution (Wesley and Brunsell 2019). These structural attributes influence the energy balance of the land surface, thereby affecting LST patterns at local, regional, and global scales. Canopy cover, or the proportion

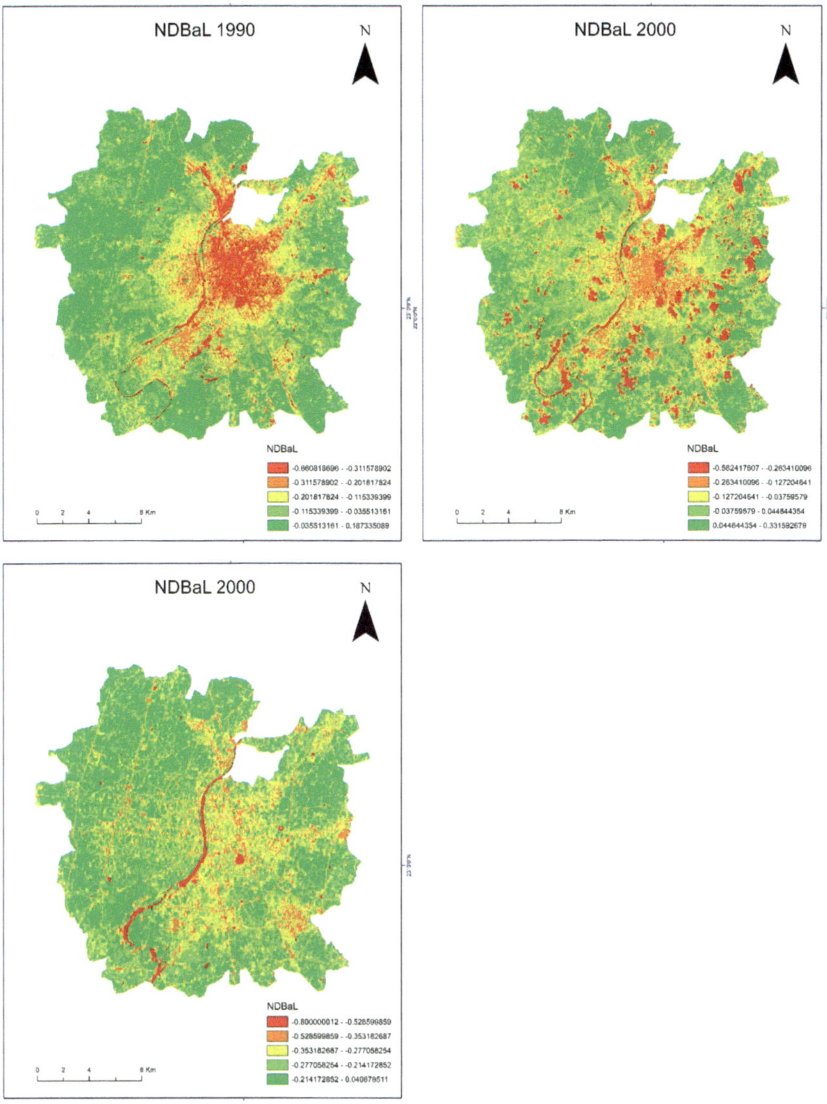

Fig. 4.6 Variation in Normalized Difference Built-up and Bareness Index (NDBaL) was calculated for the city of Ahmedabad spanning the period between 1990, 2000, and 2019

of ground covered by vegetation canopy, plays a significant role in modulating LST by influencing the amount of incoming solar radiation absorbed or reflected by the land surface. Dense vegetation canopies intercept a larger fraction of solar radiation, leading to lower surface temperatures through shading and evapotranspiration processes. In contrast, sparse vegetation cover allows more solar radiation to reach

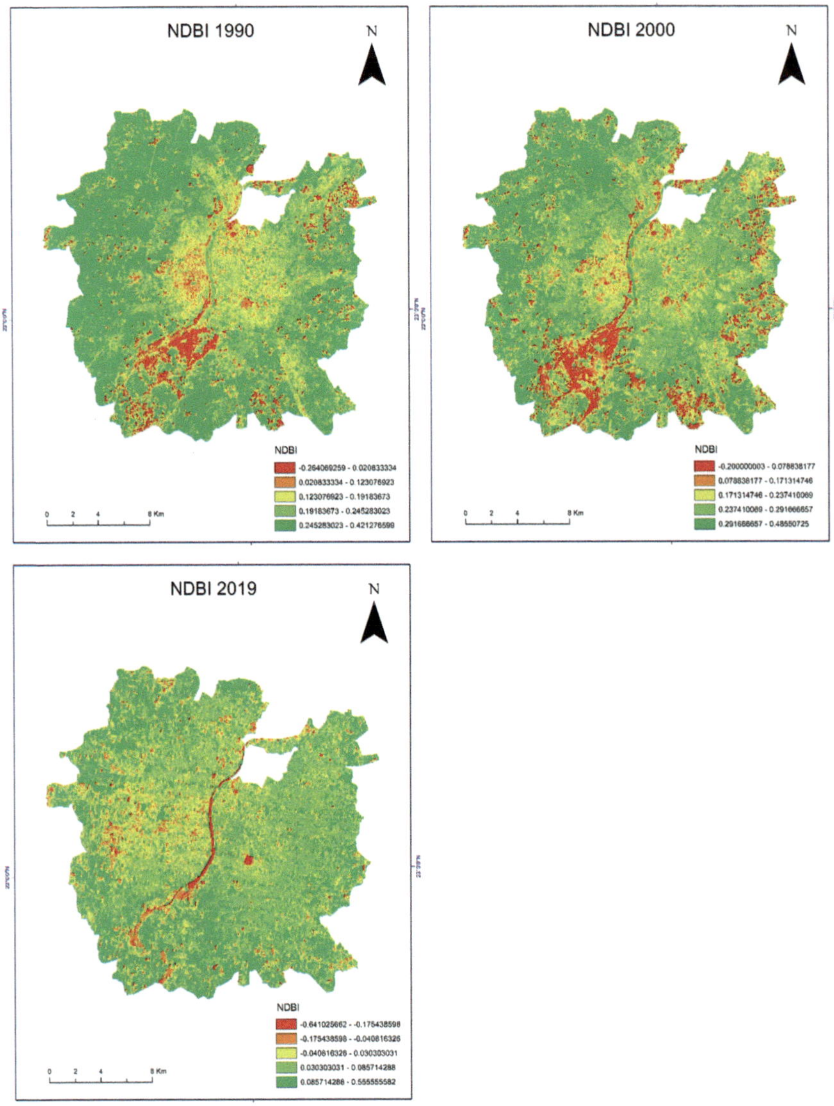

Fig. 4.7 Analysis focuses on the Normalized Difference Built-up Index (NDBI) over the city of Ahmedabad spanning the period between 1990, 2000, and 2019

the land surface, resulting in higher LST due to increased absorption of solar energy (Nature Cities 2024). Leaf area index (LAI) is another important vegetation structural parameter that influences LST dynamics. LAI is the total leaf area per unit ground area and is strongly correlated with canopy density and photosynthetic activity. High LAI values indicate dense vegetation canopies with abundant leaf cover, which can reduce LST by enhancing evaporative cooling through transpiration. Conversely, low LAI values are associated with sparse vegetation cover and higher LST due to reduced shading and transpiration (Weng 2009). Canopy height, or the vertical extent of vegetation, also influences LST by affecting the exchange of energy and moisture between the land surface and the atmosphere. Tall vegetation canopies intercept more solar radiation and promote greater turbulence within the boundary layer, leading to enhanced cooling through convective heat transfer and increased evapotranspiration rates. Shorter vegetation, on the other hand, may have limited cooling effects and can contribute to higher LST by allowing more solar radiation to reach the land surface. Biomass distribution within vegetation canopies can further modulate LST patterns by altering the partitioning of energy and water fluxes within the ecosystem (Abdullah-Al-Faisal et al. 2022). Biomass distribution affects the distribution of surface roughness elements, which in turn influences aerodynamic resistance and heat exchange processes between the land surface and the atmosphere (Colombo 2003). Moreover, biomass distribution impacts the spatial variability of evapotranspiration rates and surface albedo, both of which play critical roles in regulating LST. Several studies have investigated the relationship between vegetation structure and LST across different ecosystems and land cover types and have examined the impact of canopy cover and LAI on LST in urban areas using remote sensing data. In conclusion, vegetation structure plays a critical role in shaping land surface temperature patterns through its influence on energy balance processes and surface–atmosphere interactions (Fatemi and Narangifard 2019). Understanding the complex interactions between vegetation structure and LST is essential for ecosystem monitoring, climate modeling, and land management practices. Incorporating vegetation structural parameters into LST modeling frameworks can improve our ability to predict and mitigate the impacts of environmental changes on terrestrial ecosystems and human communities (Yue et al. 2007).

4.4.5 Risk Reduction Strategies

Ahmedabad, like many other urban areas, experiences the urban heat island effect, where built-up areas become significantly warmer than surrounding rural regions due to factors like concrete surfaces, reduced vegetation, and heat emissions from vehicles and industries. To mitigate UHI, the city has adopted several strategies. Below are some existing UHI reduction techniques in Ahmedabad.

4.4.5.1 Urban Greening Initiatives

Urban greening plays a crucial role in mitigating the urban heat island (UHI) effect by enhancing vegetation cover, improving air quality, and reducing surface temperatures. In Ahmedabad, various urban greening initiatives have been implemented to counteract rising temperatures and create a more sustainable urban environment.

Large-scale tree plantation drives have been carried out along roads, parks, and open spaces to provide shade and lower surface heat. Additionally, the city has developed green corridors and urban forests, ensuring that natural vegetation is preserved and expanded within the urban fabric. The promotion of vertical gardens and rooftop gardens has further contributed to reducing heat absorption by buildings while improving air quality. These initiatives not only enhance thermal comfort but also support biodiversity, improve public health, and contribute to overall climate resilience. By integrating green infrastructure into urban planning, Ahmedabad aims to create a more livable and cooler city for its residents.

4.4.5.2 Cool Roofs and Reflective Surfaces

Cool roofs and reflective surfaces are effective strategies for mitigating the urban heat island effect in Ahmedabad. These technologies work by reflecting more sunlight and absorbing less heat compared to conventional roofing materials, thereby reducing surface temperatures and lowering indoor cooling demands. One notable initiative in the city is the ModRoof project, which promotes modular roofing made from recycled materials like coconut husk and paper waste, providing better insulation and heat resistance. Additionally, Ahmedabad has encouraged the use of white roofs and reflective paints, particularly in low-income housing and government buildings, to enhance thermal comfort. The adoption of high-albedo materials for roads and pavements further helps in reducing heat absorption from solar radiation. By integrating these cooling solutions, Ahmedabad not only minimizes heat stress but also reduces energy consumption for air conditioning, contributing to both environmental sustainability and improved urban livability.

4.4.5.3 Water-Based Cooling Strategies

Water-based cooling strategies play a crucial role in mitigating the urban heat island effect by leveraging the cooling properties of water bodies to regulate urban temperatures. In Ahmedabad, several initiatives have been undertaken to restore and enhance water features to improve the local microclimate. The revitalization of lakes, such as Chandola Lake and Kankaria Lake, helps in temperature moderation by increasing humidity and facilitating evaporative cooling. Additionally, the city has integrated fountains, artificial ponds, and water features in public spaces, parks, and urban squares to provide localized cooling effects. These measures help in reducing heat buildup in highly urbanized areas, improving thermal comfort for residents. Furthermore, sustainable stormwater management practices, such as rainwater harvesting and the creation of urban wetlands, not only contribute to cooling but also aid in groundwater recharge. By incorporating water elements into urban design, Ahmedabad enhances its resilience to rising temperatures while promoting ecological balance and sustainable urban living.

Urban Planning and Sustainable Design play a crucial role in mitigating the urban heat island (UHI) effect by integrating climate-responsive strategies into city development. In Ahmedabad, sustainable urban planning focuses on creating well-ventilated layouts, ensuring adequate green cover, and optimizing land use to minimize heat accumulation. The city has adopted compact urban planning with designated ventilation corridors to facilitate natural airflow and reduce localized heating. Additionally, energy-efficient building orientation and design codes have been implemented to enhance indoor thermal comfort by reducing direct sun exposure and promoting passive cooling techniques. Mixed-use developments and pedestrian-friendly infrastructure further contribute to reducing heat emissions from vehicles, encouraging sustainable mobility. By prioritizing green infrastructure, water-sensitive urban design, and climate-adaptive architecture, Ahmedabad aims to create a resilient urban environment that enhances livability while combating the adverse effects of rising temperatures.

4.4.5.4 Use of Renewable Energy in Mitigating UHI in Ahmedabad

The adoption of renewable energy plays a crucial role in reducing the urban heat island (UHI) effect in Ahmedabad by minimizing heat emissions from conventional energy sources and promoting sustainable urban development. One of the key initiatives in this regard is the widespread installation of rooftop solar panels under the Gujarat Solar Power Policy, which encourages households, commercial buildings, and industries to shift toward solar energy. This not only reduces the dependency on fossil fuels but also lowers the heat generated by power plants. Additionally, the city has implemented solar-powered street lighting, which helps in reducing electricity consumption while preventing unnecessary heat buildup from traditional lighting systems. The integration of solar cooling systems in buildings is another effective measure, enhancing indoor thermal comfort while lowering overall energy demand. By investing in renewable energy solutions, Ahmedabad is making significant progress in creating a more energy-efficient and climate-resilient urban environment, helping to mitigate the long-term effects of UHI.

4.4.5.5 Community Engagement and Heat Action Plan (HAP)

Ahmedabad has implemented the Heat Action Plan (HAP) as a proactive measure to reduce the impact of extreme heat events on public health. Launched in 2013, it was India's first city-wide initiative to combat rising temperatures and is now a model for other cities. The HAP focuses on early warning systems, public awareness campaigns, and policy interventions to protect vulnerable populations, such as outdoor workers, elderly individuals, and children. The plan includes heatwave alerts, advisories on hydration and cooling strategies, and the provision of cooling shelters. Additionally, Ahmedabad's municipal authorities actively engage with local communities, NGOs, and healthcare providers to spread awareness about heat-related illnesses and preventive measures. Public outreach efforts involve educational programs, media campaigns, and community workshops to ensure citizens are well-prepared to cope with extreme temperatures. Through continuous monitoring and adaptation, the HAP has significantly improved the city's resilience to heatwaves, reducing mortality rates and enhancing overall urban climate resilience.

4.4.5.6 Sustainable Transportation

Ahmedabad has implemented several sustainable transportation initiatives to mitigate the urban heat island (UHI) effect and reduce heat emissions from vehicular traffic. One of the key strategies is the expansion of the Bus Rapid Transit System (BRTS) and Ahmedabad Metro, which provide efficient public transport options, reducing the dependence on private vehicles and lowering overall carbon emissions. Additionally, the city has developed pedestrian-friendly walkways and dedicated cycling lanes, encouraging non-motorized transport and minimizing the heat generated by asphalt roads and motor vehicles. Shaded pathways and green cover along transport routes further contribute to localized cooling. The adoption of electric buses and energy-efficient street lighting also plays a role in reducing heat emissions. By prioritizing sustainable transport solutions, Ahmedabad not only enhances mobility but also contributes to a cooler and more climate-resilient urban environment.

4.5 Spatial Distribution of LST, NDVI, NDWI, NDBaL, and NDBI

4.5.1 Association Between LST and NDVI

Figure 4.8 shows a linear regression model where the coefficient of determination in 1990, R^2, is 0.7986. This indicates that approximately 79.86% of the variability in NDVI can be explained by its linear relationship with LST. The regression equation suggests a negative relationship between LST and NDVI, meaning that higher LST values are associated with lower NDVI values. This inverse relationship is commonly observed in vegetation studies, as high temperatures can stress vegetation, resulting in decreased photosynthetic activity and, consequently, lower NDVI values. The coefficient of determination in 2000, R^2, indicates the strength of the linear relationship between LST and NDVI. In this case, an R^2 value of 0.7986 suggests a relatively strong correlation between the two variables, indicating that the model provides a good fit to the observed data points. However, it is important to consider other factors that may influence NDVI variability, as the model might not capture all sources of variation. The R^2 value of 2019, 0.9578, suggests that about 95.78% of the variability in NDVI, the dependent variable, can be attributed to changes in LST, the

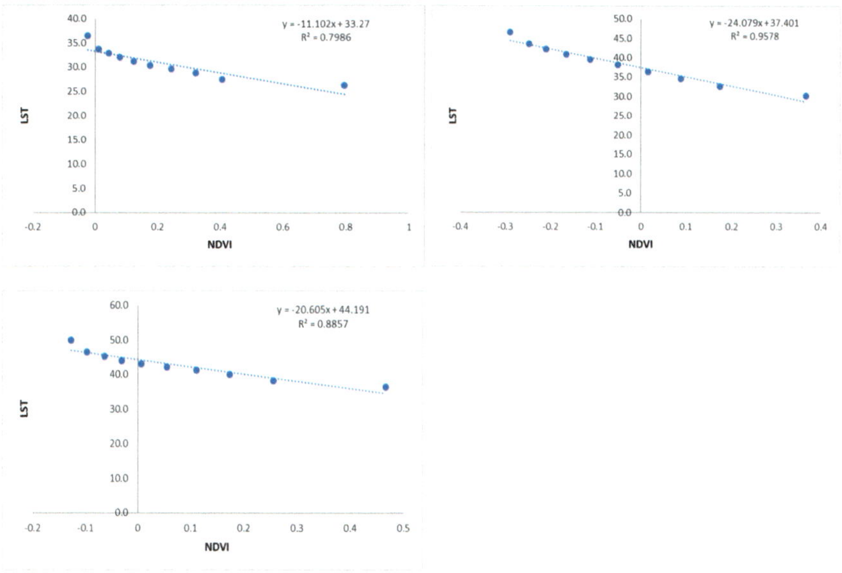

Fig. 4.8 LST and NDVI correlation charts for 1990, 2000, and 2019

independent variable, in this linear regression model. This high R^2 indicates a robust fit of the model to the data, demonstrating its ability to accurately predict NDVI based on LST. This underscores the significant and strong relationship observed between NDVI and LST in the year 2000, emphasizing temperature's crucial role in influencing vegetation health and dynamics.

An R^2 value of 0.8857 indicates that approximately 88.57% of the variation in land surface temperature (LST) can be explained by Normalized Difference Vegetation Index (NDVI) using the linear regression model. This strong R^2 suggests a robust relationship between NDVI and LST for the year 2019, highlighting how vegetation density influences surface temperature dynamics. This insight is valuable for applications such as environmental monitoring, ecosystem analysis, and agricultural management, offering significant understanding into land surface processes based on vegetation indices.

4.5.2 Association Between LST and NDWI

Figure 4.9 provided correlation coefficients (R^2 values) between Normalized Difference Water Index (NDWI) and land surface temperature (LST) for the years 1990, 2000, and 2019 indicate the strength and direction of their relationship. In 1990, the R^2 value of 0.9653 suggests a very strong positive correlation between NDWI and LST. This indicates that as the water presence, as indicated by NDWI, increases,

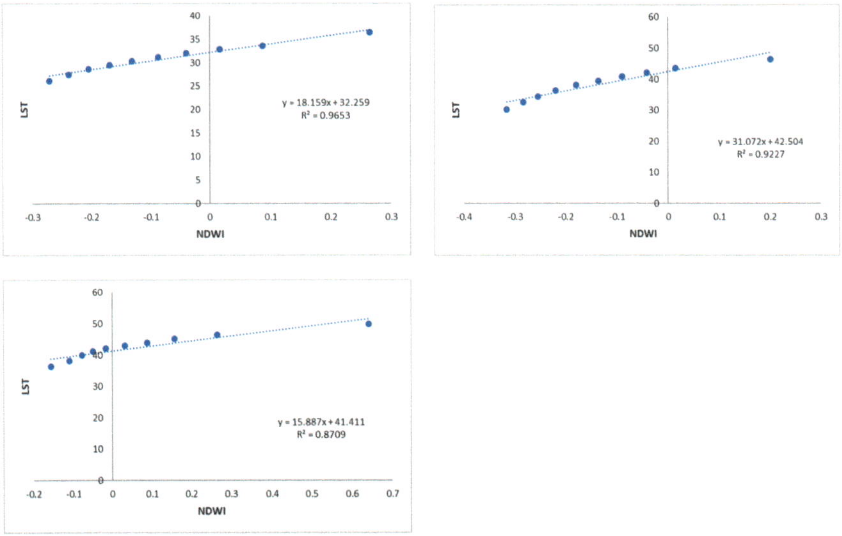

Fig. 4.9 LST and NDWI correlation charts for 1990, 2000, and 2019

the land surface temperature tends to increase as well. This could be attributed to the fact that water bodies, being good absorbers of solar radiation, can contribute to higher temperatures in their vicinity. In 2000, the R^2 value of 0.9227 indicates a strong positive correlation between NDWI and LST, similar to the trend observed in 1990. Again, this suggests that areas with higher water presence tend to have higher land surface temperatures. In 2019, the R^2 value of 0.8709 suggests a slightly weaker but still significant positive correlation between NDWI and LST compared to the previous years. However, the correlation remains strong, indicating that the relationship between water presence and land surface temperature persists over time. Overall, these findings underscore the consistent positive relationship between NDWI and LST over the examined years, highlighting the influence of water bodies on land surface temperatures in the respective years.

4.5.3 Association Between LST and NDBI

This research explores the relationship between the Normalized Difference Built-up Index (NDBI) and land surface temperature (LST) in Ahmedabad city, revealing a strong positive correlation between these variables. The coefficient of determination (R^2) for the relationship between NDBI and LST is presented for the years 1990, 2000, and 2019. For 2019, the R^2 value is 0.8975, indicating a strong positive correlation. This high R^2 value, close to 1, implies that 89.75% of the variation in LST can be attributed to the variation in NDBI for that year (Fig. 4.10).

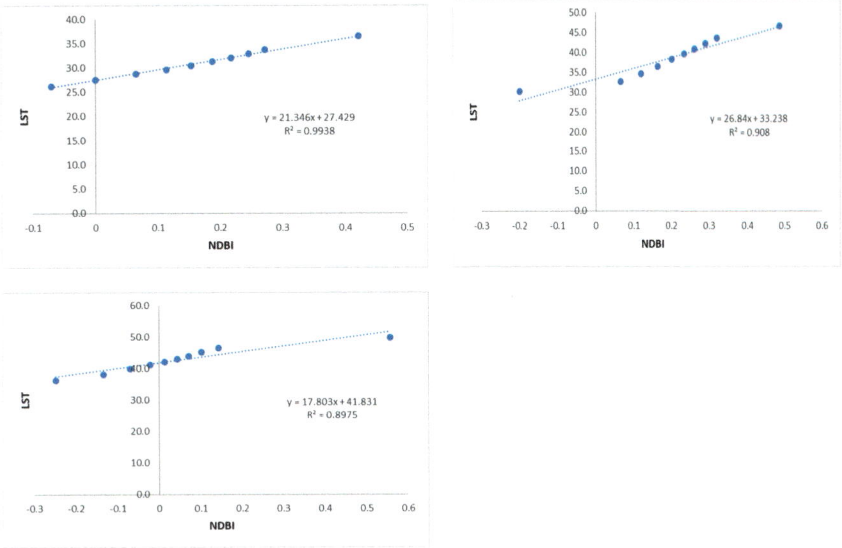

Fig. 4.10 LST and NDBI correlation charts for 1990, 2000, and 2019

The Normalized Difference Built-up Index (NDBI) is used to measure the extent of built-up areas in satellite images, while land surface temperature (LST) indicates the temperature of the Earth's surface as detected by satellite sensors. A strong positive correlation between NDBI and LST implies that regions with higher concentrations of built-up areas typically exhibit higher surface temperatures. This is often due to the heat-retaining characteristics of construction materials like concrete and asphalt, which result in elevated surface temperatures compared to areas covered by natural vegetation or bare soil.

Understanding the relationship between NDBI and LST is valuable for various applications, including urban heat island monitoring, land use planning, and climate change studies. By analyzing the correlation between these two variables, researchers can gain insights into the spatial distribution of urbanization and its impacts on local climate conditions (Fig. 4.10).

4.5.3.1 Association Between NDBaL and LST

The coefficient of determination (R^2) values obtained for the relationship between the Normalized Difference Built-up Index (NDBaI) and land surface temperature (LST) across different years demonstrate a strong correlation between these variables. In 1990, the R^2 value of 0.9982 indicates that approximately 99.82% of the variability in LST is explained by variations in NDBaI. Similarly, in 2000 and 2019, the R^2 values of 0.9866 and 0.9825, respectively, show that around 98.66 and 98.25% of the variability in LST can be attributed to changes in NDBaI.

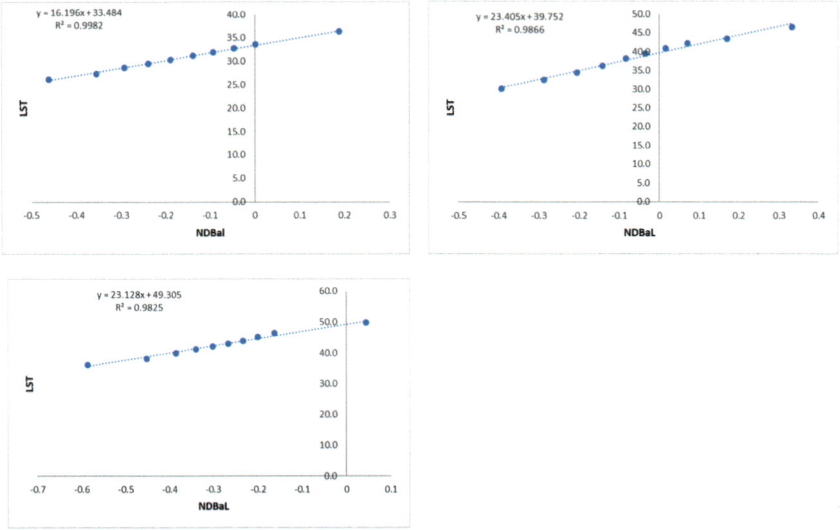

Fig. 4.11 LST and NDBaL correlation charts for 1990, 2000, and 2019

The high R^2 values demonstrate a strong relationship between NDBaI and LST over the three years examined. NDBaI, a metric derived from satellite imagery that indicates the presence and density of built-up areas, proves to be a reliable predictor of land surface temperature. This strong correlation implies that urbanization and the density of built-up areas significantly impact land surface temperature patterns.

The high R^2 values demonstrate a strong relationship between NDBaI and LST over the three years examined. NDBaI, a metric derived from satellite imagery that indicates the presence and density of built-up areas, proves to be a reliable predictor of land surface temperature. This strong correlation implies that urbanization and the density of built-up areas significantly impact land surface temperature patterns.

In conclusion, the analysis underscores the strong relationship between NDBaI and LST across multiple years, as evidenced by the consistently high R^2 values. This highlights the importance of considering built-up density as a key factor influencing land surface temperature patterns, and it emphasizes the potential of NDBaI as a valuable tool for assessing urban heat island effects and informing urban planning strategies aimed at mitigating temperature-related impacts in urban areas (Fig. 4.11).

4.5.3.2 Association Between Biophysical Parameter and LST for the Year 1990

Table 4.1 shows a correlation analysis for the year 1990 between land surface temperature (LST) and several spectral indices: the Normalized Difference Vegetation Index (NDVI), the Normalized Difference Built-up Index (NDBI), the Normalized Difference Bareness Index (NDBaI), and the Normalized Difference Water Index (NDWI).

Table 4.1 Correlation matrix of environmental parameters for the year 1990

1990 year	LST	NDVI	NDBI	NDBaI	NDWI
LST	1				
NDVI	– 0.89362	1			
NDBI	0.996885	– 0.90986	1		
NDBaI	0.9991	– 0.90136	0.997836	1	
NDWI	0.9825	– 0.80393	0.969206	0.980026	1

The correlation coefficients indicate the strength and direction of the relationships between these variables.

The correlation coefficient between LST and NDVI is – 0.89362, showing a strong negative relationship. This means that as NDVI rises (reflecting denser vegetation), LST generally falls, which is consistent with the expected cooling effect of vegetation on surface temperatures. On the other hand, the correlation coefficients between LST and urban-related indices (NDBI and NDBaI) are significantly high, indicating strong positive correlations. This suggests that as urbanization or built-up areas increase, LST tends to increase as well, aligning with the urban heat island effect where urban areas have higher temperatures than their surroundings.

Additionally, the correlation between LST and NDWI, which indicates water presence, is moderately negative. This implies that areas with higher water content generally have lower surface temperatures, likely due to the cooling effect of water bodies. These correlation values offer important insights into the relationships between LST and various land-cover types, helping to understand surface temperature dynamics and informing urban planning, environmental management, and climate change mitigation efforts (Fig. 4.12).

4.5.3.3 Association Between Biophysical Parameter and LST for the Year 2000

Table 4.2 showcases a correlation analysis examining the relationship between land surface temperature (LST) and several spectral indices, such as the Normalized Difference Vegetation Index (NDVI), Normalized Difference Built-Up Index (NDBI), Normalized Difference Bareness Index (NDBaI), and Normalized Difference Water Index (NDWI). The correlation coefficients highlight the extent of the linear association between these variables.

Land surface temperature (LST) shows a strong negative correlation with the Normalized Difference Vegetation Index (NDVI) (– 0.97869), indicating that areas with more vegetation cover tend to have lower temperatures. On the other hand, there are positive correlations between LST and the Normalized Difference Built-up Index (NDBI) (0.952897), Normalized Difference Bareness Index (NDBaI) (0.993283), and Normalized Difference Water Index (NDWI) (0.960556). This suggests that built-up areas, bare surfaces, and water bodies generally have higher temperatures.

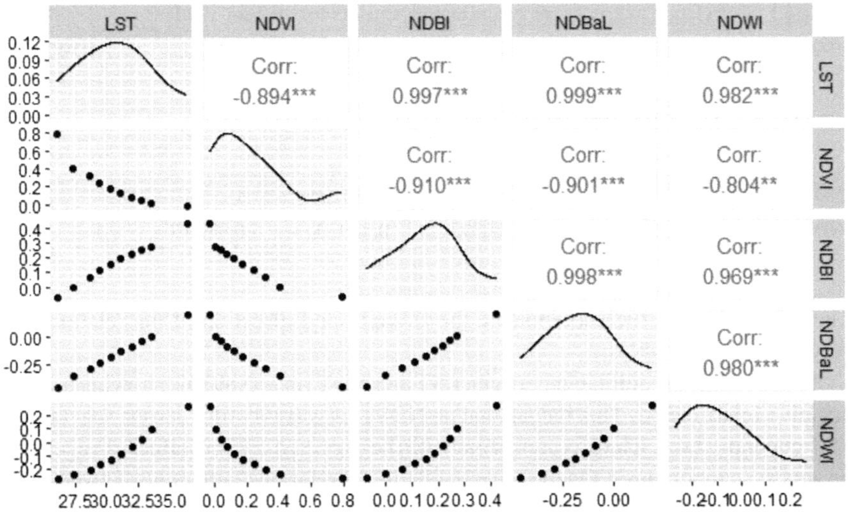

Fig. 4.12 Grid map of correlation between LST, NDVI, NDBI, NDBaL, and NDWI for the year 1990

Table 4.2 Correlation matrix of environmental parameters for the year 2000

2000 year	LST	NDVI	NDBI	NDBal	NDWI
LST	1				
NDVI	– 0.97869	1			
NDBI	0.952897	– 0.96596	1		
NDBal	0.993283	– 0.95756	0.952648	1	
NDWI	0.960556	– 0.88821	0.897225	0.980234	1

Additionally, the strong negative correlations between NDVI and the other indices (NDBI, NDBal, and NDWI) indicate that densely vegetated areas typically have lower levels of built-up surfaces, bare ground, and water bodies. This aligns with ecological principles, where vegetated areas provide cooling effects through transpiration and shading. Overall, the correlation analysis highlights the complex relationships between land surface temperature and various land cover types, underscoring the importance of considering multiple spectral indices to understand the thermal behavior of different land surfaces (Fig. 4.13).

4.5.3.4 Association Between Biophysical Parameter and LST for the Year 2019

Table 4.3 presents the correlation analysis of environmental parameters for the year 2019 and reveals significant interrelationships among land surface temperature

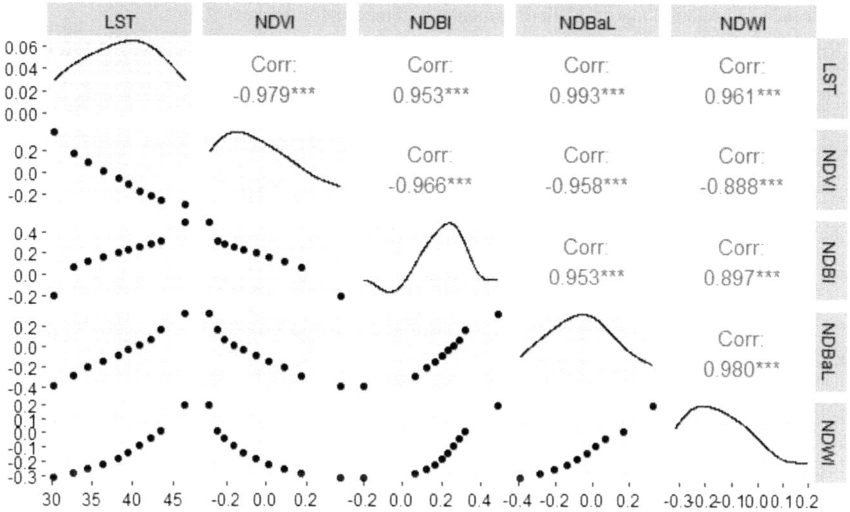

Fig. 4.13 Grid map of correlation between LST, NDVI, NDBI, NDBaL, and NDWI for the year 2000

(LST), vegetation cover, built-up density, and surface moisture. A strong negative correlation ($r = -0.9411$) was observed between LST and the Normalized Difference Vegetation Index (NDVI), indicating that higher LST values are associated with lower vegetation density, thereby highlighting the role of vegetation in mitigating urban heat. Conversely, LST exhibited strong positive correlations with the Normalized Difference Built-up Index (NDBI) ($r = 0.94735$) and the Normalized Difference Bareness Index (NDBaI) ($r = 0.991199$), suggesting that increased built-up density and bare surfaces contribute significantly to thermal intensification due to their higher heat retention and emission properties. Furthermore, NDVI demonstrated moderate negative correlations with NDBI (r = -0.8239), NDBaI ($r = -0.93799$), and the Normalized Difference Water Index (NDWI) ($r = -0.76708$), indicating that areas with higher vegetation cover tend to have lower built-up density, reduced surface bareness, and comparatively lower water presence. A strong positive correlation was identified between NDBI and NDBaI ($r = 0.968742$), suggesting a high degree of spatial co-occurrence between built-up expansion and barren land. These findings emphasize the intricate interactions among LST, vegetation cover, built-up intensity, and surface moisture, providing critical insights into the thermal and environmental dynamics of the study area. Such insights are crucial for informing sustainable urban planning and climate adaptation strategies (Fig. 4.14).

Table 4.3 Correlation matrix of environmental parameters for the year 2019

2019 year	LST	NDVI	NDBI	NDBal	NDWI
LST	1				
NDVI	– 0.9411	1			
NDBI	0.94735	– 0.8239	1		
NDBal	0.991199	– 0.93799	0.968742	1	
NDWI	0.933233	– 0.76708	0.975519	0.934193	1

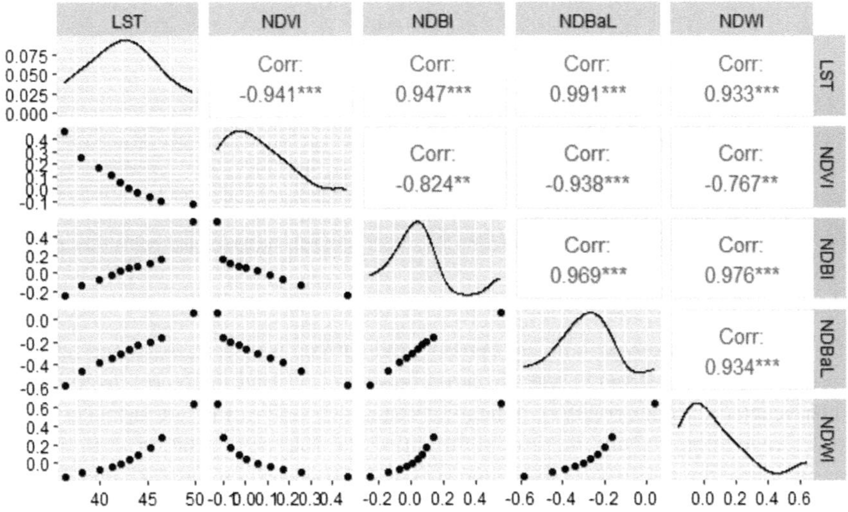

Fig. 4.14 Grid map of correlation between LST, NDVI, NDBI, NDBaL, and NDWI for the Year 2019

4.6 Conclusion

This could suggest a shift in land-use dynamics or changes in environmental factors that influenced vegetation patterns independently of urbanization. Alternatively, it could also indicate limitations in the applicability of NDBI and NDVI indices for capturing the complexity of land cover changes in more recent years. Overall, the analysis of NDBI versus NDVI provides valuable insights into the dynamics of urbanization and vegetation cover over time. The variations in R^2 values across different years highlight the changing relationships between built-up areas and vegetation, reflecting the dynamic nature of land use and environmental processes. These findings are essential for urban planning, environmental management, and sustainable development initiatives, as they contribute to a better understanding of the interactions between urbanization and ecosystem dynamics. The outcomes of this study underscore the intensification of urban heat island (UHI) effects, attributed to rapid urbanization between 1990 and 2019. The positive correlation observed between land surface temperature (LST) and Normalized Difference Built-up Index (NDBI) further validates the impact of urbanization on temperature rise. While infrastructural enhancements have led to vegetation losses in some areas, there were also instances of vegetation improvement noted in certain regions. This indicates a complex interplay between urban development and greenery preservation. In conclusion, the proliferation of man-made constructions in urbanized regions and the reduction in vegetation cover due to infrastructural enhancements significantly contribute to temperature rise in the study region. To effectively mitigate the effects of UHIs, proactive measures should be implemented both before and during the development phases of townships. These measures could include the integration of green spaces, such as parks and urban forests, into urban planning initiatives, the implementation of green infrastructure like green roofs and permeable pavements, and the adoption of sustainable land-use practices to preserve and enhance vegetation cover. By prioritizing these strategies, urban areas can mitigate the impacts of UHIs, enhance thermal comfort, and promote overall environmental sustainability.

Authors Contribution All authors have contributed equally to this work.

Funding The research was not supported by any funding.

Data Availability Statement The datasets during and/or analyzed during the current study available from the corresponding author on reasonable request.

Potential Conflicts of Interest On behalf of all authors, the corresponding author states that there is no conflict of interest.

Informed Consent Informed consent was obtained from all individual participants included in the study.

AI Discloser No AI tool has been used for writing the manuscript.

References

Aflaki A et al (2017) Urban heat island mitigation strategies: a state-of-the-art review on Kuala Lumpur, Singapore and Hong Kong. Cities 62:131–145. https://doi.org/10.1016/j.cities.2016.09.003

Anguelovski I, Connolly JJ, Garcia-Lamarca M, Cole H, Pearsall H (2018) New scholarly pathways on green gentrification: What does the urban 'green turn' mean and where is it going? Prog Hum Geogr 43(6):1064–1086. https://doi.org/10.1177/0309132518803799

Ahern J (2012) Urban landscape sustainability and resilience: The promise and challenges of integrating ecology with urban planning and design. Landscape Ecol 28(6):1203–1212. https://doi.org/10.1007/s10980-012-9799-z

Abdullah-Al-Faisal, Rahman, MS, Kafy, AA, Roy, S, Khan, AR, Rahman, MM (2022) Impact on land surface temperature and urban heat island, due to land use/land cover change in Dhaka Metropolitan Area, using remote sensing and GIS techniques. In Re-envisioning advances in remote sensing. CRC Press, pp 69–102. https://doi.org/10.1201/9781003224624-6

Bokaie M, Zarkesh MK, Arasteh PD, Hosseini A (2016) Assessment of urban heat island based on the relationship between land surface temperature and land use/land cover in Tehran. Sustain Cities Soc 23:94–104. https://doi.org/10.1016/j.scs.2016.03.009

Booth P (1997) Newman, Peter and Thornley, Andy, 'urban planning in Europe: international competition, national system and planning projects' (book review). Town Planning Review 68(2):279. https://doi.org/10.3828/tpr.68.2.4012561447335407

Carlson TN (1986) Regional-scale estimates of surface moisture availability and thermal inertia using remote thermal measurements. Remote Sens Rev 1(2):197–247. https://doi.org/10.1080/02757258609532069

Chen D et al (2022) Landscape and vegetation traits of urban green space can predict local surface temperature. Sci Total Environ 825:154006. https://doi.org/10.1016/j.scitotenv.2022.154006

Cervero R, Murakami J (2009) Rail and property development in Hong Kong: experiences and extensions. Urban Studies 46(10):2019–2043. https://doi.org/10.1177/0042098009339431

Colombo R (2003) Retrieval of leaf area index in different vegetation types using high resolution satellite data. Remote Sens Environ 86(1):120–131. https://doi.org/10.1016/S0034-4257(03)00094-4

Engel RA, Millard-Ball A, Turner VK (2023) Contributions of roads to surface temperature: evidence from Southern California. Environ Res Commun 5(1):015004. https://doi.org/10.1088/2515-7620/acabb8

Fatemi M, Narangifard M (2019) Monitoring LULC changes and its impact on the LST and NDVI in District 1 of Shiraz City. Arab J Geosci **12**(4). https://doi.org/10.1007/s12517-019-4259-6

Gamon JA, Serrano L, Surfus JS (1997) The photochemical reflectance index: an optical indicator of photosynthetic radiation use efficiency across species, functional types, and nutrient levels. Oecologia 112(4):492–501. https://doi.org/10.1007/s004420050337

Geng X, Li C, Yu ZW, Fan Z, Wang X (2022) Natural-anthropogenic environment interactively causes the surface urban heat island intensity variations in global climate zones. Environ Int 170:107574. https://doi.org/10.1016/j.envint.2022.107574

Harlan SL, Brazel AJ, Prashad L, Stefanov WL, Larsen L (2006) Neighborhood microclimates and vulnerability to heat stress. Soc Sci Med 63(11):2847–2863. https://doi.org/10.1016/j.socscimed.2006.07.030

Khandelwal S, Goyal R, Kaul N, Mathew A (2018) Assessment of land surface temperature variation due to change in elevation of area surrounding Jaipur, India. Egypt J Remote Sens Space Sci 21(1):87–94. https://doi.org/10.1016/j.ejrs.2017.01.005

Li Z, et al (2023) Satellite remote sensing of global land surface temperature: definition, methods, products, and applications. Rev Geophys 61(1) https://doi.org/10.1029/2022rg000777

Mustafa EK, et al (2020) Study for predicting land surface temperature (LST) using Landsat data: a comparison of four algorithms. Adv Civil Eng 2020(1). https://doi.org/10.1155/2020/7363546

Nelson N (2021) Smart urbanism not smart cities: takeaways from Bike City Amsterdam. Urban Notes. https://doi.org/10.21428/2bdcbd0f.8ad4512e

Nature Cities (2024) Neighboring rural areas provide cities with 'green clothes' that reduce urban heat islands. Nat Cities **1**(8), 499–500. https://doi.org/10.1038/s44284-024-00092-y

Ouyang W, Morakinyo TE, Ren C, Ng E (2020) The cooling efficiency of variable greenery coverage ratios in different urban densities: a study in a subtropical climate. Build Environ 174:106772. https://doi.org/10.1016/j.buildenv.2020.106772

Ozelkan E, Bagis S, Ozelkan EC, Ustundag BB, Ormeci C (2014) Land surface temperature retrieval for climate analysis and association with climate data. Eur J Remote Sens 47(1):655–669. https://doi.org/10.5721/eujrs20144737

Peng J, Cheng X, Hu Y, Corcoran J (2022) A landscape connectivity approach to mitigating the urban heat island effect. Landscape Ecol 37(6):1707–1719. https://doi.org/10.1007/s10980-022-01439-3

Voogt JA, Oke TR (2003) Thermal remote sensing of urban climates. Remote Sens Environ 86(3):370–384. https://doi.org/10.1016/S0034-4257(03)00079-8

Vergara Perucich JF (2019). Florida, Richard (2017). The new urban crisis: how our cities are increasing inequality, deepening segregation, and failing the middle class—And what we can do about it. Documents d'Anàlisi Geogràfica 65(1):202–203. https://doi.org/10.5565/rev/dag.535

Weng Q, Lu D, Schubring J (2004) Estimation of land surface temperature–vegetation abundance relationship for urban heat island studies. Remote Sens Environ 89(4):467–483. https://doi.org/10.1016/j.rse.2003.11.005

Wesley EJ, Brunsell NA (2019) Greenspace pattern and the surface urban heat island: a biophysically-based approach to investigating the effects of urban landscape configuration. Remote Sens 11(19):2322. https://doi.org/10.3390/rs11192322

Weng Q (2009) Thermal infrared remote sensing for urban climate and environmental studies: methods, applications, and trends. ISPRS J Photogramm Remote Sens 64(4):335–344. https://doi.org/10.1016/j.isprsjprs.2009.03.007

Yuan Y, Santamouris M, Xu D, Geng X, Li C, Cheng W, Su L, Xiong X, Fan Z, Wang X, Liao C (2025) Surface urban heat island effects intensify more rapidly in lower income countries. npj Urban Sustain 5(11). https://doi.org/10.1038/s42949-025-00198-9

Yue W, Xu J, Tan W, Xu L (2007) The relationship between land surface temperature and NDVI with remote sensing: application to Shanghai Landsat 7 ETM+ data. Int J Remote Sens 28(15):3205–3226. https://doi.org/10.1080/01431160500306906

Zhang Y, Ge J, Wang S, Dong C (2025) Optimizing urban green space configurations for enhanced heat island mitigation: a geographically weighted machine learning approach. Sustain Cities Soc 119:106087. https://doi.org/10.1016/j.scs.2024.106087

Chapter 5
Innovative Strategies and Solutions for Mitigating the Urban Heat Island (UHI) Effect

Mohammad Fahad Aslam, Tahir Abdul Rahman Siddiquee, S. V. P. Najid, Umair Ali, and Zulnoorain Khairoowala

Abstract The Urban Heat Island (UHI) effect poses significant challenges to cities globally, exacerbating energy consumption, public health risks, and environmental degradation. Traditional mitigation strategies, while beneficial, often fall short in densely populated urban environments where UHI intensity is increasing. This chapter provides a comprehensive analysis of both traditional and innovative UHI mitigation approaches, emphasizing emerging technologies and integrated frameworks that enhance urban resilience. Key solutions explored include IoT-based temperature monitoring, AI-driven predictive cooling, advanced materials like phase-change and reflective nano coatings, and nature-based blue-green infrastructure. Through comparative analysis and case studies from cities such as Barcelona, Singapore, New York City, Copenhagen, Tokyo, and Seoul, the chapter highlights the effectiveness of adaptive, data-driven approaches in cooling urban spaces. Furthermore, the chapter discusses the importance of ongoing monitoring, community participation, and policy frameworks that foster public–private collaboration, ensuring the long-term success of UHI strategies. This synthesis of innovative methods and traditional practices provides actionable insights for creating resilient, sustainable urban environments in response to rising temperatures.

M. F. Aslam (✉) · U. Ali · Z. Khairoowala
Faculty of Architecture, Planning and Design, Integral University, Lucknow, India
e-mail: ar.fhd@outlook.com

U. Ali
e-mail: ar.umair.ali@gmail.com

Z. Khairoowala
e-mail: khairoowala5291@gmail.com

T. A. R. Siddiquee
Architecture Section, University Polytechnic, Aligarh Muslim University, Aligarh, India
e-mail: tahir.aligarh89@gmail.com

S. V. P. Najid
Bearys Enviro-Architecture Design School, Mangalore, India
e-mail: najid@beads.edu.in

K. Mukherjee et al. (eds.), *Urban Climate*, The Urban Book Series,
https://doi.org/10.1007/978-3-032-17895-4_5

Keywords Urban Heat Island (UHI) · Mitigation strategies · Urban resilience · Sustainable urban planning · Adaptive policy frameworks · Long-term monitoring

5.1 Introduction

The Urban Heat Island (UHI) effect remains a critical concern for cities worldwide, leading to increased energy consumption, health risks, and environmental challenges. Traditional mitigation strategies, while effective to some extent, often fall short in addressing the increasing intensity of UHI, especially in densely populated urban areas. The growing urgency of the issue has spurred the exploration of innovative approaches and advanced, integrated solutions to mitigate UHI effects and enhance urban resilience.

Recent studies emphasize that addressing UHI requires not only improving existing solutions but also incorporating advanced cooling technologies, data-driven urban planning, and multi-stakeholder policy frameworks to create sustainable urban environments. Technological advancements such as IoT-based temperature monitoring systems and AI-driven predictive models are proving to be promising tools in understanding and managing UHI dynamics. According to Salmanian and Bayat, it is essential for urban planners to systematically evaluate these advancements and integrate them into effective policy frameworks that prioritize climate resilience and environmental sustainability (Salmanian and Bayat 2023). Furthermore, recent research by Keith et al. indicates the need for proactive urban governance, suggesting that urban planners should engage more actively in policy-making processes to address heat-related inequities and enhance climate risk management (Keith et al. 2023). As urbanization intensifies, city planners and policymakers must adopt climate-sensitive urban design principles and nature-based solutions, such as blue-green infrastructure, to reduce urban heat and improve thermal comfort. This entails promoting green infrastructure, reflective pavements, and energy-efficient solutions that align with broader sustainability goals (Irfeey et al. 2023).

To achieve these objectives, this chapter explores a wide range of emerging technological solutions, including IoT-driven cooling strategies, innovative materials, and nature-based approaches. It emphasizes the integration of smart technologies and community-driven frameworks to effectively mitigate UHI effects and create resilient urban spaces. Additionally, it highlights the importance of adaptive policy frameworks and long-term monitoring to support sustainable urban transformation and maintain the effectiveness of UHI mitigation strategies over time.

5.2 Research Methodology

This section outlines the research methodology used to assess the effectiveness of traditional and innovative strategies for Urban Heat Island (UHI) mitigation. A mixed-methods approach is employed, combining a systematic literature review, case study analysis, and comparative assessment to ensure a robust and comprehensive evaluation.

5.2.1 Research Approach

The chapter utilizes a mixed-methods approach to analyze UHI mitigation strategies. This approach is structured around three primary components:

a. **Systematic Literature Review**: Analyzes both traditional and innovative strategies for UHI mitigation.
b. **Comparative Case Study Analysis**: Examines real-world applications and results of these strategies in selected cities.
c. **Comparative Framework Application**: Evaluates strategies based on specific key performance indicators (KPIs).

This multi-layered approach allows for an in-depth assessment of UHI mitigation strategies in diverse urban contexts.

5.2.2 Data Collection and Literature Review

A systematic review of peer-reviewed literature from reputable databases including ScienceDirect, SpringerLink, IEEE Xplore, Scopus, etc., was conducted, focusing on publications from 2015 to 2024. The keywords used for the search included *Urban Heat Island, mitigation strategies, innovative technologies, green infrastructure, climate-responsive design*, and *policy frameworks*.

The literature review is divided into two main sections:

a. **Traditional Strategies**: Overview of commonly applied methods like green roofs, urban parks, and reflective pavements, emphasizing their effectiveness and limitations in reducing UHI.
b. **Innovative Strategies**: Examination of contemporary approaches, including IoT-driven cooling solutions, advanced materials like reflective nano coatings, and integrative frameworks that combine technology with nature-based solutions.

5.2.3 Case Study Selection and Analysis

The case study analysis was designed to provide practical insights into the application and effectiveness of UHI mitigation strategies in specific cities. Cities were selected based on the following criteria:

a. **Demonstrated use of innovative UHI mitigation solutions** (e.g., smart technologies, green–blue infrastructure).
b. **Incorporation of integrated policy frameworks** for urban resilience and sustainability.
c. **Availability of empirical data and research findings** on implemented UHI strategies.

Selected cities for the case studies include Barcelona, Singapore, Tokyo, Copenhagen, Seoul, and New York City. These cities are known for their proactive approaches to UHI mitigation, each showcasing unique methods of integrating advanced technologies, policy frameworks, and community-based solutions.

5.2.4 Comparative Analysis Framework

A comparative framework was developed to evaluate traditional and innovative UHI mitigation strategies based on key performance indicators (KPIs) that are crucial for understanding their efficacy, feasibility, and overall impact. This framework provides a structured method for assessing each strategy, considering the following KPIs:

a. **Temperature Reduction and Energy Savings**: Measures the degree to which each strategy reduces urban temperatures and its impact on energy consumption. This is evaluated by comparing quantitative data on temperature reductions and energy savings for both traditional and advanced solutions.
b. **Economic Feasibility and Scalability**: Assesses the initial costs, long-term economic viability, and scalability of each strategy. Traditional methods are compared against contemporary solutions to evaluate cost-effectiveness and suitability for broader application across urban contexts.
c. **Community Participation and Policy Integration**: Examines how each strategy engages the community in implementation and aligns with existing policies. Innovative strategies often require more complex policy support, while traditional methods may have easier community engagement pathways.

The framework allows for a clear comparison of each strategy's strengths and weaknesses, highlighting where innovative approaches address gaps left by traditional methods. This comparative assessment is instrumental in identifying the most sustainable and scalable options for UHI mitigation in urban areas.

5.2.5 Data Analysis Techniques

The data from literature reviews and case studies were analyzed using qualitative content analysis to identify patterns, themes, and insights. Quantitative findings, such as temperature reductions and energy savings, were integrated into the analysis to provide an evidence-based understanding of the relative effectiveness of each strategy. Additionally, triangulation was employed as an alternative validation approach, comparing findings with outcomes from previous meta-analyses and high-impact studies to enhance reliability. This approach ensures that the conclusions drawn align with established research and evidence-based practices.

5.3 Literature Review

5.3.1 Traditional UHI Mitigation Strategies

Traditional strategies for UHI mitigation have primarily focused on green roofs, urban parks, and reflective pavements, which have been widely recognized for their capacity to reduce surface temperatures and improve urban microclimates. Green roofs, for instance, can reduce rooftop temperatures significantly and mitigate the UHI effect by increasing vegetation cover in densely built areas. A study by Cipolla et al. demonstrated that green roofs in Bologna reduced temperatures by as much as 18.9 °C compared to traditional roofs, with a projected decrease in air temperature of up to 3 °C when extensively implemented (Cipolla et al. 2018).

Reflective pavements are another widely used method, effective in increasing the albedo of urban surfaces and reflecting more solar radiation back into the atmosphere. Wang et al. reviewed cool pavements and found that reflective materials contribute significantly to UHI mitigation by reducing surface temperatures and improving thermal comfort during extreme heat events (Wang et al. 2021). However, despite their benefits, these traditional methods often fall short in heavily urbanized environments due to limited scalability and maintenance requirements, particularly in the case of green roofs which necessitate consistent irrigation and upkeep (Akbari and Kolokotsa 2016).

5.3.2 Innovative UHI Mitigation Strategies

In response to the limitations of traditional methods, cities and researchers have explored innovative strategies leveraging advanced materials and technology to enhance the resilience of urban areas to UHI. Smart urban cooling technologies like IoT-based monitoring systems and AI-driven predictive models have been integrated into urban environments to provide real-time temperature monitoring and adaptive cooling. Such technologies allow for responsive adjustments in urban cooling based on real-time data, which improves efficiency and reduces energy demands (Irfeey et al. 2023).

Another area of innovation includes cool materials and advanced green infrastructure. Cool surfaces with reflective nano coatings or phase-change materials (PCMs) are capable of reflecting more sunlight and storing heat energy efficiently, reducing surface temperatures even in dense urban centers (Grdenić and Tomšić 2017). Furthermore, next-generation green roofs that integrate hydroponic systems or solar panels demonstrate a hybrid approach, combining vegetation with renewable energy to enhance cooling and support sustainable energy production (Degirmenci et al. 2021).

Nature-based solutions have also been expanded to incorporate blue-green infrastructure, where interconnected water bodies and green spaces jointly contribute to lowering temperatures and increasing air humidity. This approach has shown to be particularly effective in cities facing prolonged summer heat, offering additional environmental and recreational benefits (Moulai et al. 2019).

5.3.3 Comparative Analysis of Traditional and Contemporary Strategies

By employing a comparative framework, the strengths and limitations of traditional and innovative UHI mitigation strategies can be evaluated against key performance indicators (KPIs) such as temperature reduction, economic feasibility, community engagement, and policy integration.

a. **Temperature Reduction and Energy Savings**: Traditional strategies like green roofs and reflective pavements are effective in moderate cooling but may be insufficient in high-density urban centers. In contrast, advanced materials like PCMs and reflective nano coatings offer higher efficiency in temperature reduction and are better suited for extreme conditions (Rossi et al. 2015).
b. **Economic Feasibility and Scalability**: While traditional methods like parks and green roofs are cost-effective initially, their scalability is often limited by space and water availability. Innovative solutions, although more costly, tend to offer higher ROI over time due to energy savings and reduced maintenance needs. This scalability advantage makes them viable options for large urban areas (Aleksandrowicz et al. 2017).
c. **Community Participation and Policy Integration**: Traditional approaches generally align well with existing urban policies and are easily adopted by local communities. However, newer technologies, such as smart cooling systems, require more complex policy support, including data-sharing frameworks and privacy considerations for IoT-based systems (Akbari and Kolokotsa 2016). This complex integration can pose challenges but also provides an opportunity for cities to modernize and improve urban resilience.

d. **Implementation Challenges and Contextual Constraints:** While innovative strategies such as AI-driven cooling and IoT-based monitoring offer significant potential, they often require substantial initial investment and advanced technical infrastructure, limiting their adoption in low-resource settings. Studies by Morales-Inzunza et al. (2023) emphasize that the scalability of cool materials and smart systems can be hindered by economic and regulatory barriers. Furthermore, green infrastructure approaches—though environmentally beneficial—demand continuous maintenance, including irrigation, pruning, and replacement of plant material, particularly in arid or rapidly urbanizing regions (Ladan et al. 2023). Cultural acceptability and community buy-in also affect the success of these interventions. As Boateng et al. (2023) note, participatory planning and localized design are crucial for ensuring long-term viability and equitable outcomes.

The visual matrix below (Table 5.1) summarizes the comparative analysis, showing the performance of each strategy across KPIs:

This matrix provides a clear illustration of where innovative strategies outperform traditional approaches, especially in areas like scalability and temperature reduction. Each approach's performance within this framework emphasizes that an integrated mix of traditional and contemporary methods is essential for effective UHI mitigation across diverse urban settings.

5.4 Emerging Technological Solutions

5.4.1 Smart Urban Cooling Technologies

Smart urban cooling technologies leverage IoT-based systems, sensor networks, and AI-driven heat prediction models to provide real-time temperature monitoring and adaptive cooling solutions in urban environments. IoT sensors, for instance, can capture detailed temperature data across various city zones, enabling efficient tracking of heat distribution and UHI intensity in real time. This data is invaluable for predictive models that adjust cooling strategies based on current and forecasted temperatures, enhancing response effectiveness and energy efficiency (Chen et al. 2022). For instance, AI-driven models allow for predictive cooling adjustments in high-traffic areas, which helps reduce peak UHI effects and improve urban livability (Grdenić and Tomšić 2017).

Furthermore, combining IoT-based data with cloud analytics enables urban managers to engage in dynamic cooling management, reducing energy costs associated with air conditioning and helping to prevent overuse of energy resources in heavily populated areas (Akbari and Kolokotsa 2016). Such integrations not only enhance temperature management but also align with broader sustainability goals.

Table 5.1 Performance of each strategy across KPIs

Strategy	Temperature reduction	Energy savings	Economic feasibility	Scalability	Community engagement	Policy integration	Implementation constraints
Traditional green roofs	Moderate	Moderate	High	Limited	High	High	High maintenance and water demand
Reflective pavements	Moderate	Moderate	Moderate	High	Moderate	Moderate	Effective in hot-dry climates; limited visual appeal
IoT smart cooling systems	High	High	Moderate	High	Low	Moderate	High setup cost; requires digital infrastructure
Phase-change materials (PCMs)	High	High	High	High	Low	High	Material cost; performance varies by climate
Blue-green infrastructure	High	Moderate	Moderate	Moderate	High	Moderate	Requires land and maintenance; regional feasibility varies

5.4.2 Innovative Building Materials and Cool Surfaces

In recent years, there has been significant progress in innovative materials designed to mitigate UHI effects, with particular focus on reflective nano coatings and phase-change materials (PCMs). Reflective nano coatings have proven highly effective in reducing surface temperatures by reflecting solar radiation away from buildings and pavement surfaces. These materials are especially valuable in densely built environments where traditional green infrastructure might be impractical (Malet-Damour et al. 2023; Pigliautile et al. 2020).

PCMs offer another promising solution by absorbing and releasing thermal energy based on ambient temperatures, thereby maintaining cooler surface temperatures. Morales-Inzunza et al. highlighted PCMs' effectiveness in urban cooling, especially in warm climates where temperature fluctuations can be intense (Morales-Inzunza et al. 2023). PCMs are often used in advanced cool roofs, where they can reduce rooftop temperatures, decreasing energy demand for air conditioning within buildings.

5.4.3 Advanced Green Infrastructure and Urban Vegetation

The development of advanced green infrastructure has also contributed significantly to UHI mitigation, with designs integrating vegetation with renewable energy systems, like photovoltaic (PV) panels and adaptive urban parks. These approaches extend beyond traditional green roofs by combining greenery with technology that enhances urban cooling while generating renewable energy. For example, green roofs with integrated hydroponic systems or PV panels are increasingly being adopted to optimize both energy efficiency and cooling in urban environments (Mat Santamouris and Yun 2020).

Moreover, vertical gardens and adaptive urban parks are effective in dense urban areas where horizontal space for greenery is limited. Ibrahim noted that such green infrastructure, combined with reflective materials and cooling technologies, can significantly lower temperatures and increase urban resilience against heat waves, especially in arid and semi-arid regions (Ibrahim 2021). These structures not only contribute to UHI mitigation but also enhance urban biodiversity and air quality.

5.5 Integrated Urban Planning and Policy Frameworks

5.5.1 *Climate-Sensitive Urban Design*

Climate-sensitive urban design integrates principles of shading, natural ventilation, and water bodies to mitigate the Urban Heat Island (UHI) effect effectively. Shading is an essential design element, achieved through tree canopies, green facades, and building structures that block direct sunlight, thereby lowering surface and ambient temperatures. Studies by Banerjee et al. demonstrate that urban layouts with optimized shading can reduce air temperatures by up to 5 °C during peak summer, significantly enhancing urban comfort levels (Banerjee et al. 2022). In addition, natural ventilation allows air to circulate freely within urban spaces, reducing stagnant hot air pockets, a crucial aspect for hot, densely populated cities. This effect can be enhanced by orienting streets and building facades in line with prevailing winds, as shown in case studies in Singapore and Hong Kong (Ng et al. 2012).

The integration of water bodies such as fountains, ponds, and misting systems provides additional cooling through evaporative processes. Studies have shown that urban parks with water features can reduce local temperatures and provide thermal comfort in high-density urban settings (Armson et al. 2012). These climate-responsive design strategies contribute to creating more resilient and comfortable cities, essential for adapting to intensifying urban heat.

5.5.2 *Policies and Regulatory Approaches for UHI Mitigation*

Effective Urban Heat Island mitigation requires comprehensive policy frameworks that align with broader urban sustainability and climate resilience goals. Many cities have adopted policies to regulate building materials, enforce green roofing, and mandate urban greening in response to UHI challenges. For example, Tokyo's Green Building Program mandates that all new construction projects include green roofs and facades, aiming to lower surface temperatures and enhance biodiversity (Shimpo 2022). Similarly, New York City's Local Law 92 and 94 requires green or reflective roofs on new and significantly renovated buildings, supporting both energy efficiency and UHI mitigation (NYC Dep. Build. 2019).

Multi-stakeholder frameworks are essential in urban climate policy, as they facilitate collaborative efforts among local governments, private developers, community organizations, and environmental agencies. These frameworks are increasingly seen in European and Asian cities, where policies encourage public–private partnerships to implement green infrastructure and cooling technologies. For instance, Copenhagen's climate adaptation strategy emphasizes collaboration between municipal authorities and local residents to promote green and blue infrastructure, showcasing how collective efforts can bolster UHI mitigation (Kabisch et al. 2016).

The integration of UHI mitigation within sustainability policies allows cities to address multiple urban challenges simultaneously. By aligning with goals for carbon reduction, biodiversity, and air quality improvements, UHI mitigation policies contribute to comprehensive urban resilience strategies. However, studies indicate that successful policy implementation requires consistent community engagement, transparent communication, and adequate funding to ensure long-term effectiveness (Lindberg and Grimmond 2011).

5.6 Nature-Based Solutions (NBS) for Urban Resilience

5.6.1 Integration of Blue and Green Infrastructure

Nature-based solutions (NBS), especially the integration of blue and green infrastructure, have shown significant promise in mitigating the Urban Heat Island (UHI) effect by enhancing urban cooling, supporting biodiversity, and improving air quality. Green infrastructure, such as parks and vegetation, combined with water features like ponds, fountains, and urban wetlands, contributes to cooling through evapotranspiration and provides natural shade, lowering surface, and ambient temperatures (Hassan et al. 2023). Studies have highlighted the success of blue-green infrastructure in cities like London and Copenhagen, where strategically placed greenery and water bodies create cooler microclimates and contribute to enhanced urban resilience (Mutafoglu et al. 2017).

For instance, in Mediterranean cities where summer temperatures soar, green roofs combined with water-retentive landscapes effectively reduce the UHI effect, while simultaneously addressing stormwater management and reducing flood risks (Leone et al. 2020; Pelorosso et al. 2017). Such integration has been particularly effective in high-density urban areas, where traditional green spaces are limited, and maximizing the cooling effect is essential for improving public health and comfort.

5.6.2 Community-Driven and Participatory Approaches

Community engagement plays a crucial role in the design, implementation, and maintenance of NBS, particularly as cities strive to create resilient and livable urban environments. Community-driven approaches foster a sense of ownership among residents, leading to improved care and sustainability of green spaces, which ultimately supports the long-term effectiveness of these solutions. Participatory strategies, such as involving local communities in the planning of urban parks and green roofs, have been instrumental in cities like Barcelona and Singapore, where active public involvement has enhanced the ecological and social impact of UHI mitigation efforts (Boateng et al. 2023).

Research shows that urban greening initiatives with community involvement not only reduce urban heat but also provide social benefits, such as improved mental health, recreational opportunities, and stronger community ties. For example, initiatives in New York City that involve residents in park upkeep and tree-planting programs have improved both environmental conditions and local quality of life, showcasing the importance of participatory approaches in enhancing urban resilience (Tauhid 2018).

These participatory strategies not only improve long-term effectiveness but also ensure that solutions are tailored to local social and climatic contexts, minimizing the risk of underuse or neglect. However, widespread adoption of such approaches requires adequate funding, maintenance planning, and social inclusivity—factors that are often underdiscussed in early-phase implementation strategies.

5.7 Case Studies of Innovative UHI Solutions

5.7.1 Smart Cities and Cooling Technologies

The integration of IoT and AI-driven cooling solutions in smart cities has become a vital approach for mitigating the Urban Heat Island (UHI) effect. In this section, we examine how Barcelona, Singapore, and New York City have utilized advanced cooling technologies, emphasizing real-time temperature monitoring, data-driven decision-making, and adaptive strategies that promote urban resilience.

5.7.1.1 Barcelona: IoT-Enabled Cooling at La Plaça De Les Glòries

Barcelona has been at the forefront of smart city initiatives, particularly in addressing the UHI effect. One notable project is the smart cooling initiative at La Plaça de les Glòries, where the city has implemented a network of IoT sensors to monitor temperature and humidity levels continuously. These sensors provide real-time data to inform targeted cooling interventions, optimizing cooling efficiency based on immediate conditions. By tracking heat patterns, city planners can deploy mobile cooling units or activate misting systems in areas where temperatures exceed predetermined thresholds, reducing thermal discomfort in public spaces (Calderó Gràcia 2015).

The project has contributed to reduced energy costs associated with cooling while improving pedestrian comfort in high-traffic areas. Additionally, Barcelona's smart cooling project at La Plaça de les Glòries serves as a model for integrating technology with public spaces, where real-time data can support timely responses to temperature fluctuations, thus enhancing urban livability.

5.7.1.2 Singapore: AI-Powered Cooling Solutions Under the Smart Nation Initiative

Singapore's Smart Nation Initiative represents one of the most comprehensive smart city frameworks globally, leveraging AI and IoT technologies to enhance urban cooling and combat the UHI effect. Singapore's strategy includes AI-powered predictive models, which anticipate temperature peaks and automatically trigger cooling mechanisms in areas with high foot traffic, especially in dense urban zones (Aflaki et al. 2017). For instance, the city employs sensors to detect temperature, humidity, and even crowd density, which informs when and where cooling measures are needed most.

The use of AI-driven predictive cooling aligns well with Singapore's urban greening efforts, as it integrates seamlessly with green infrastructure like rooftop gardens and vertical greenery. By strategically cooling spaces in advance of peak temperature periods, Singapore has managed to achieve significant reductions in ambient temperature, which not only improves public comfort but also reduces the energy burden on buildings for air conditioning. This proactive, technology-enabled approach illustrates how AI can transform urban cooling from a reactive to a predictive process, maximizing efficiency and minimizing environmental impact.

5.7.1.3 New York City: Cool Neighborhoods NYC and IoT Temperature Monitoring

New York City has implemented Cool Neighborhoods NYC, a program designed to reduce heat exposure in vulnerable communities through IoT-driven temperature monitoring and localized cooling interventions. The program focuses on areas with high heat exposure and limited access to green spaces, leveraging IoT sensors to track temperatures across various neighborhoods. Data collected through these sensors enable the city to deploy temporary cooling measures, such as cooling stations and tree canopies, in neighborhoods identified as heat islands (Moulai et al. 2019).

In addition to short-term cooling interventions, New York City's program includes a long-term vision that integrates IoT data into urban planning to increase vegetation and implement reflective materials in high-heat areas. The use of IoT technology in Cool Neighborhoods NYC demonstrates how data-driven urban cooling strategies can address specific vulnerabilities, ensuring that the most affected areas receive adequate attention and resources. This approach highlights the value of smart technologies in enhancing community resilience, especially in urban environments with diverse and densely populated neighborhoods.

5.7.2 *Green–Blue Infrastructure in European and Asian Cities*

In recent years, several cities across Europe and Asia have successfully integrated green and blue infrastructure as a dual approach to mitigate the Urban Heat Island (UHI) effect, improve urban resilience, and enhance quality of life. This section examines the cases of Copenhagen, Tokyo, and Seoul, highlighting their unique approaches and successful implementations.

5.7.2.1 Copenhagen: Climate-Resilient Blue-Green Infrastructure

Copenhagen is widely recognized for its proactive approach to climate adaptation, particularly through its integration of blue-green infrastructure. The city's Cloudburst Management Plan focuses on creating resilient urban spaces that incorporate water retention systems, green corridors, and parks to manage stormwater and mitigate UHI simultaneously. One of the key projects under this plan is the Tåsinge Plads, Denmark's first climate-resilient neighborhood park, designed with features such as rain gardens, bioswales, and permeable surfaces to control excess rainwater while providing cooling benefits to the surrounding area (Kabisch et al. 2016).

Through these efforts, Copenhagen has demonstrated that blue-green infrastructure not only mitigates urban heat but also supports flood management, biodiversity, and recreational opportunities for residents. This integrated approach to UHI mitigation showcases how multifunctional infrastructure can contribute to overall urban resilience and sustainable urban development.

5.7.2.2 Tokyo: Integrating Green Infrastructure in a High-Density Urban Environment

Tokyo's dense urban landscape presents unique challenges for implementing green spaces. However, the city has been successful in leveraging vertical greenery and rooftop gardens as part of its UHI mitigation strategy. The Tokyo Green Plan mandates the inclusion of green spaces on rooftops and encourages the use of wall vegetation on high-rise buildings. This strategy not only reduces surface temperatures but also improves air quality and enhances urban aesthetics (Shimpo 2022).

Additionally, Tokyo has developed urban cooling islands in areas like Shibuya, which include water features, shaded seating areas, and green roofs to create cooler microclimates. These green spaces have been shown to reduce temperatures by several degrees in densely populated areas, providing a respite for pedestrians and reducing the energy needed for air conditioning. The integration of both green and blue infrastructure within a high-density city like Tokyo highlights the feasibility of UHI mitigation even in space-constrained environments.

5.7.2.3 Seoul: Revitalizing Streams as Cooling Corridors

Seoul has undertaken ambitious projects to restore natural waterways and create blue-green corridors to combat UHI and improve urban resilience. The most famous project, Cheonggyecheon Stream Restoration, transformed a former concrete roadway into a flowing stream surrounded by green spaces. This restored stream corridor has reduced nearby temperatures, enhancing cooling for pedestrians and improving overall air quality. Studies indicate that temperatures around the Cheonggyecheon Stream are 3–5 °C lower compared to surrounding areas, illustrating the significant impact of integrating water and greenery into the urban fabric (Han and Huh 2008; Hwang 2015; Kim et al. 2024; Kim and Jung 2019; Lee et al. 2022).

Seoul's success with the Cheonggyecheon project has inspired similar initiatives throughout the city, where green and blue infrastructure continues to play a vital role in sustainable urban planning. The project highlights how restoring natural waterways in urban settings can provide effective and lasting UHI mitigation, while also offering ecological, social, and recreational benefits for city residents.

5.8 Evaluating the Long-Term Impact of Innovative UHI Strategies

5.8.1 Importance of Ongoing Monitoring and Assessment

For Urban Heat Island (UHI) mitigation strategies to be effective, continuous monitoring and assessment are essential. Long-term evaluation allows cities to assess the efficacy of implemented strategies, adapt them based on changing urban climates, and measure their contribution to overall urban resilience. Effective UHI mitigation goes beyond immediate cooling and aims for sustained impact over time, which requires comprehensive data on variables such as temperature reduction, energy savings, and urban comfort levels (Banerjee et al. 2022).

Monitoring the long-term effectiveness of UHI mitigation strategies is particularly important due to climate change's evolving nature and its direct impact on urban temperatures. According to researchers, even slight increases in global temperatures can intensify the UHI effect, particularly in cities with dense urban structures and limited green spaces. Therefore, assessing the sustained impact of UHI solutions can provide insights into their adaptation to different climates and their resilience to increased heat exposure over time (Morales-Inzunza et al. 2023; Santamouris 2015).

5.8.2 Tools and Metrics for Periodic Evaluation

Periodic evaluation is crucial for assessing the long-term effectiveness of UHI mitigation strategies, allowing for timely adaptations based on real-world data. The following Table 5.2 highlights easy-to-implement tools and metrics that cities can use for ongoing assessment without extensive validation procedures.

Table 5.2 lists tools that support evaluation with minimal resource requirements, making them accessible for a range of urban settings. These include:

- **Temperature Tracking Systems**: Tools such as IoT-based temperature sensors and thermal imaging can be deployed to track surface and air temperatures in real time. For example, temperature monitoring stations and thermal cameras can provide precise data on temperature variations across different urban areas, highlighting areas where UHI mitigation strategies may need to be reinforced (Chen et al. 2022).
- **Energy Consumption Analysis**: Analyzing energy usage, particularly for air conditioning in buildings near UHI mitigation sites, can provide indirect data on the cooling effectiveness of green roofs, cool pavements, and shaded areas. Reduced energy consumption indicates a decrease in heat accumulation and a corresponding reduction in the need for mechanical cooling, which can be quantified and compared over time (Akbari and Kolokotsa 2016).
- **Community Feedback and Surveys**: Gathering feedback from residents through surveys or digital platforms can offer valuable insights into perceived improvements in comfort levels and the overall quality of life. Community feedback can also serve as an indicator of social acceptance and engagement with UHI strategies, which is crucial for their long-term success (Aflaki et al. 2017).

Table 5.2 Easy-to-implement tools and metrics for long-term UHI mitigation evaluation

Tool/metric	Description	Purpose	Ease of implementation
Temperature tracking systems	Real-time monitoring using IoT sensors, thermal cameras, or temperature stations	Measures variations in surface and air temperatures across urban areas to assess cooling effectiveness	High—requires minimal installation and remote data collection
Energy consumption analysis	Tracks energy usage in nearby buildings, focusing on air conditioning usage	Provides indirect data on UHI reduction by evaluating energy savings in air-conditioned spaces	Moderate—requires periodic data collection from energy meters
Community feedback and surveys	Collects resident feedback on perceived thermal comfort, urban comfort, and quality of life	Engages the community, providing qualitative insights on social acceptance and overall satisfaction	High—simple survey or digital feedback, low-cost and effective

By incorporating these metrics, urban planners can make informed decisions on where and how to adapt UHI mitigation measures. Periodic assessments using these tools allow for low-cost evaluations without the need for extensive validation processes, making them accessible for cities with varying resources and technical expertise.

5.9 Recommendations and Future Directions

Based on the insights from the comparative analysis of Urban Heat Island (UHI) mitigation strategies, several recommendations emerge to optimize the integration of innovative technologies, urban planning, and policy support. The integration of smart cooling technologies such as IoT-enabled monitoring and AI-driven predictive models should be prioritized in dense urban centers to allow for adaptive, data-driven responses to temperature fluctuations. These technologies have proven effective in cities like Singapore and New York City, where real-time monitoring supports targeted interventions that maximize cooling with minimal resource usage (Aflaki et al. 2017).

Furthermore, urban planning should incorporate climate-sensitive design principles that encourage natural shading, ventilation, and the strategic use of green–blue infrastructure. Policymakers should mandate the inclusion of these elements in urban development guidelines, as seen in Tokyo's Green Plan and Copenhagen's cloudburst strategy. Integrating these approaches in building codes and land-use policies can enhance resilience to urban heat, while also supporting biodiversity and stormwater management (Kabisch et al. 2016).

Policy support is essential for scaling UHI mitigation efforts. Municipalities should create incentives for private developers to adopt reflective materials, green roofs, and advanced cooling technologies. Additionally, establishing multi-stakeholder frameworks that include government agencies, community groups, and private entities can foster collaborative implementation, as demonstrated in New York City's Cool Neighborhoods program. Policy frameworks that support public–private partnerships can extend the reach of UHI mitigation strategies, creating synergies between climate resilience and urban development (Santamouris 2015; Mat Santamouris 2013; Mat Santamouris and Yun 2020).

Future research in UHI mitigation should focus on advancing cooling technologies, innovative policy frameworks, and real-time adaptive management tools. Emerging materials, such as phase-change materials and reflective nano coatings, could enhance cooling efficiency while reducing energy demand (Leone et al. 2020). Research on flexible policies, including incentives for green–blue infrastructure and community-driven projects, will be essential for diverse urban contexts (Kabisch et al. 2016). Additionally, exploring IoT-enabled sensors and predictive analytics can support real-time monitoring, allowing cities to dynamically adjust UHI strategies based on immediate temperature data, thus boosting urban resilience to heat (Chen et al. 2022).

To support more effective implementation, planners should prioritize interventions based on regional climate characteristics and urban form. For example, reflective coatings and phase-change materials are ideal for dry, compact cities with limited green space, while blue-green infrastructure is more appropriate for tropical or temperate climates with sufficient rainfall. This contextual prioritization ensures that UHI strategies remain both technically effective and economically viable.

5.10 Conclusion

This chapter has highlighted the critical importance of innovative and integrated approaches for effective Urban Heat Island (UHI) mitigation. Traditional UHI strategies, while beneficial, often lack the adaptability and efficiency required for today's rapidly urbanizing and warming cities. The chapter emphasized how emerging technologies—such as IoT-enabled monitoring systems, AI-driven predictive models, and advanced cooling materials—provide cities with precise and adaptive tools for reducing urban temperatures. Case studies from cities like Barcelona, Singapore, New York City, Copenhagen, Tokyo, and Seoul demonstrate the tangible benefits of combining green–blue infrastructure, smart technologies, and community engagement, leading to greater resilience and environmental sustainability.

The findings reinforce that sustainable urban planning must incorporate climate-sensitive design principles, policy frameworks, and technological advancements. Cities must prioritize shading, natural ventilation, and cooling infrastructure in urban layouts, supported by policies that encourage public–private partnerships and stakeholder collaboration. By focusing on resilient and adaptable UHI mitigation strategies, cities can better protect residents from extreme heat, reduce energy demand, and enhance quality of life. As climate challenges intensify, adopting these comprehensive approaches will be essential to creating livable and sustainable urban environments for the future.

While this study outlines a broad suite of strategies, it is essential to recognize certain limitations. High upfront costs, data privacy concerns in smart systems, and maintenance demands for green infrastructure may constrain adoption. Additionally, the scalability and cultural acceptability of these solutions vary significantly across cities. Future research should explore cost–benefit analyses, flexible regulatory frameworks, and mechanisms to tailor UHI solutions to local socio-environmental conditions.

Potential Conflicts of Interest On behalf of all authors, the corresponding author states that there is no conflict of interest.

AI Discloser ChatGPT has been used for grammar correction and sentence restructuring.

Authors Contribution All authors have contributed equally to this work.

Funding The research was not supported by any funding.

References

Aflaki A, Mirnezhad M, Ghaffarianhoseini A, Ghaffarianhoseini A, Omrany H, Wang Z-H, Akbari H (2017) Urban heat island mitigation strategies: a state-of-the-art review on Kuala Lumpur, Singapore and Hong Kong. Cities 62:131–145

Akbari H, Kolokotsa D (2016) Three decades of urban heat islands and mitigation technologies research. Energy Build 133:834–842

Aleksandrowicz O, Vuckovic M, Kiesel K, Mahdavi A (2017) Current trends in urban heat island mitigation research: observations based on a comprehensive research repository. Urban Clim 21:1–26

Armson D, Stringer P, Ennos AR (2012) The effect of tree shade and grass on surface and globe temperatures in an urban area. Urban Forestr Urban Green 11(3):245–255

Banerjee S, Yik SK, Dzyuban Y, Crank PJ, Yi RPX, Chow WTL (2022) Analysing impacts of urban morphological variables and density on outdoor microclimate for tropical cities: a review and a framework proposal for future research directions. Build Environ 225:109646

Boateng EA, Asibey MO, Cobbinah PB, Adutwum IO, Blija DK (2023) Enabling nature-based solutions: Innovating urban climate resilience. J Environ Manage 332:117433

Calderó Gràcia M (2015) Projecte de millora de les propietats d'illa de calor per la plaça de les Glòries de Barcelona i el seu entorn. Universitat Politècnica de Catalunya

Chen Y, Wang Y, Zhou D, Gu Z, Meng X (2022) Summer urban heat island mitigation strategy development for high-anthropogenic-heat-emission blocks. Sustain Cities Soc 87:104197

Cipolla S, Maglionico M, Semprini G, Villani V, Bonoli A (2018) Green roofs as a strategy for urban heat island mitigation in Bologna (Italy). Acta Horticulturae 295–300. https://doi.org/10.17660/ActaHortic.2018.1215.54

Degirmenci K, Desouza KC, Fieuw W, Watson RT, Yigitcanlar T (2021) Understanding policy and technology responses in mitigating urban heat islands: a literature review and directions for future research. Sustain Cities Soc 70:102873

Grdenić G, Tomšić Ž (2017) Renewable energy sources and other energy technologies as a measure for mitigating the impact of urban heat islands. J Energy: Energija 66(1–4)

Han S-G, Huh J-H (2008) Estimate of the heat island and building cooling load changes due to the restored stream in Seoul Korea. Int J Urban Sci 12(2):129–145

Hassan Z, Shahbaz B, Gallardo Lopez F (2023) Enhancing blue/green infrastructure for resilient urban environments: smart solutions and nature-based strategies. In: 3rd international congress on engineering and life science ICELIS, pp 50

Hwang SW (2015) Restoration of urban streams and transforming lifestyle patterns: two cases in Seoul, South Korea. J Architect Plann Res 131–144

Ibrahim MS (2021) Mitigation strategies of the urban heat island over Greater Cairo Metropolitan Area, Egypt utilizing ENVI-met model. Catrina: Int J Environ Sci 24(1):35–47

Irfeey AMM, Chau H-W, Sumaiya MMF, Wai CY, Muttil N, Jamei E (2023) Sustainable mitigation strategies for urban heat island effects in urban areas. Sustainability 15(14):10767

Kabisch N, Strohbach M, Haase D, Kronenberg J (2016) Urban green space availability in European cities. Ecol Ind 70:586–596

Keith L, Gabbe CJ, Schmidt E (2023) Urban heat governance: examining the role of urban planning. J Environ Planning Policy Manage 25(5):642–662. https://doi.org/10.1080/1523908X.2023.2244446

Kim A-R, Seol J, Lim B-S, Lim C-H, Kim G-S, Lee C-S (2024) Response of plant phenology on microclimate change depending on land use intensity in Seoul Central Korea. Forests 15(4):718

Kim H, Jung Y (2019) Is Cheonggyecheon sustainable? a systematic literature review of a stream restoration in Seoul, South Korea. Sustain Cities Soc 45:59–69

Ladan T, Ibrahim M, Ismail M, Mukhtar A (2023) The potential of urban green infrastructure in mitigating urban heat islands in the semi-arid regions. Int J Acad Res Bus Soc Sci 13. https://doi.org/10.6007/IJARBSS/v13-i6/17392

Lee K, Kim Y, Sung HC, Kim SH, Jeon SW (2022) Surface urban heat island in South Korea's new towns with different urban planning. Environ Monit Assess 194(5):360
Leone A, Gobattoni F, Pelorosso R, Calace F (2020) Nature-based climate adaptation for compact cities: green courtyards as urban cool islands. Plurimondi 18:83–110
Lindberg F, Grimmond CSB (2011) The influence of vegetation and building morphology on shadow patterns and mean radiant temperatures in urban areas: model development and evaluation. Theoret Appl Climatol 105:311–323
Malet-Damour B, Bigot D, Rivière G (2023) Experimental and numerical analysis on a thermal barrier coating with nano-ceramic base: a potential solution to reduce urban heat islands? Eng 4(3):2421–2442
Morales-Inzunza S, González-Trevizo ME, Martínez-Torres KE, Luna-León A, Tamayo-Pérez UJ, Fernández-Melchor F, Santamouris M (2023) On the potential of cool materials in the urban heat island context: scalability challenges and technological setbacks towards building decarbonization. Energy Build 296:113330
Moulai M, Kharvari F, Shahhosseini G, Esmailpour Zanjani N (2019) Study of the urban heat island mitigation strategies: the case of two cities. J Urban Manag Energy Sustain 1(2):34–40
Mutafoglu K, Ten Brink P, Dekker S, Woollard J, Schweitzer JP (2017) Adapting to climate change and improving urban resilience: the role of nature and biodiversity protection in cities. In: Building a climate resilient economy and society. Edward Elgar Publishing, pp 59–77
Ng E, Chen L, Wang Y, Yuan C (2012) A study on the cooling effects of greening in a high-density city: an experience from Hong Kong. Build Environ 47:256–271
NYC Department of Buildings (2019) NYC Department of Buildings. https://www1.nyc.gov/assets/buildings/local_laws/ll92-94_2019.pdf
Pelorosso R, Gobattoni F, Leone A (2017) Green courtyards as urban cool islands: towards nature-based climate adaptation plans of compact cities. CSE-City Safety Energy 1:27–36
Pigliautile I, Chàfer M, Pisello AL, Pérez G, Cabeza LF (2020) Inter-building assessment of urban heat island mitigation strategies: field tests and numerical modelling in a simplified-geometry experimental set-up. Renew Energy 147:1663–1675
Rossi F, Castellani B, Presciutti A, Morini E, Filipponi M, Nicolini A, Santamouris M (2015) Retroreflective façades for urban heat island mitigation: experimental investigation and energy evaluations. Appl Energy 145:8–20
Salmanian M, Bayat A (2023) Urban heat island: a primary guide for urban designers. Fut Energy 2(4):10–23
Santamouris M (2015) Analyzing the heat island magnitude and characteristics in one hundred Asian and Australian cities and regions. Sci Total Environ 512:582–598
Santamouris M (2013) Using cool pavements as a mitigation strategy to fight urban heat island—a review of the actual developments. Renew Sustain Energy Rev 26:224–240
Santamouris M, Yun GY (2020) Recent development and research priorities on cool and super cool materials to mitigate urban heat island. Renew Energy 161:792–807
Shimpo N (2022) Urban ecological life in a metropolitan area—an insight from Satoyama conservation activities in the Greater Tokyo Area. Landscape Ecol Eng 18(1):109–119
Tauhid FA (2018). Urban green infrastructure for climate resilience: a review. Nat Nat Acad J Architect 5(1):58–65
Wang C, Wang Z-H, Kaloush KE, Shacat J (2021) Cool pavements for urban heat island mitigation: a synthetic review. Renew Sustain Energy Rev 146:111171

Chapter 6
Application of Remote Sensing and GIS for Developing Heat Action Plans

Mohit Kumar and Sarika Keswani

Abstract The world is urbanizing rapidly. By 2050, more than two-thirds of the world's population will reside in cities. Increased urbanization has modified many biophysical processes which often lead to a rapid increase in local ambient temperatures. Such increase in intra-urban temperatures results in the formation of urban heat islands (UHI), raising concerns over the health of people and the environment. In the last few decades, the incidences of heat waves have increased, apparently because of climate change. Therefore, the city authorities across the globe are working hard to develop city-specific heat action plans (HAPs). These HAPs are targeted at saving the lives of people during heat wave episodes, especially those living in UHIs, through early adoption of preventive measures. Remote sensing and Geographic Information System (GIS) can greatly aid in developing HAPs. In this paper, we have provided an example of how city-level HAPs can be made more robust using remote sensing and GIS, and discuss the way forward.

Keywords Urban Heat Islands (UHIs) · Heat action plans (HAPs) · Land surface temperature (LST) · Remote sensing · GIS

6.1 Introduction

In the past few decades, the world has become more urban. For example, in 2018, approximately half of the global population was urban and by 2050 more than two-thirds of the world will reside in cities (UN 2018). Increasingly urbanizing world will see many modifications in basic life processes which will affect the living beings on air, water and land. One of the consequences of urbanization is rapid increase in

M. Kumar (✉)
Department of Geography, Adamas University, Barasat, India
e-mail: mohit6kumar@gmail.com

S. Keswani
Symbiosis Centre for Management Studies (SCMS), Symbiosis International (Deemed University), Nagpur, India

K. Mukherjee et al. (eds.), *Urban Climate*, The Urban Book Series,
https://doi.org/10.1007/978-3-032-17895-4_6

local ambient temperature (both air temperature as well as land surface temperature), which has significantly impacted the lives of citizens across all the major global cities. The cities are experiencing the formation of Urban Heat Islands (UHIs) with temperatures higher than the surrounding relatively rural areas. Such UHIs are mostly characterized by heat-absorbing land forms built mostly of impervious materials like concrete and asphalt (Tyagi et al. 2021). These UHIs make daily livelihood activities difficult for human beings as our optimum physiological range of temperature for proper functioning is exceeded. Many of these human health impacts of elevated temperatures are manifested in the form of cardiovascular, cerebrovascular and respiratory ailments (Sharma et al. 2019). These UHIs when persist for few days in same area lead to formation of heat waves. A constant upsurge in temperature during the heat wave events can also lead to power outages as the energy consumption and energy demands rise exponentially. This is because as the temperatures soar on the higher side, artificial cooling requirements (like Air Conditioners, Fans etc.) for the human beings will also increase. For example, during April 2012 to March 2017, UHI formation caused an increased electricity consumption of 2600 GWh (i.e. 11.4%) annually on an average over eight districts of Delhi (Kumari et al. 2021). Such power outages in high temperatures can be uncomfortable, inconvenient and even dangerous, especially for children and the elderly. The UHI incidents, and in turn the incidences of heat waves, have increased in recent years, possibly due to anthropogenic climate change (IPCC 2013; Sharma et al. 2019). As predicted by Intergovernmental Panel on Climate Change (IPCC), the incidences of heat waves are supposed to increase in future (IPCC 2013; Sharma et al. 2019). IPCC report also suggests an increase in the frequency and duration of the heat waves in near future (IPCC 2013; Sharma et al. 2019). The incidences of heat waves are also on the rise in India. During the summer months of March–May, the country faces heat waves in one or more areas almost every year (Mishra et al. 2017). Eleven of India's 15 warmest years have occurred recently during 2008–2022 (IMD 2022). The annual mean surface air temperature over the country was reportedly 0.51 °C above the average for 1981–2010 period in 2022 (IMD 2022). The year 2022 was the fifth warmest year on record since nationwide records commenced in 1901 (IMD 2022). The increased events of heat waves over the years have caused many fatal incidents in India. In India, heat waves have been recognized as one of the most important natural threats in terms of number of people died in cities. During 1992 and 2015, more than 24,000 deaths were attributed to heat waves across the country (NDMA 2019). In 2016, more than 1000 deaths were attributed to heat wave in India (Guleria and Gupta 2018). It is therefore increasingly felt that a heat action plan (HAP) should be developed at city scale, which will guide the administrators to target their efforts at vulnerable locations and vulnerable population during the intense period of heat waves. The identification of vulnerable areas becomes challenging because meteorological locations within the Indian cities are very scarce and therefore do not provide an accurate representation of intra-urban temperature variability. In fact, many of the Indian cities even today either do not have any meteorological station or have just one station. Satellite remote sensing provides a cost-effective and reliable way to map and monitor the land surface temperatures across the city and identify locations with increased temperatures. The thermal urban maps provide a pictorial representation of the distribution of vulnerable locations (often referred to as 'hotspots') within the cities. Identifying hotspots within a city is the first step in dealing with heat waves as this exercise can help direct and prioritize interventions to areas where they are most needed during heat waves. These 'hotspots' exhibit relatively higher temperatures than that of surrounding non-urban areas and the thermal maps help identify the areas showing accumulation of hotspots. The population living in such high-temperature areas, especially those living with minimum or no basic amenities, lives under high physiological and socio-economic risks due to thermal stress. Using Geographic Information System (GIS), the population and the land use/land cover (LULC) maps can be used in sync with the land surface temperature (LST) maps to properly analyse the situation and develop appropriate solutions. In addition, a comparative study of thermal hotspot maps for each season will help to identify the months in which most cautions need to be taken. For ambient air temperature, one can rely on measurements provided by meteorological stations, or taken by the researchers using hand-held portable devices. However, these measurements are mostly taken at a point and therefore do not sufficiently characterize the temperature of entire city. Satellite remote sensing can be used to overcome this limitation. Satellites offer a vantage point in the space as they cover a large swath of earth's surface. The LST data provided by satellites by capturing the information in thermal infrared region of electromagnetic spectrum has been used by various researchers across the globe (Zhang and Cheng 2019; Wang et al. 2017; Jin 2012; Li et al. 2012; Yuan and Bauer 2007; Chen et al. 2006). In addition, these satellite-derived LSTs have been correlated with ambient air temperatures in some studies to obtain a complete picture of the study area (Mutiibwa et al. 2015; Unger 2009; Vadasz 1994). However, general public and people in administration rarely read such papers, and even if they do, they find it very difficult to comprehend and implement the suggested solutions. In the case of India, many cities' HAP do not contain any LST map of the area and as such, the administrators are unaware of the potentials of remotely sensed data for developing a robust HAP. Therefore, we are writing this discussion paper with a view to demonstrate how satellite data can be used to identify hotspots during heat wave periods and target the efforts more towards vulnerable locations. We have provided an example of Delhi, the capital city of India, and discussed how satellite-derived LST can be used to better plan for adaptation and resilience during hot summer days. In particular, our objectives were to demonstrate the feasibility of satellite data for developing a robust HAP by providing information about vulnerable locations at the finest administrative level, to showcase the spatial expansion of LST rise across the city as the summer progresses from March to June, and to suggest potential solutions for reducing the impact of heat stress on vulnerable population in identified hotspots.

6.2 Literature Review

There have been many studies since 1900s on UHI covering the various urban areas of the world and employing a variety of techniques (Rasul et al. 2017; Stewart and Oke 2009). Though remote sensing data have been used by many researchers to study UHI (Tran et al. 2006, Chakraborty et al. 2013), some researchers also combined satellite-based information with specific physical models for the purpose (Hafner and Kiddner 1999). Some studied temporal temperature variations (Mochida et al. 1997), others analysed temperature differences between cities and surrounding rural areas (Rizwan et al. 2008).

The study of UHIs using remote sensing data started in 1970s. Matson et al. (1978) employed 0.9 km resolution NOAA-5 VHRR data for UHI studies in 11 US cities. Price (1979) used 500 m Heat Capacity Mapping Mission (HCMM) satellite data to study the UHI pattern in various American cities. He found that satellite-sensed temperatures in many large cities are 10–15 °C warmer than the temperatures of rural areas surrounding those cities. In addition, he advised that for a thorough interpretation of the increased surface temperatures in cities, it is essential to study the relationship between remotely sensed surface temperatures and locally measured air temperatures. Brest (1987) examined visible and near-IR Landsat data for fourteen land cover types from urban and rural environments. Bailing and Brazel (1988) used AVHRR thermal data to examine the radiant surface temperature of the Phoenix, Arizona and reported a direct relationship between temperatures and land-use patterns. Gallo et al. (1993) studied the relationship between various satellite-derived parameters such as LST and vegetation index for the Seattle city. They found a declining relationship between Normalized Difference Vegetation Index (NDVI) and the surface temperature. Urban areas reported low values of NDVI compared to rural areas. Chen et al. (2006) studied the relationship between land use/land cover change and land surface temperature in Pearl River Delta (PRD) region of China. They used Landsat TM/ETM + images of the duration 1990 to 2000 and calculated brightness temperature as well as land use/land cover changes. They found that UHI effect became more prominent in the areas which experienced more urbanization. Srivastava et al. (2010) estimated land surface temperatures (LST) and spectral emissivities in the eastern part of India using several remote sensing data which included data from ASTER, MODIS and Landsat satellites. The satellite-derived LSTs were validated using ground recorded temperature data collected through Telatemp Model AG-42D Portable Infrared Thermometer. In thermally homogenous areas, satellite-derived LSTs showed good agreement with ground measurements. However, in less homogenous areas such as agricultural fields they differed by up to 2 °C. Sobrino et al. (2012) carried out an experiment to determine 'optimal characteristics in terms of spatial resolution and the time of acquisition that a satellite must possess to properly monitor the UHI effect'. The authors generated LST maps at different spatial resolution using the Airborne Hyperspectral Scanner (AHS) imagery for Madrid, Spain. The authors concluded that for properly estimating the surface urban heat island effect at district level, spatial resolutions greater than 50 m are required.

Sharma and Joshi (2013) employed Landsat images of 2010–11 period to study UHI in Delhi, India. For the year 2010, LST maps were prepared for all the months of the year except March. For March, the data of 2011 was used, as the cloud-free data for 2010 was not available. The authors reported that areas with high temperature during winter months were located towards interior parts of the city whereas during summer months, UHI pattern was seen towards south-western parts. The authors attributed this trend to the presence of large agricultural fields around the peripheral northern, western and south-western parts of the city. Rajeshwari and Mani (2014) employed Landsat-8 data for estimating LST of Dindigul district of Tamil Nadu, India for 2014. The authors used Split-Window (SW) algorithm to calculate LST. This algorithm is based on computation of spectral radiance and emissivity. The authors reported that LST was high in the barren regions whereas it was low in hilly areas. The impact of urbanization and climate change on urban temperatures was reviewed by Chapman et al. (2017). The authors reported that there has been little attention given on how climate change and urbanization pattern will influence the heat stress in future. They also cautioned that such lethargy and ignorance may make it difficult to adapt to the changing climatic conditions. Dutta et al. (2018) assessed the spatial and temporal variation of urban heat island in two areas of India using Landsat data. The authors used building and vegetation indices to characterize the spatio-temporal pattern of temperature in different land-use conditions. For the National Capital Region (NCR), LST, Normalized Difference Built-up Index (NDBI) and Normalized Difference Vegetation Index (NDVI) maps, prepared for 1998, 2009 and 2018, revealed the increase in thermal hotspots as well as built-up area during the two decades. The authors reported that in 1998, thermal hotspots were observed in small patches over Southwest Delhi, Delhi-Gurgaon-Faridabad border and eastern fringe of Delhi, whereas in 2018 Meerut and Ghaziabad were also included as thermal hotspot. The authors also reported a decline in vegetation in NCR at the rate of 2.09% in the first decade and 0.05% in the second decade. The second region they investigated was West Bengal for which LST, NDBI and NDVI maps were generated for 1998, 2009 and 2018. The authors reported that area under thermal hotspots has expanded during the two decades. In 1998, Kolkata was observed as the thermal hotspot, with few sandy patches in southern part. In 2018, the heat island had reportedly expanded to east and north of Howrah, south-east of North 24 Parganas and north of South 24 Parganas. Dissanayake et al. (2019) investigated the impact of land use/land cover changes on surface urban heat island in Kandy, Sri Lanka. They used Landsat data for the years 1996, 2006 and 2017, and employed several methods such as pixel-based supervised classification, change detection and grid-based density analysis for the purpose. The authors used area occupied by impervious surfaces (IS) as the proxy for urbanization and reported that area occupied by IS increased from 2.3% in 1996 to 23.9% in 2017. This increase in IS area was accompanied by gradual decrease in forest cover in the study area. Pramanik and Punia (2019) used satellite data from Landsat and Sentinel platforms to study the cooling effects of urban greenery in Delhi, India. The LST of the city was mapped using Landsat-8 data, whereas vegetation was mapped using Sentinel-2 data. The cooling effects of urban vegetation was characterized using statistical methods such as spatial auto-regression model, ordinary least square regression and multiple linear regression. They reported a fragmented vegetated area or a green patch with complex shape can increase the LST whereas the dense vegetation has the capability to effectively reduce the LST. Chung et al. (2020) analysed the correlation between surface air temperature (SAT) and land surface temperature (LST). The correlation study was done among the following parameters: the Terra and Aqua Moderate Resolution Imaging Spectroradiometer (MODIS) daytime and nighttime LST and maximum, minimum and mean SAT measured at 79 weather stations of the Korea Meteorological Administration (KMA) from 2008 to 2018. The results showed that the maximum SAT (TMX) had a good correlation with the daytime LST of Terra MODIS, with a Pearson's Correlation Coefficient (R) of 0.92 and root mean square error (RMSE) of 4.8 °C and the minimum SAT (TMN) showed a good correlation with the nighttime LST of Terra MODIS, with an R of 0.93 and RMSE of 4.2 °C. Tyagi et al. (2021) synthesized global literature related to the physics of UHI phenomena, and its impact on various sectors. They emphasized that mitigation and adaptation of UHI would require developing a plan which brings together efforts taken at various levels of administration.

Looking at the above literature, there appears two prominent gaps. One, the assessment of spatio-temporal distribution of temperature and humidity (two very important factors resulting in heat stress) has not been done to the finest administrative levels, possibly because of non-availability of high spatio-temporal resolution satellite data. This also is the major lacuna in various HAPs. And the other gap we noticed is the lack of synergy between climate change studies and the impact of urbanization studies which makes it difficult to understand and prepare for future heat waves. Unless we get a clear picture of spatio-temporal distribution of temperature and other climate parameters at the finest administrative scale, it will be very difficult to appropriately develop HAPs for saving lives and livelihoods during heat wave events.

6.3 Heat Action Plans (HAPs) Developed in India

The HAP developed for the city of Ahmedabad in 2013 was the first such initiative in the entire South Asia (https://www.exemplars.health/stories/ahmedabad-indias-heat-action-plan, page accessed on 24 July 2023). Since then its updated versions have been released in 2016 and 2019 (https://www.nrdc.org/sites/default/files/ahmedabad-heat-action-plan-2016.pdf, page accessed on 24 July 2023; https://www.c40knowledgehub.org/s/article/Ahmedabad-Heat-Action-Plan-2019?language=en_US, page accessed on 24 July 2023). The 2013 HAP focused on developing a coordination between different agencies and communities to increase preparedness, information-sharing and response coordination to protect vulnerable populations from extreme heat. It was implemented by the Ahmedabad Municipal Corporation (AMC) in association with other agencies. AMC carried out a cross-sectional randomized cluster sample survey in 2011 to understand the vulnerability of the population with respect to parameters such as age, work, health conditions, access to water, sanitation, health care, information and social connectedness (Tran et al. 2013). The vulnerable section of the population identified in that study included children, the elderly, slum dwellers and the outdoor workers. AMC adopted a strategy to raise mass awareness about the impacts of heat waves in the city by disseminating information through radio, television, newspapers, email, text and WhatsApp messages, along with distributing pamphlets and displaying hoardings in the city before a heatwave hits. The plan was later evaluated also to assess its impact and it was claimed that in the two years after the launch of HAP there was at least a 25% decrease in May's excess all-cause mortality (https://www.nrdc.org/sites/default/files/ahmedabad-hap-evaluation.pdf, page accessed on 24 July 2023). However, the analysis was criticized for lack of data on whose lives were saved (e.g. were they among the most vulnerable and disadvantaged groups?), and to what extent this statistic is the result of the HAP, rather than other factors (Nastar 2020). None-the-less, the Ahmedabad HAP remained the most popular HAP at least in India, and since then HAPs have been developed for 17 Indian States and over 130 Indian cities (Golecha et al. 2021). The basic strategy followed for developing HAPs revolved around the following three pillars: identifying vulnerable populations and the causes of their higher risk to heat waves, developing strategies and coordinated responses, and activating heat alerts. A study by Pillai and Dalal (2023) analysed 37 HAPs developed in India and identified the following shortcomings:

1. Most HAPs are not built for local context and have an oversimplified view of the hazard.
2. Nearly all HAPs are poor at identifying and targeting vulnerable groups.
3. HAPs are underfunded.
4. HAPs have weak legal foundations.
5. HAPs are insufficiently transparent.
6. Capacity building is sectorally targeted.

6.4 How Remote Sensing and GIS Can Help Develop More Robust HAPs?

The first prerequisite to develop a HAP is to understand the spatial and temporal distribution of temperature. As explained earlier, air temperature data provided by meteorological stations are far and scanty to appropriately characterize the intra-urban variations. Satellite data can be used in this case to obtain city-wide representation of land surface temperature in a cost-effective and repetitive manner. In addition, satellite data can also provide information related to land use factors like vegetation, water and built-up at city scale to help understand the causes of higher intra-urban temperature variability. Many researchers have also correlated the LST obtained from satellite data with the air temperature recorded by meteorological stations (Price 1979; Gallo et al. 1993) and suggested that for a thorough interpretation of the increased surface temperatures in cities, the relationship between remotely sensed surface temperatures and locally measured air temperatures should also be studied (Price 1979). None-the-less, with paucity of air temperature data, satellite observed LST data has been and is being widely utilized by scientists and researchers across the globe to study heat waves. It is pertinent to explain here that LST refers to direct measurement of earth's surface/'skin' temperature whereas the air temperature recorded by meteorological stations are mostly observed at some distance above the ground surface (mostly 2 m above the ground surface). Thus, LST allows for a relatively more detailed analysis (Almeida et al. 2021). It can help determine the radiative load on earth's surface by analysing longwave radiation and turbulent heat fluxes between the earth and the atmosphere thereby providing an understanding of warming trend on earth's surface (Almeida et al. 2021).

Among the various remote sensors used for studying heat waves, most commonly used are the thermal sensors which operate in the infrared electromagnetic range of 8–14 μm. This is the range through which earth's emission is characterized, and the thermal sensors record this emission as Digital Numbers (DN), which are later converted into temperature data according to sophisticated mathematical equations. However, a disadvantage of thermal sensors is their incapability to penetrate through clouds. In particular, for studying tropical Indian cities, obtaining a cloud-free data becomes a significant challenge, which restricts their full utilization for the purpose. To overcome the inaccessibility of cloud-free data, researchers have resorted to the data received from passive microwave sensors; however, these sensors do not provide detailed spatial resolution as is possible with thermal sensors (Almeida et al. 2021). Some researchers have emphasized on using both these methods together to complement the inability of each sensor (Nguyen and Henebry 2016). Here, we are focusing more on LST retrieval using thermal data as this is the most used method across the remote sensing community.

LST retrieval from thermal data requires input on atmospheric effects, sensor parameters, i.e. spectral range and viewing angle, and surface parameters such as emissivity and geometry (Sekertekin and Bonafoni 2020). There are different approaches developed for LST retrieval based on factors such as emissivity and atmospheric effects. These algorithms are named depending on the number of thermal bands used. For instance, single-channel or mono-window algorithms use one thermal band. However, split-window or multi-channel methods include more than one Thermal Infrared (TIR) band.

For the purpose of developing the city-level HAPs, one needs to map the LST at the finest administrative resolution. For Indian cities, the finest administrative intra-city units are 'Wards' and therefore the HAPs should incorporate the spatial and temporal details of LST and the vulnerable population at ward levels. At the same time, it is important to identify the satellite datasets which can be used for ward-level mapping. Through literature review, we found that thermal infrared satellite data having low spatial resolution could not accurately account for surface temperature variations and heat fluxes for intra-urban heterogeneous landscapes (Mohamed et al. 2017; Weng 2009; Pu et al. 2006). In addition, for studying temperature variations within the city, medium-resolution satellite data proved to be more useful (Mohamed et al. 2017; Zhang et al. 2013; Mallick et al. 2013; Gao et al. 2011). Sobrino et al. (2012) suggested that sensors with < 50 m pixel size may be most ideal for accurate

Table 6.1 Different satellite datasets and their properties

Satellite/sensor	Spectral bands	Special resolution	Temporal resolution
Landsat 8 OLI/TIRS	11 bands	30 m	16 days, equatorial crossing time around 10 AM
Sentinel-3 SLSTR	11 bands	1 km/0.5 km	Twice a day/Daily
Kalpana-1	3 bands	8 km (for thermal band), 2 km for rest	23–33 min
MODIS	36 bands	1 km	1–2 days, equatorial crossing time for Terra 10:30 AM and for Aqua 1:30 PM
ASTER	14 bands	90 m (for thermal band), 15–30 m for rest	16 days, equatorial crossing time around 10:30 AM
AVHRR	6 bands	1–4 km	Twice a day (one in afternoon 1:30–2:30 PM; and other in night 1:30–2:30 AM)

intra-urban thermal mapping. With this background, we evaluated different thermal satellite datasets with respect to their properties (Table 6.1).

After screening, Landsat satellite data, in particular Landsat-8 data, seems most suitable among the available options. Landsat-8 data are openly available at a spatial resolution of 30 m. This satellite has two instruments: Operational Land Imager (OLI) and Thermal Infrared Sensor (TIRS). Landsat 8 provides data in 11 spectral bands, out of which two are thermal bands, viz. band 10 and 11 (https://landsat.usgs.gov/what-are-band-designations-landsat-satellites). Though the data in thermal bands are collected at 100 m resolution, it is resampled to 30 m by the United States Geological Survey (USGS) before distributing to users. As band 11 has more uncertainty than band 10 (Yu et al. 2014), therefore, here we are presenting an example of LST retrieval from band 10 of Landsat 8 data.

6.5 Data and Methodology

We downloaded satellite data of Delhi, the capital city of India, for the summer months of 2020 from the Earth Explorer portal of United States Geological Survey (USGS) (https://earthexplorer.usgs.gov/). For each of the four summer months, viz., March, April, May and June, cloud-free satellite data were downloaded. Thus, we got satellite data of 13 March 2020, 30 April 2020, 16 May 2020 and 17 June 2020. To compute LST from satellite data, we followed the procedure adopted by Rajeshwari and Mani (2014). All the satellite datasets have energy recorded in the form of Digital Numbers (DN), which were converted to Top of Atmosphere (TOA) spectral radiance (L) for further analysis as per Eq. 6.1. The L was determined by multiplying multiplicative rescaling factor (0.000342) of Thermal Infrared (TIR) bands with its corresponding TIR band and adding additive rescaling factor (0.1) to it.

$$L = M * Q + A \tag{6.1}$$

where L—Top of Atmospheric Radiance in watts/(m^2 * srad * μm)

M—Band specific multiplicative rescaling factor.

Q—band 10 image.

A—Band specific additive rescaling factor.

The spectral radiance values thus computed were then converted to brightness temperature (BT) through following equation (Rajeshwari and Mani 2014):

$$BT = K2/L\ [(K1/L) + 1] \tag{6.2}$$

where K1 and K2—thermal conversion constant and L—Top of Atmospheric spectral radiance. Next, we computed Land surface emissivity (LSE). The land surface emissivity is indispensable for LST computation from satellite data. The emissivity of land, unlike that of oceans, can differ significantly from unity and vary with vegetation, surface moisture, roughness and viewing angles. There are several methods to compute LSE. In this case, Normalized Difference Vegetation Index (NDVI)-based approach was used to compute LSE because of its simplicity and wider acceptability. The equation to compute LSE is (Rajeshwari and Mani 2014):

$$\text{LSE}\ (\varepsilon) = \varepsilon\text{s}\ (1 - \text{FVC}) + \varepsilon\text{v} * \text{FVC} \tag{6.3}$$

where εs and εv—soil and vegetative emissivity values of the corresponding bands,

FVC—Fractional Vegetation Cover.

FVC for an image was calculated by

$$\text{FVC} = \text{NDVI—NDVIs}/\ \text{NDVIv} - \text{NDVIs} \tag{6.4}$$

where NDVIs—NDVI reclassified for soil

NDVIv—NDVI reclassified for vegetation.

NDVI is computed from red and near-infrared bands of Landsat 8 OLI data as follows:

$$\text{NDVI} = (\text{NIR} - \text{Red})/\ (\text{NIR} + \text{Red}) \tag{6.5}$$

The LST is now retrieved using the following equation (Yu et al. 2014):

$$LST = \gamma[\varepsilon - 1(\Psi 1BT + \Psi 2) + \Psi 3] + \delta \tag{6.6}$$

The symbols γ, Ψ and δ are coefficients and the equations used to compute these coefficients can be found in Yu et al. (2014). All these computations were carried out in ArcGIS 10.5 software, though one can also use other software such as QGIS, ENVI, ERDAS IMAGINE, GEOMATICA.

6.6 Results and Discussion

The LST maps of Delhi for the summer season of 2020 are shown in Fig. 6.1. As we can see, the temperature increases from the month of March to the month of May, and more and more areas within the city are experiencing higher temperatures and are represented in red colour. The southeastern regions get heated first in March, and then the western and northern regions showed elevated temperatures during April and May. By June, the entire city is seen experiencing elevated temperature. As we have developed these maps at ward level, we can also understand ward-wise distribution of temperatures in the city. For example, on 13 March 2020, majority of the wards in Shahdara North and Shahdara South showed temperature above 30 °C (Fig. 6.2). In particular, wards like Saboli (054-E), Harsh Vihar (055-E), Joharipur (052-E), Karawal Nagar West (061-E) in Shahdara North experienced temperature in excess of 32 °C (Fig. 6.2). In Shahdara South, most of the areas in the wards such as Dallupura (005-E), Gharoli (006-E) and I.P. Extension (019-E) experienced temperatures in excess of 33 °C (Fig. 6.2). On 30 April 2020, much of the areas in northern and western side of Delhi experienced relatively warmer temperatures. This does not mean that the areas which showed higher temperatures on 13 March 2020 have cooled down on 30 April 2020. Wards of Shahdara North and Shahdara South experienced temperatures in excess of 42 °C on 30 April 2020. However, the northern and western parts recorded temperatures in excess of 45 °C, therefore these areas were depicted in bright red colour and the other areas were depicted in lighter colours. The sudden increase of temperatures in northern and western parts of the city in April relative to March in 2020 is worth noticing. In particular, wards such as Narela and Najafgarh were heated more than the other wards. Incidentally, Najafgarh and Narela, both areas, have built-up structures as well as barren lands, with very less vegetation and water bodies, as seen in the NDVI map of Delhi (Fig. 6.3). This may be the reason for more increase in temperatures in these areas in April compared to March in 2020. Chakraborty et al. estimated an average increase in LST of nearly 20 °C over urban built-up areas in Delhi during 2000–2010. This increase was mainly seen in southern part of the city as satellite cities like Gurgaon and NOIDA came up in the vicinity, with urban structures like multinational office buildings, multiplexes, shopping malls, apartments and residential areas. Much of the vegetation in Delhi is around central part and to some extent in southern part. This may be the reason that the areas around the central part of Delhi are relatively cooler.

By May and June, almost all of the city experienced high temperatures (Fig. 6.1). This shows that preparedness for the heatwaves should be started before March and areas in south-eastern wards such as Shahdara North and Shahdara South

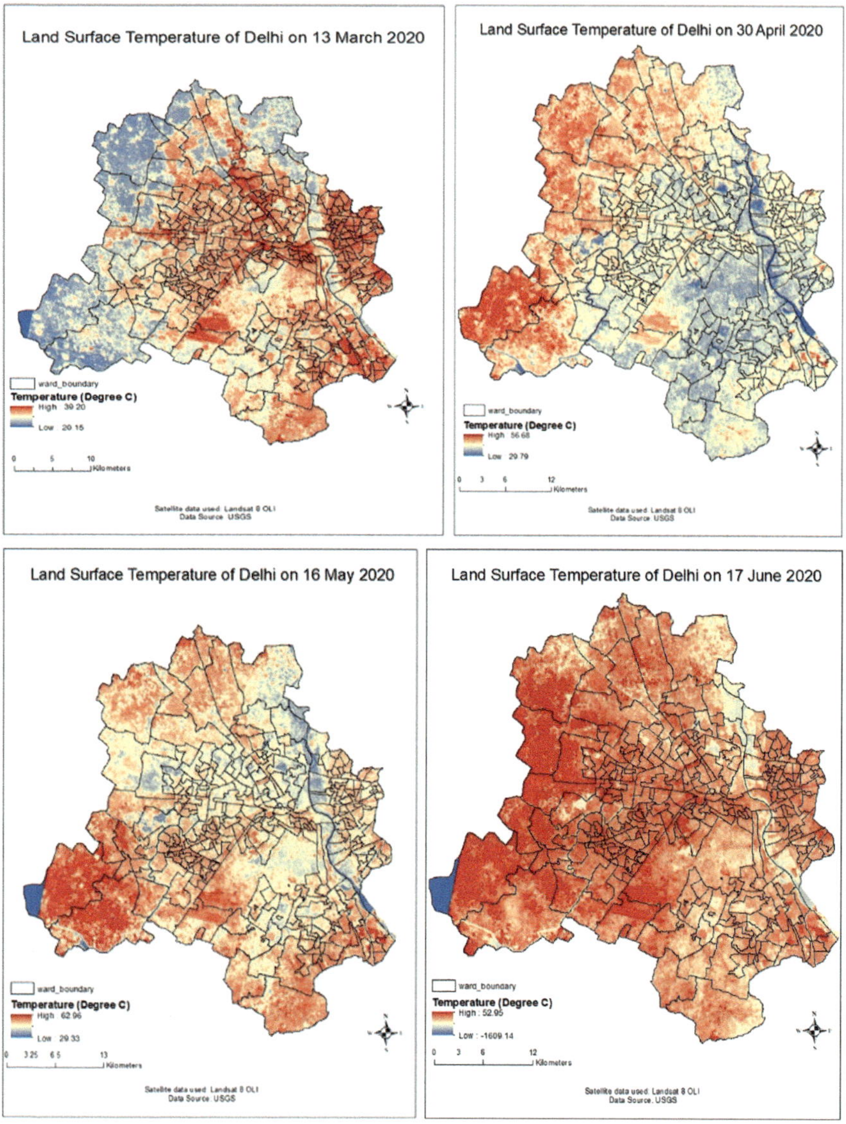

Fig. 6.1 LST maps of Delhi for summer months of 2020

should be given special attention during these preparations. There are many slums in these areas. The slum dwellers in these areas constitute a significant portion of vulnerable populations, and they should be sensitized to the impacts of heatwaves. Areas with less vegetation cover should be vegetated fast. In particular, recent efforts like Miyawaki Forests have shown promising results in greening an area very quickly and the Municipal Corporation of Delhi (MCD) has plans to grow

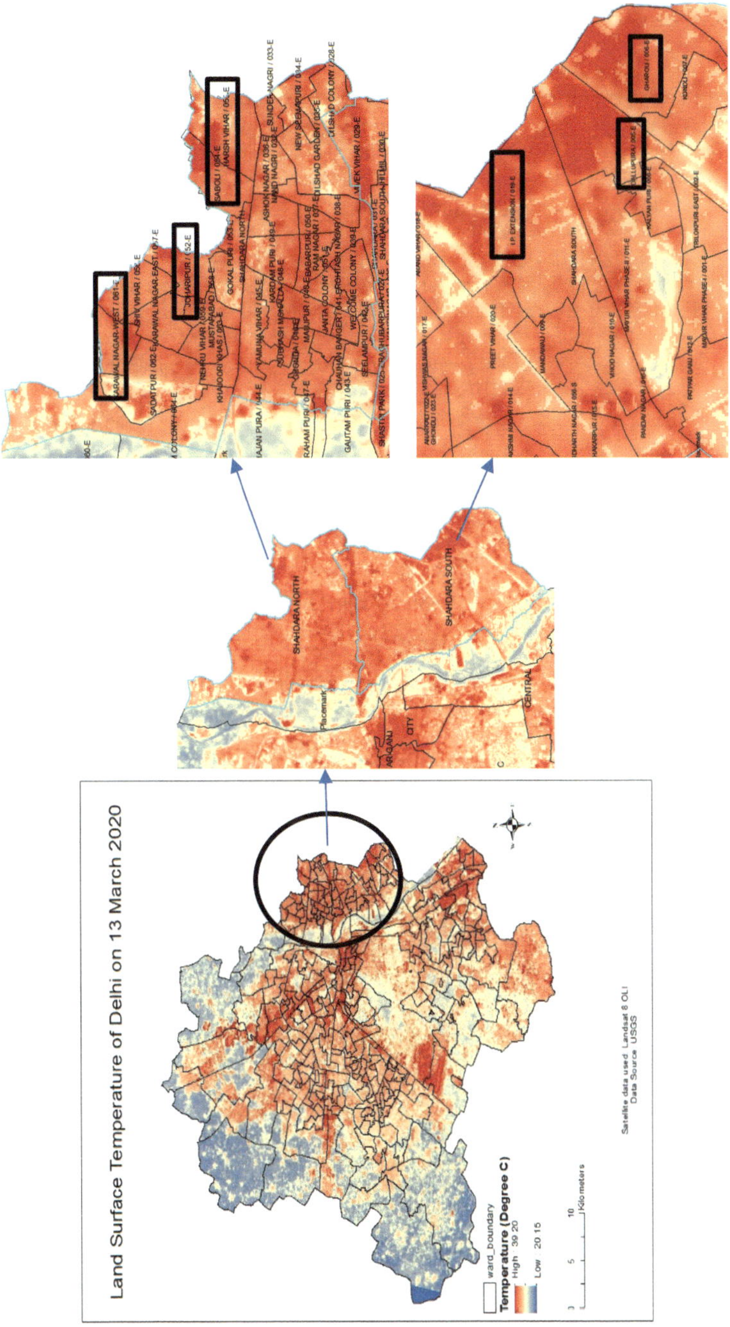

Fig. 6.2 Spatial distribution of thermal hotspots in Delhi on 13 March 2020

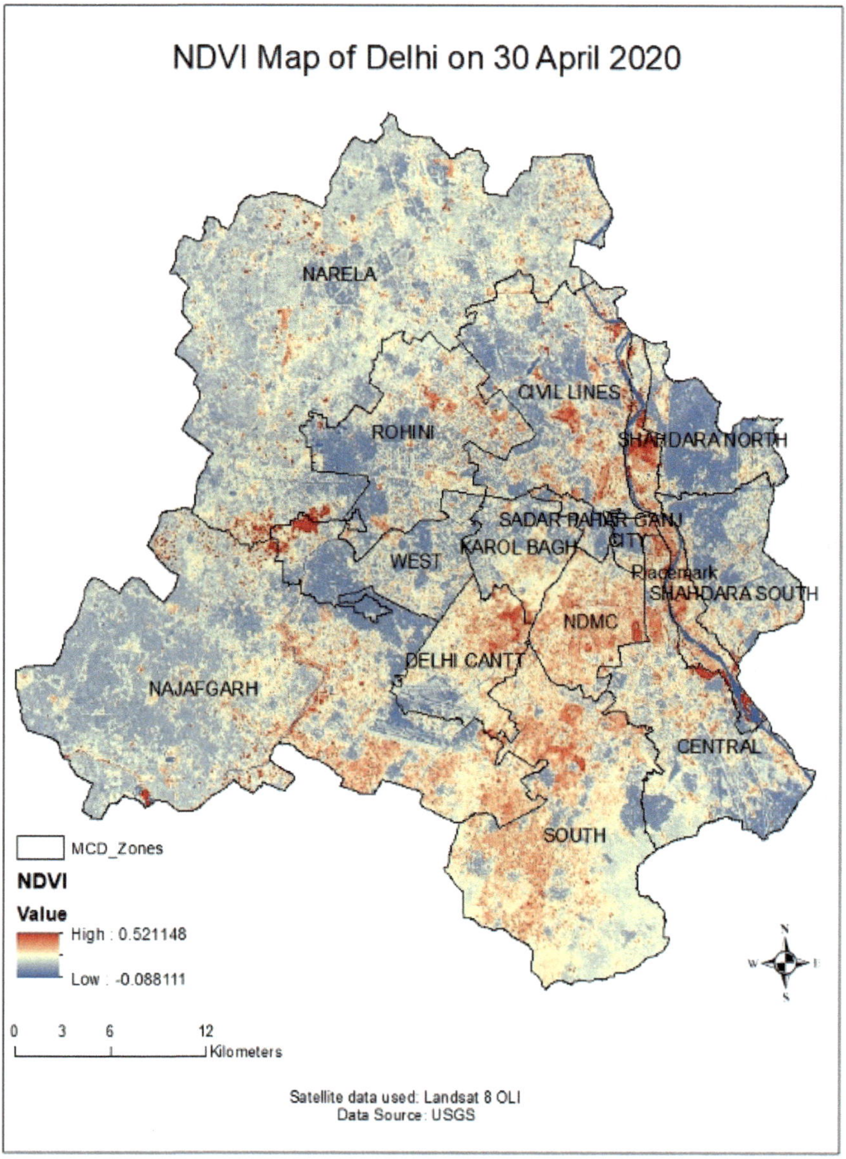

Fig. 6.3 NDVI Map of Delhi on 30 April 2020

micro-urban forest shades using this technique in Delhi (https://www.indiatvnews.com/delhi/what-is-miyawaki-technique-how-it-will-be-implemented-in-delhi-mcd-latest-news-2023-03-13-854152#:~:text=Locations%20distributed%20across%2010%20acres,month%2Dold%20park%20in%20Dallas., page accessed on 24 July, 2023).

6.7 Limitations of Present Work

Though satellite-derived LST maps have their utility in explicitly explaining its spatial distribution, one of the biggest challenges associated with this approach lies in its validation. Ideally, ground measurements available from thermocouples, infrared thermometers or other instruments at different locations in the study area should be used as ground truth LST data for validation. However, this data is not available in many locations. Many researchers have compared the LST derived from satellite data with air temperature observed by meteorological stations. However, we did not find it suitable to do the validation of satellite-derived LST maps with ground-based air temperature data. For ambient air temperature, one can rely on measurements provided by meteorological stations, or taken by the researchers using hand-held portable devices. However, these measurements are mostly taken at a point and therefore do not sufficiently characterize the temperature of entire city. Many cities in India do not have a single meteorological station even today. Many researchers have also argued against validating the satellite-derived LST with air temperature data. For example, Yu et al. (2012) stressed that due to spatial resolution mismatches and the physical differences between LST and ground-based temperature measurements, validation of former using latter can introduce significant errors. Further Reiner et al. (2023) observed that there can be a strong agreement between long-term trends of LST and air temperature; however, the difference between instantaneous LST and air temperature can be high. Satellites like Landsat, which are polar orbiting satellites, provide instantaneous LST [Reiner et al. (2023)] and therefore comparison with air temperature for validation purposes is not recommended. Reiner et al. (2023) also mentioned, 'While validation studies are necessary, especially from the point of view of climate-change-related studies, which require a high accuracy and stability of LST [148], the use of existing remote sensing LST datasets to study global change processes should be extended'. Moreover, the administrators will require a relative knowledge of spatial distribution of temperature in his/her administrative territory for planning during summers, and these maps can be used for these purposes. Further, the paper has not included computation of a vulnerability index for different wards by employing socio-economic and/or demographic data along with remotely sensed information. We believe a ward-level vulnerability index will further strengthen the HAPs and aid a lot in planning during heat-wave periods.

6.8 Conclusions and Way Forward

In this paper, we have demonstrated how remote sensing can help develop HAPs at city levels. The application of remote sensing and GIS was demonstrated in identifying vulnerable locations at the finest administrative levels, citing an example of Delhi, India. Our approach will address some of the lacunae identified in HAPs developed for Ahmedabad and other cities by Pillai and Dalal (2023). For example, our approach will help prepare HAPs tailored to local conditions and land use, highlighting locations which are more vulnerable to heat stress. In addition, such an approach will help the administrators in visualizing the temperature distribution across the wards during summer months. This, in turn, will help in understanding the sharp rise in temperature in some of the wards across the different summer months. All these points become crucial for planning for heat waves. Further work based on this approach may involve spatial vulnerability modelling and/or quantitative socio-economic risk assessment of different sections of the society such as traffic personnel, labourers, daily-wage workers, auto/tri-rickshaw drivers, hawkers etc. It is important that we refine the available information at the finest granular level so that benefits of advanced techniques like remote sensing and GIS can be shared with most down-trodden sections of the society. Going ahead, the information provided by remote sensing can also be used to model and predict future temperature changes. Such type of information will be an invaluable resource for planners and city administrators and can help save lives.

HAPs based on satellite-derived LST maps can be powerful tools for identifying heat-vulnerable zones. However, their integration into policy requires specific guidelines for integrating such types of information in the urban planning and public health management. There is dire need to embed LST-based vulnerability maps into municipal by-laws and disaster management protocols. Operationalizing remote sensing and GIS-based outputs will require training municipal staff and health workers to understand and use GIS tools. For maintaining smooth and effective coordination across different municipal/civic bodies, disaster management agencies and scientists will need to establish formal channels between GIS experts, urban planners and emergency responders. The municipal/government authorities can approach private agencies, including private universities, to provide help with data infrastructure, analysis and visualization. In addition, civic bodies should also involve local communities in such planning exercises and collect their feedback to improve the overall system.

The need of the hour is to incorporate the information available from different satellites into the city-level HAPs and prioritize mitigation and adaptation efforts.

Potential Conflicts of Interest On behalf of all authors, the corresponding author states that there is no conflict of interest.

Informed Consent Informed consent was obtained from all individual participants included in the study.

AI Discloser No AI tool has been used for writing the manuscript.

6.9 Data Availability Statement

The datasets used are landsat satellite datasets, which are openly available at USGS Earthexplorer website (https://earthexplorer.usgs.gov/).

Authors Contribution All authors have contributed equally to this work.

Funding The research was not supported by any funding.

References

Almeida CRd, Teodoro AC, Gonçalves (2021) Study of the Urban Heat Island (UHI) using remote sensing data/techniques: a systematic review. Environments 8(10):105. https://doi.org/10.3390/environments8100105

Bailing RC Jr, Brazel SW (1988) High-resolution surface temperature patterns in a complex urban terrain. Photogramm Eng Rem Sens 54:1289–1293

Brest CL (1987) Seasonal albedo of an urban/rural landscape from satellite observations. J Appl Meteorol 26:1169–1187

Chakraborty SD, Kant Y, Mitra D (2013) Assessment of land surface temperature and heat fluxes over Delhi using remote sensing data. J Environ Manage 148:143–152

Chapman S, Watson JEM, Salazar A, Thatcher A, McAlpine CA (2017) The impact of urbanization and climate change on urban temperatures: a systematic review. Landscape Ecol 32:1921–1935

Chen XL, Zhao HM, Li PX, Yin ZY (2006) Remote sensing image based analysis of the relationship between urban heat island and land use/cover changes. Rem Sens Environ 104(2):133–146

Chung J, Lee Y, Jang W, Lee S, Kim S (2020) Correlation analysis between air temperature and land surface temperature and prediction of air temperature using tensorflow long short-term memory for the period of occurrence of cold and heat waves. Rem Sens 12(3231):1–25

Dissanayake DMSLB, Morimoto T, Ranagalage M, Murayama Y (2019) Land-use/land-cover changes and their impact on surface urban heat islands: case study of Kandy City Sri Lanka. Climate 7(99):1–20

Dutta K, Basu D, Agarwal S (2018) Temporal and spatial analysis of urban heat island using Landsat satellite data: two Indian case studies. In: ISPRS annals of the photogrammetry, remote sensing and spatial information sciences, vol IV-5, ISPRS TC V Mid-term Symposium "Geospatial Technology–Pixel to People", 20–23 November 2018, Dehradun, India

Gallo KP, McNab AL, Karl TR, Brown JF, Hood JJ, Tarpley JD (1993) The use of a vegetation index for assessment of the urban heat island effect. Int J Rem Sens 14:2223–2230

Gao ZQ, Gao W, Chang NB (2011) Integrating temperature vegetation dryness index (TVDI) and regional water stress index (RWSI) for drought assessment with the aid of Landsat TM/ETM plus images. Int J Appl Earth Obs Geoinf 13:495–503

Golechha M, Mavalankar D, Bhan SC (2021) India: heat wave and action plan implementation in Indian cities. In: Ren C, McGregor G (eds) Urban climate science for planning healthy cities. biometeorology, vol 5. Springer, Cham. https://doi.org/10.1007/978-3-030-87598-5_13

Guleria S, Gupta AK (2018) Heat wave in India documentation of state of Telangana and Odisha (2016). National Institute of Disaster Management, New Delhi, pp 124

Hafner J, Kidder SQ (1999) Urban heat island modeling in conjunction with satellite-derived surface/soil parameters. J Appl Meteorol Climatol 38:448–465

IMD (2022) Statement on Climate of India during 2022. India Meteorological Department (IMD). Downloaded from https://mausam.imd.gov.in/Forecast/marquee_data/Statement_climate_of_india_2022_final.pdf, Accessed on 23 July 2023

IPCC (2013) Climate Change 2013: the physical science basis. In: Contribution of working group I to the fifth assessment report of the intergovernmental panel on climate change. Cambridge University Press, Cambridge, United Kingdom and New York, NY, USA

Jin MS (2012) Developing an index to measure urban heat island effect using satellite land skin temperature and land cover observations. J Clim 25(18):6193–6201

Kumari P, Garg V, Kumar R, Kumar K (2021) Impact of urban heat island formation on energy consumption in Delhi. Urban Clim 36:100763. https://doi.org/10.1016/j.uclim.2020.100763

Li Y, Zhang H, Kainz W (2012) Monitoring patterns of urban heat islands of the fast-growing Shanghai Metropolis, China: using time-series of Landsat TM/ETM+ data. Int J Appl Earth Obs Geoinf 19(10):127–138

Mallick J, Rahman A, Singh CK (2013) Modeling urban heat islands in heterogeneous land surface and its correlation with impervious surface area by using night-time ASTER satellite data in highly urbanizing city Delhi-India. Adv Space Res 52:639–655

Matson M, Mcclain E, McGinnis D Jr, Pritchard J (1978) Satellite detection of urban heat islands. Mon Weather Rev 106:1725–1734

Mishra V, Mukherjee S, Kumar R, Stone DA (2017) Heat wave exposure in India in current, 1.5, and 2.0 ◦C worlds. Environ Res Lett 12:1–9

Mochida A, Murakami S, Ojima T, Kim S, Ooka R, Sugiyama H (1997) CFD analysis of mesoscale climate in the Greater Tokyo area. J Wind Eng Ind Aerodyn 67:459–477

Mohamed AA, Odindi J, Mutanga O (2017) Land surface temperature and emissivity estimation for Urban Heat Island assessment using medium- and low-resolution space-borne sensors: a review. Geocarto Int 32:455–470

Mutiibwa D, Strachan S, Albright T (2015) Land surface temperature and surface air temperature in complex terrain. IEEE J Select Topics Appl Earth Observ Rem Sens. https://doi.org/10.1109/JSTARS.2015.2468594

Nastar M (2020) Message sent, now what? a critical analysis of the heat action plan in Ahmedabad, India. Urban Sci 4:53. https://doi.org/10.3390/urbansci4040053

NDMA (2019) National guidelines for preparation of action plan–prevention and management of heat wave. National Disaster Management Authority, New Delhi, India, p 62

Nguyen LH, Henebry GM (2016) Urban heat islands as viewed by microwave radiometers and thermal time indices. Rem Sens 8:831

Pillai AV, Dalal T (2023) How is India adapting to heatwaves? an assessment of heat action plans with insights for transformative climate action. CPR report

Pramanik S, Punia M (2019) Assessment of green space cooling effects in dense urban landscape: a case study of Delhi India. Modell Earth Syst Environ 5:867–884

Price JC (1979) Assessment of the urban heat island effect through the use of satellite data. Mon Weather Rev 107:1554–1557

Pu R, Gong P, Michishita R, Sasagawa T (2006) Assessment of multi-resolution and multi-sensor data for urban surface temperature retrieval. Rem Sens Environ 104:211–225

Rajeshwari A, Mani ND (2014) Estimation of land surface temperature of Dindigul district using Landsat 8 data. Int J Res Eng Technol 3(5):122–126

Rasul A, Balzter H, Smith C, Remedios J, Adamu B, Sobrino JA, Srivanit M, Weng Q (2017) A review on remote sensing of urban heat and cool islands. Land 6:38

Reiners P, Sobrino J, Kuenzer C (2023) Satellite-derived land surface temperature dynamics in the context of global change—a review. Rem Sens 15:1857. https://doi.org/10.3390/rs15071857

Rizwan AM, Dennis LYC, Liu C (2008) A review on the generation, determination and mitigation of Urban Heat Island. J Environ Sci 20:120–128

Sekertekin A, Bonafoni S (2020) Land surface temperature retrieval from landsat 5, 7, and 8 over rural areas: assessment of different retrieval algorithms and emissivity models and toolbox implementation. Rem Sens 12(2):294. https://doi.org/10.3390/rs12020294

Sharma R, Joshi PK (2013) Analyzing monthly dynamics of SUHI in Delhi using satellite information. Joint Urban Rem Sens Conferen. https://doi.org/10.1109/jurse.2013.6550714

Sharma R, Hooyberghs H, Lauwaet D, De Ridder K (2019) Urban heat islands and future climate change—implications for Delhi's heat. J Urban Health 96(2):235–251

Sobrino JA, Oltra-Carrio R, Soria G, Bianchi R, Paganini M (2012) Impact of spatial resolution and satellite overpass time on evaluation of the surface urban heat island effects. Rem Sens Environ 117:50–56

Srivastava PK, Majumdar TJ, Bhattacharya AK (2010) Study of land surface temperature and spectral emissivity using multi-sensor satellite data. J Earth Syst Sci 119(1):67–74

Stewart I, Oke T (2009) Newly developed "thermal climate zones" for defining and measuring urban heat island magnitude in the canopy layer. In: Proceedings of the T.R. Oke Symposium: Urban Scales, Urban Systems and the Urban Heat Island (Joint between the Timothy R. Oke Symposium and the Eighth Symposium on the Urban Environment), Garmisch-Partenkirchen Germany, 12 January 2009Srivastava PK, Majumdar TJ, Bhattacharya AK (2010) Study of land surface temperature and spectral emissivity using multi-sensor satellite data. J Earth Syst Sci 119(1):67–74

Tran H, Uchihama D, Ochi S, Yasuoka Y (2006) Assessment with satellite data of the urban heat island effects in Asian mega cities. Int J Appl Earth Obs Geoinf 8:34–48

Tran KV, Azhar GS, Nair R, Knowlton K, Jaiswal A, Sheffield PE, Mavalankar D, Hess JJ (2013) A cross-sectional, randomized cluster sample survey of household vulnerability to extreme heat among slum dwellers in Ahmedabad, India. Int J Environ Res Public Health 10:2515–2543

Tyagi A, Kumar M, Bhan SC et al (2021) Review of urban heat islands: monitoring, forecast and impacts. Vayumandal 47(2)

Unger J, Gal T, Rakonczai J, Mucsi L, Szatmari J, Tobak Z, van Leeuwen B, Fiaala K (2009) Air temperature versus surface temperature in urban environment. Paper presented at The Seventh International Conference on Urban Climate, 29 June- 3 July 2009, Yokohama, Japan

United Nations (2018) World Urbanization Prospects: 2018 revision. https://www.un.org/development/desa/publications/2018-revision-of-world-urbanization-prospects.html. Accessed on 23 Oct 2019

Vadasz V (1994) On the relationship between surface temperature, air temperature and vegetation index. Adv Space Res 14(3):41–44

Wang Y, Zhan Q, Ouyang W (2017) Impact of urban climate landscape patterns on land surface temperature in Wuhan, China. Sustainability 9(1700), 1–16

Weng Q (2009) Thermal infrared remote sensing for urban climate and environmental studies: methods, applications, and trends. ISPRS J Photogramm Rem Sens 64:335–344

Yu Y, Tarpley D, Privette JL, Flynn LE, Xu H, Chen M, Vinnikov KY, Sun D, Tian Y (2012) Validation of GOES-R satellite land surface temperature algorithm using SURFRAD ground measurements and statistical estimates of error properties. IEEE Trans Geosci Remote Sens 50(3):704–713

Yu X, Guo X, Wu Z (2014) Land surface temperature retrieval from landsat 8 tirs—comparison between radiative transfer equation-based method, split window algorithm and single channel method. Rem Sens 6:9829–9852. https://doi.org/10.3390/rs6109829

Yuan F, Bauer ME (2007) Comparison of impervious surface area and normalized difference vegetation index as indicators of surface urban heat island effects in Landsat imagery. Rem Sens Environ 106(3):375–386

Zhang H, Qia ZF, Ye XY, Cai YB, Ma WC, Chen MN (2013) Analysis of land use/land cover change, population shift, and their effects on spatiotemporal patterns of urban heat islands in metropolitan Shanghai China. Appl Geograph 44:121–133

Zhang Y, Cheng J (2019) Spatio-temporal analysis of urban heat island using multisource remote sensing data: a case study in Hangzhou, China. IEEE J Select Topics Appl Earth Observ Rem Sens. https://doi.org/10.1109/JSTARS.2019.2926417

Part II
Urban Policy, Governance and Sustainable Development

Chapter 7
Transforming Energy and Emission-Intensive Industry Towards Zero Emissions—A Policy Framework

Baidyanath Mukherjee, Arun Kr. Singh, Anil Kumar Sharma, and Shalu Tyagi

Abstract With the increasing urgency to achieve net-zero emissions and stabilise the climate, there is a growing recognition of the need for a comprehensive understanding of the key factors driving the green industry. The shift towards zero emissions as a target represents a significant paradigm shift for industry and industrial policy. Sectors like steel, cement and chemicals, known as hard-to-abate sectors, have largely remained unaffected by climate policies. To address this disparity, a fundamental change is required, moving away from traditional industrial policies that protect the industry towards transformative strategies that reshape the industry itself. By embracing a comprehensive industrial policy framework aligned with the zero-emissions target, we can drive transformative changes, promote sustainable practise, and create a greener future for industries and the environment. This research proposes a framework for industrial policy to address the complexity of the climate challenge effectively.

Keywords Green Industry · Net zero · Climate change · Decarbonization · Zero emission · Technology innovation

B. Mukherjee (✉)
AURO University, Surat, India
e-mail: baidyanath.mukherjee@aurouniversity.edu.in

A. Kr. Singh · A. K. Sharma · S. Tyagi
Greater Noida College of Law, Greater Noida, India

A. K. Sharma
Vidya University, Meerut, India

K. Mukherjee et al. (eds.), *Urban Climate*, The Urban Book Series,
https://doi.org/10.1007/978-3-032-17895-4_7

7.1 Introduction

As decarbonisation becomes a priority, greener industry practices have progressed from basic approaches to more sophisticated systems. This includes moving beyond traditional pollution control methods to focus on cleaner production techniques, material substitution, minimising resource usage, transforming waste into valuable inputs, and adopting industrial ecology principles. By mimicking natural ecosystems, industries aim to integrate their production systems sustainably and reduce their environmental impact (Organisation for Economic Co-operation and Development 2009). In 2022, global energy-related carbon dioxide (CO_2) emissions rose by 0.9%, amounting to an increase of 321 million tonnes and reaching a record high of over 36.8 gigatonnes (Gt). Industries such as steel, chemicals, cement and aluminium, which are significant contributors to CO_2 emissions, account for nearly half of all energy-related CO_2 emissions. These sectors are also responsible for more than 70% of direct industrial emissions (CO_2 Emissions in 2022—*Analysis—IEA* 02 March 2023). Based on the most recent data evaluated by the Intergovernmental Panel on Climate Change (IPCC) for the United Nations Framework Convention on Climate Change (UNFCCC) negotiations, it is essential for developed countries to achieve a reduction of 58–93% in CO_2 emissions from the industrial sector by 2050, relative to 2010 levels. This target aligns with the range spanning from the 10th to the 90th percentile of all 1.5 °C scenarios, covering a wide range of potential outcomes (*IAMC 1.5 °C Scenario Explorer and Data Hosted by IIASA (Release 1.X)* 2018; Masson-Delmotte et al.2021). Due to the pressing need to reduce carbon emissions and the serious consequences of failing to do so, it is vital to strive for zero emissions across all sectors, including industry. However, the computational models used in the latest IPCC report do not provide enough detailed information about the specific opportunities and challenges for cutting emissions in the industrial sector. Therefore, there is a necessity to delve deeper into understanding the complexities of emission reduction within industry. This exploration is essential for developing effective strategies and overcoming potential obstacles (Edelenbosch et al. 2017). In the scenarios being considered, there is a significant emphasis on negative emission technologies (NETs) to offset greenhouse gas emissions. However, relying heavily on NETs can divert attention away from the importance of exploring alternative methods and strategies to rapidly reduce industrial emissions and ultimately achieve zero emissions. While NETs can play a role in mitigating emissions, it is crucial to also prioritise and discuss other approaches that directly address the reduction of industrial emissions. By focusing on a broader range of solutions and considering alternative means, we can ensure a more comprehensive and effective approach to tackling industrial emissions and working towards the goal of zero emissions (Smith et al. 2016).

Achieving zero emissions in global industrial systems will require substantial changes within and across current value chains, economic sectors and policy domains (C. G. F. Bataille 2020). Given the interconnectedness of material, product and service production and consumption systems within modern economies, there is no singular solution to effectively transform energy-intensive and emission-heavy industries. These industries face distinctive challenges characterised by their significant energy and emission levels during production processes, a persistent and strong dependence on carbon-intensive technologies and infrastructures due to substantial capital investments, and lengthy investment cycles that allow for major process changes only over extended periods, often spanning decades. Additional obstacles include their deep-rooted historical presence, their symbolic and economic importance to local communities, and the global competition they face in commodity markets, where largely undifferentiated products are traded (Energy Technology Perspectives 2020—Analysis—IEA 10 September 2020; Sci-Hub | The Transition of Energy Intensive Processing Industries towards Deep Decarbonization: Characteristics and Implications for Future Research. Renewable and Sustainable Energy Reviews, 79, 1303–1313 n.d.; Seto et al. 2016). Hence, it is necessary to pursue multiple complementary solutions simultaneously across value chains to address these challenges. These solutions encompass technological advancements, organisational changes and shifts in behaviour. To achieve decarbonisation, various approaches need to be pursued, such as enhancing energy efficiency, optimising materials usage, decreasing overall demand, establishing circular material flows, adopting electrification, transitioning to alternative fuels and implementing carbon capture technologies for either utilisation or storage purposes. By employing a combination of these strategies, a comprehensive and effective path towards decarbonisation can be established (Bataille et al. 2018). Achieving short-term reductions in emissions within existing processes typically involves improving energy efficiency and transitioning to cleaner fuels. However, achieving zero emissions requires substantial changes to entire production processes. This necessitates proactive efforts from companies in energy and emissions-intensive industries, which have been slow to adopt effective carbon management practices (Dietz et al. 2018). Additionally, governments must take an active role in identifying and implementing tailored strategies specific to each region to achieve zero emissions (Waisman et al. 2019).

To implement these solutions effectively, industrial policy, encompassing all policy initiatives that influence industrial development, must recognise that achieving zero emissions is a fundamental objective within this policy domain (Springer Science + business Media and Springer 2020). Modern industrial policy originated from the intention to safeguard industries that were considered crucial to state interests during the post-war period. Various policy instruments, such as state ownership, direct subsidies and tax breaks, were employed to support and protect these industries. Additionally, trade-related policies, including the imposition of tariffs, were used as instruments within industrial policy to regulate trade and protect domestic industries (Grabas and Nützenadel 2014). In subsequent years, the emphasis of industrial policy shifted towards promoting the growth of high-tech sectors like information and communication technology (ICT) and biotechnology. Additionally, industrial policy aimed to provide support and assistance to small- and medium-sized enterprises (SMEs) due to their vital role in fostering entrepreneurship, job creation and local economic development.

During the 1970s, environmental policy primarily focused on implementing end-of-pipe solutions to address local and regional pollution. However, in recent times, there has been a shift in industrial policy towards promoting sustainable development by emphasising a systems-oriented approach rather than solely prioritising economic growth (Aiginger 2015; Rodrik 2014; Warwick 2013). This change recognises the need to transform modern economies to achieve environmental sustainability. This shift reflects a parallel progression in innovation policy, transitioning from a broad focus on supporting various forms of economic development to adopting more targeted and transformative objectives driven by specific challenges (Schot and Steinmueller 2018). Considering that the objective of these policies extends beyond merely supporting innovation development, but rather aims to initiate and propagate targeted changes within the systems delivering essential societal services like transportation, housing or food, it becomes imperative to embrace strategic policymaking that incorporates a portfolio of policies rather than relying on isolated instruments (Edmondson et al. 2018; *Green Industry Development in Different Types of Regions* 2019). While carbon pricing plays a significant role in climate policy, it alone is inadequate to bring about the transformative change necessary to address climate challenges effectively. To drive transformative change, a combination of strategies is essential (Ball 2018; Rosenbloom et al. 2020).

In the past, approaches to green industrial policy and clean technology innovation did not give explicit consideration to the crucial importance of achieving zero emissions. They also overlooked the significant transformations required in the production, utilisation and recycling of emissions-intensive basic materials. However, since the adoption of the Paris Agreement, there has been a growing recognition of this challenge. In recent years, various stakeholders, including non-governmental organisations (NGOs), industrial actors and academic scholars, have increasingly focused on the need for low-carbon industrial development. This shift in attention is reflected in the publication of numerous policy briefs and recommendations addressing the transition towards sustainable and low-carbon industrial practices (Busch et al. 2018; *Policy Briefing: Forging a Carbon–Neutral Heavy Industry by 2050: How Europe Can Seize the Opportunity|Corporate Leaders Groups* 3 October 2019). In the ongoing debate on industrial policy for transitioning energy and emission-intensive industries towards zero emissions, it is contended that an effective industrial policy should adopt a structured and comprehensive approach (Meckling and Allan 2020). Drawing from recent insights, this research emphasises that successful climate policy requires the use of a diverse range of instruments and initiatives, strategically implemented in combination and sequence (Meckling et al. 2017).

This paper presents a comprehensive framework for industrial policy aimed at guiding the transformation of energy and emission-intensive industries towards zero emissions. The framework consists of eight key pillars. Firstly, it emphasises the importance of establishing a clear and ambitious policy, setting specific targets for emissions reduction aligned with international agreements like the Paris Agreement. Secondly, it advocates for a supportive regulatory environment that promotes the adoption of clean technologies and sustainable practices through policies such as emissions limits and energy efficiency mandates. Thirdly, financial incentives and support are proposed to encourage investments in clean technologies, renewable energy and research and development, alongside affordable financing and infrastructure development.

The fourth pillar, Collaborative Partnerships, stresses the need for cooperation among governments, industry stakeholders, research institutions and civil society to foster innovation and address technical challenges. Fifthly, Technology Innovation and Adoption are highlighted, promoting research into clean technologies and facilitating their commercialisation through pilot projects and technology transfer programmes. The sixth pillar focuses on Skill Development and Workforce Transition, advocating for training programmes to equip the workforce with the skills needed for zero-emissions industries, including reskilling and job placement services.

Circular Economy Approaches constitute the seventh pillar, encouraging resource efficiency, waste reduction and sustainable consumption patterns through eco-design and circular supply chains. Lastly, the framework emphasises the importance of Monitoring, Reporting and Evaluation to track progress, assess policy effectiveness and ensure accountability, guiding continuous improvement. While interconnected, each pillar is presented distinctly to clarify its specific goals and strategies within the broader context of industrial policy for achieving sustainability goals.

7.2 The Path Towards Zero-Emission Industrial Policy

7.2.1 Clear and Ambitious Policy

Green industrial policy catalyses advancing green industries. Existing literature on green industrial policy emphasises its focus on sustainability, advocating for increased investment in research and development (R&D) and the adoption of green technologies to achieve decarbonisation across all economic sectors. Future industrial policy needs to expand its scope beyond mere economic efficiency and growth, embracing a broader approach that combines efficiency with a clear sense of purpose, aiming to achieve both environmental and socio-economic objectives. Public policies play a crucial role in shaping the pace and direction of change by establishing specific incentive systems. Recognising directionality as a core aspect of industrial policy involves understanding the importance of fostering innovations that not only maximise efficiency but also align with the desired developmental goals (Weber and Rohracher 2012). A direction-oriented industrial policy does not solely advocate for centralised state projects (Mowery et al. 2010). Instead, it necessitates collaborative partnerships between private and state actors (Aiginger and Rodrik 2020) from the bottom up to facilitate the advancement of not only new technologies but also new techno-economic paradigms (Mathews 2013). Setting the desired direction of progress involves not only supporting sustainable solutions but also restricting less sustainable alternatives. Thus, the policy mix should include initiatives that encourage the creation of sustainable options and those that facilitate the removal of unsustainable ones (Kivimaa and Kern 2016; Rogge and Johnstone 2017). The interplay of creation and destruction results in both winners and losers, which may require various measures such as reskilling, job creation and compensation to ensure legitimacy and acceptance.

To guide development effectively, overarching policy frameworks like climate laws are crucial in establishing the ambition of policies (Eskander and Fankhauser 2020). However, it is important to complement these efforts with detailed plans tailored to the specific needs of various industries. Trade associations and individual companies have a critical role in developing roadmaps that outline feasible pathways to achieve zero emissions. Directionality can also be strengthened through strategic infrastructure investments, giving governments opportunities to support system-building and speed up the transformation process. Comprehensive approaches that include policy frameworks, industry-specific plans and infrastructure investments are crucial for steering development towards sustainable objectives (Cass et al. 2018). To successfully achieve the desired direction of development, it is essential to have a comprehensive policy toolbox that includes a range of direct and indirect policy measures (Andersson and Karpestam 2014). However, a critical component of this policy mix is achieving alignment and coherence across various areas. This alignment has two aspects: first, policies should align with the intended direction of development, ensuring they are consistent with the desired goals and outcomes. Second, coherence within the policy mix is necessary, meaning the different policies must work together harmoniously and complement each other to drive the desired changes. This alignment and coherence are essential for maximising the effectiveness and impact of the policy measures.

7.2.2 Supportive Regulatory Environment

The shift towards fully decarbonising industry presents a new policy challenge for governments, requiring them to develop the managerial and bureaucratic capacity necessary to effectively oversee this transformation. Previously, with short-term and incremental emission reduction targets, as seen in the Kyoto Protocol, industrial emissions could be decreased through relatively low-cost measures such as energy efficiency improvements and the switch from coal to gas. These efforts were supported by traditional energy efficiency policies. However, as the focus moves from short-term, low-cost solutions to achieving zero emissions and fundamentally transforming the industrial sector, new challenges arise that demand the collective attention of governments, industry and other stakeholders. Innovating policy processes is essential to effectively addressing these challenges (Upham et al. 2014). Climate policy has been integrated into energy and transport policies for many years, which has led to the development of institutional capacity in these areas. This includes establishing goals, fostering policy learning and building expertise within government ministries and agencies, as well as in academic research. However, the focus in these sectors has traditionally been on short-term, incremental changes rather than long-term, transformational strategies and capabilities (Johansson et al. 2020). However, achieving zero industrial emissions remains a relatively new and underdeveloped area in policy and governance. This is particularly evident in the plastics industry, where early policy measures have been fragmented and do not adequately address the heavy reliance on fossil feedstock and energy, along with the consequent greenhouse gas emissions (Nielsen et al. 2020). Looking ahead, it is crucial to address the structural obstacles present in the integrated oil, gas, petrochemical and plastic industries, particularly their interconnected reliance on fossil resources. Establishing a robust institutional capacity that comprehensively understands these complexities is essential for effectively tackling emerging challenges. Independent organisations, like the UK Climate Change Committee, can play a valuable role by providing greater transparency, incorporating diverse perspectives and ensuring political accountability. This approach can foster a more holistic and informed strategy to navigate these industries' intricacies and drive meaningful progress towards sustainability.

Creating stable conditions is essential for businesses to confidently invest in new technologies. However, it is equally important to maintain the flexibility to adapt and customise policies in response to evolving technological advancements, changing demand patterns and geopolitical factors. Balancing this policy reform process is challenging, as there is a risk of regulatory capture by influential businesses aiming to maintain the status quo. Additionally, there is a concern that proposed policies might create unfair conditions for some businesses or generate excessive profits for certain actors. For example, the EU Emissions Trading System (ETS) has resulted in significant windfall profits for several highly emissions-intensive industries due to the overallocation of free emission rights (*Additional Profits of Sectors and Firms from the EU ETS: 2008–2019—CE Delft—EN* 2021). It is essential to consider the democratic implications of creating stable investment conditions and ensure that mechanisms for policy adaptation and responsiveness are in place to accommodate future knowledge and priorities.

In pursuing transformative change, policy coherence is crucial for efficient and effective outcomes, often requiring institutional reforms. A coherent approach to industrial transformation will span various traditional policy areas, especially in a future that prioritises material efficiency, circularity and electrification with new sectoral connections. This approach will go beyond building codes, waste management, product standards, electricity market design, resource security, environmental permits, trade, carbon capture and storage infrastructure, regulations, research, development, innovation and market creation. It emphasises the need for a policy framework that adopts a systems-oriented perspective for sustainable development, moving away from a sole focus on growth (Aiginger 2015; Rodrik 2014).

7.2.3 Financial Incentives and Support

Advancing green industries and transitioning to a net-zero economy require substantial support and funding for several key areas, including human capital, research and development (R&D), policy innovation, institutions and infrastructure. However, insufficient financing, high costs and limited knowledge often hinder innovation in green technologies, impacting green productivity. Technological and production process improvements typically rely on incremental advancements and knowledge spillovers from innovative developments. While the green industry's innovation-driven potential can decouple natural resource consumption from economic growth, expecting rapid progress may be unrealistic. The pace and scale of innovation largely depend on a given society's societal and political dynamism and responsiveness. Responsible government agencies must provide the necessary support and resources to foster innovation in these areas. To drive innovation, businesses, governments and other stakeholders need to collaborate, explore and provide financial incentives. For instance, China's environmental regulations have effectively incentivised the development of green technology, especially in listed state-owned enterprises and technology- and capital-intensive sectors (Cai et al. 2020). China has seen considerable success in technological innovation, particularly in developing and manufacturing solar photovoltaic (PV) and wind turbine technologies. This success stems from institutional support, the allocation of public resources, government policies and international collaborations. These combined efforts have fostered an environment conducive to innovation, significantly advancing China's green technology sector (Nahm 2021).

7.2.4 Collaborative Partnerships

Collaboration along the value chain is crucial for the successful development and widespread adoption of zero-emission innovations in the industrial sector (Rissman et al. 2020). By involving various actors and facilitating collective learning, this approach allows for the effective combination of diverse knowledge and expertise. In the context of energy and emission-intensive industries, specialised engineering firms have traditionally played a vital role in technological advancements. Industry companies often outsource or engage in intensive collaborations with these firms to drive process development. This collaborative approach enhances the chances of achieving zero emissions by leveraging shared resources, insights and experiences across the entire value chain (Wesseling et al. 2017). In certain cases, it is essential to foster collaboration across sectors that possess complementary knowledge and capabilities, such as in the upscaling of forest biorefineries. However, engaging a diverse group of actors in these collaborations can be challenging. Despite the difficulties, such cross-sector collaborations are necessary to leverage a wide range of expertise, resources and perspectives, ultimately enhancing the potential for successful upscaling and innovation in industries like forest biorefineries, where multiple sectors have valuable contributions to make (Bauer et al. 2018; Karltorp and Sandén 2012).

Achieving zero emissions effectively requires governance to adopt a systems perspective that recognises the interdependence of various sectors. This approach is particularly vital for two key pathways that require innovative systems: the electrification of industrial processes (either directly or through hydrogen) and carbon capture and storage (CCS) (Lechtenböhmer, Nilsson, et al. 2015a, b). Successfully implementing these pathways requires access to low-carbon electricity, underpinned by strong power transmission and distribution systems. Additionally, establishing CO_2 and hydrogen distribution networks and storage facilities depends on the widespread adoption of complementary innovations. Investing in this infrastructure necessitates long-term planning and a unified vision among government, industry and civil society. Collaborative scenario development, roadmaps and action plans involving all stakeholders are essential tools for facilitating learning, communication and coordination during the transition (Lechtenböhmer et al. 2015b).

7.2.5 Technology Innovation and Adoption

Innovation plays a pivotal role in improving the efficiency of green industries and expediting the shift towards a net-zero economy (Sun et al. 2017). Research on innovation processes has highlighted the importance of government interventions in securing sufficient investments in research, development and innovation (RDI) that meet societal needs. These interventions also help firms benefit from their RDI investments by establishing intellectual property rights systems. Since the late 1980s, systemic perspectives have become increasingly influential in both research and policy, emphasising the interconnections among various actors, institutions and factors that influence innovation processes (Godin 2009; Schot and Steinmueller 2018). These perspectives emphasise the interactive nature of innovation processes, extending beyond traditional actors like firms, universities and research institutes. They recognise the essential contributions of public sector actors, intermediary organisations and demand-side participants in generating knowledge and fostering innovation. These broader viewpoints understand innovation as a collaborative and multifaceted process that involves a variety of stakeholders and the interplay between supply and demand factors (Edquist 1997). Moreover, these perspectives emphasise the impact of both formal and informal institutions linked to particular industries on the dynamics of innovation processes. Recent studies have delved into the potential of innovation policy to facilitate systemic change and transformation (*Green Industry Development in Different Types of Regions* 2019).

Energy-intensive industries often feature a structure where a few dominant players control the primary production segment of the value chain. These industries typically allocate relatively low investments in research, development and innovation (RDI), usually amounting to less than 1% of their annual revenue (Wesseling et al. 2017). These industries are defined by their high capital intensity and a primary focus on incremental process improvements rather than product innovation. Introducing new technologies in these sectors requires compatibility with existing processes. High capital utilisation is essential for recovering investment costs, leading to increasing returns and a path dependency where future developments are shaped by past decisions and investments (Arthur 1994). Industrial transformation often requires radical innovations that are seen as risky, difficult to integrate and less competitive than existing technologies (Sood and Tellis 2005). As a result, companies exploring emerging technologies encounter higher production costs and limited immediate financial returns. To overcome this challenge, public research, development and innovation (RDI) policies are essential in reducing the cost of innovation for businesses. These policies include public funding for fundamental research, pilot and demonstration facilities and educational and training support. By offering this assistance, public RDI policies help firms overcome obstacles, promoting the adoption and commercialisation of transformative technologies (Nemet 2009; Schot and Steinmueller 2018).

7.2.6 Skill Development and Workforce Transition

Today's energy workforce possesses skills above the global average. However, emerging clean energy industries will demand an even greater proportion of skilled employees. This necessitates the establishment of new educational programmes, certification processes and vocational training initiatives. Additionally, targeted training programmes for upskilling or reskilling the existing workforce will be crucial. By providing specialised training and education for both new entrants and current workers, the industry can ensure a sufficient supply of skilled individuals to meet the evolving demands of clean energy sectors.

Traditional energy sectors, especially coal, are currently undergoing significant workforce changes due to increased automation and global trends in the fuel mix. The coal industry is particularly affected by these changes. In the coming years, many other sectors, including oil and gas, heavy industry and road transport, will also experience substantial shifts in employment patterns. These changes are likely driven by factors such as technological advancements, evolving market dynamics and the growing emphasis on sustainable and low-carbon alternatives (Energy Agency 2022).

Ensuring a sufficient supply of skilled workers in these sectors is strategically crucial during the transition. Developing a suitable long-term transition plan is imperative to mitigate these risks. According to the recently published World Energy Employment report by the International Energy Agency (IEA), the energy sector employed over 65 million individuals, constituting approximately 2% of global employment in 2019. This includes workers involved in fuel supply, the power sector, vehicle manufacturing, end-use sectors and energy efficiency, (*World Energy Employment—Analysis—IEA* 2024). Despite substantial overall employment figures in the energy sector, the situation is nuanced. Workers' skill sets vary widely based on industry, job function and geographical location. This diversity means that while clean energy sectors are expected to grow rapidly in the coming years, companies may struggle to find adequately skilled workers tailored to specific regional project needs. Addressing this challenge requires establishing new educational programmes, certification processes and vocational training initiatives. Additionally, efforts to upskill or reskill the current workforce are crucial to bridging existing skill gaps. The goal is to ensure that the workforce possesses the precise skills necessary to meet the evolving demands of the clean energy industry across various locations and job functions. Meanwhile, as the coal sector faces rapid decline, governments and companies are prioritising a just transition for coal workers. Countries such as South Africa and Spain have conducted assessments of their coal workforces to better understand the labour market dynamics (Energy Agency 2022). Social dialogues, joint strategies and skills training programmes are being implemented to facilitate a smooth transition away from coal.

India aims to achieve 450 GW of renewable energy capacity by 2030, which is expected to attract substantial investments and create numerous job opportunities. With the government committed to expanding renewables in line with the Paris Agreement, employment in this sector is projected to grow by 30% by the same year. According to the International Labour Organisation (ILO), India's transition to a green economy could generate an additional 3 million jobs in renewables by 2030. To support this goal, the government is actively implementing training and educational programmes to prepare the workforce for the renewable energy industry's demands (Energy Agency 2022). The European Union is preparing for a significant skills transformation in the green and digital sectors *Transition: Targeted Training of Workers Can Turn Challenges into Opportunities | European Economic and Social Committee* n.d.). Over the next ten years, approximately 90% of jobs will require digital skills. However, only about 44% of Europeans have basic digital skills, and just one in five possess advanced skills. Acknowledging the urgency of preparing the workforce for this transition, various stakeholders including governments, companies and industry organisations are actively collaborating to develop extensive educational and skills training initiatives.

7.2.7 Circular Economy Approaches

To combat the worsening effects of climate change, it is increasingly necessary to implement strategies that promote a circular economy across all sectors. This approach minimises waste, maximises resource efficiency and prioritises recycling and reuse. The objective is to reduce carbon emissions by 45% by 2030 and achieve carbon neutrality by 2050 (Yang et al. 2023). Reducing waste in production and distribution processes is a primary objective of the circular economy. The Waste Framework Directive has strengthened the connection between the circular economy and waste management (Directive 2008/*98/EC of the European Parliament and of the C...—EUR-Lex* 22.11.2008). As a result, the circular economy offers an effective way to improve waste recovery and promote recycling efforts. The conversion of waste into reusable materials illustrates the ecological transition of traditional economic systems (Ghisellini et al. 2016), addressing the depletion of non-renewable resources and environmental pollution. The regional circular economy, driven by an eco-dynamic system, aims to enhance the organisational structure and operational mechanisms of production, circulation and consumption at the regional level. This approach maximises economic benefits while minimising resource consumption and environmental harm. By pursuing waste-to-material transformation, humanity can advance ecologically and economically while striving for sustainability in the ecological civilisation era. The circular economy model emphasises harmonious coexistence between humans and the ecosystem, promoting the exchange of materials, information and energy. We can establish a comprehensive ecological system by integrating natural ecosystems and developing an ecological civilisation. This section examines the circular economy as a waste management strategy, highlighting the global need to convert waste into reusable materials and create ecosystems that prioritise economic efficiency and environmental sustainability. Adopting circular economy principles worldwide will increase the prominence of these strategies on the international stage and amplify their global impact.

The circular economy offers a promising approach to addressing climate change by reducing carbon emissions across various sectors, including industry, waste management, energy use, construction and transportation. It also positively impacts air and water quality, energy consumption, natural resources and waste disposal. However, using bio-based materials in the circular economy may increase pressures on land use and land cover.

The circular economy provides opportunities for industrial growth and optimisation of the food system. Companies can collaborate with governments and universities to recycle materials and create consumer-friendly products. Effective circular economy strategies can reduce resource and energy inputs, as well as waste and emissions, throughout the product life cycle. In the food system, these strategies can help reduce the use of non-renewable resources, increase recycling and manage food waste. They can also streamline the supply chain, reduce food consumption and provide innovative waste treatment methods. Overall, the circular economy has the potential to create a more sustainable and efficient economic system (Yang et al. 2023).

Establishing an eco-industrial park based on a circular economy approach can provide multiple benefits. It consolidates recycling, treatment, production, packaging and marketing of waste within park businesses, improving waste management. It also facilitates trading bulk product waste by assessing quality and performance and uploading relevant data to the cloud. Moreover, implementing waste separation at the source enhances long-term sustainability outcomes. These practices allow an eco-industrial park to optimise resource utilisation, reduce waste generation and promote a more sustainable and efficient circular economy.

7.2.8 Monitoring, Reporting and Evaluation

In the institutional setting, the establishment of systems for continuous monitoring and evaluation will play a crucial role. These systems serve two important purposes: learning and political accountability. By continuously monitoring the progress and outcomes of policies and initiatives, valuable insights and lessons can be gained, allowing for ongoing learning and improvement. They enable stakeholders, including policymakers and the public, to hold decision-makers accountable for their actions and ensure that policies align with the desired objectives. Overall, these systems enhance the effectiveness and transparency of governance processes (Mickwitz 2003).

The presence of a learning effect through monitoring and evaluation processes can mitigate the risk of asymmetric knowledge between regulatory bodies and industries. In many cases, regulators may have limited information about the intricacies of specific industries or technologies, while industry actors possess specialised knowledge in these areas. By implementing robust monitoring and evaluation mechanisms, this knowledge gap can be reduced. These mechanisms can operate independently but alongside existing governmental structures, taking the form of an observatory or advisory function. The UK Climate Change Committee serves as an example of such an independent body that offers expertise and guidance in climate-related matters. The success of such bodies has inspired the establishment of similar institutions worldwide, ensuring that informed and impartial advice is available to policymakers (Weaver 2024). In addition to establishing independent bodies, another approach to organising monitoring and evaluation processes is integrating them within existing government structures, such as agencies and ministries. The specific objectives, mandate, resources, independence and organisational structure may differ between countries based on their political and institutional traditions. Nonetheless, the imperative of institutional learning remains critical.

A further challenge lies in developing an internationally effective system to track and evaluate progress in the industrial sector. Given the global nature of the industry and its impact on emissions and sustainability, a unified approach to monitoring and evaluating progress becomes essential. Coordinating efforts across nations and establishing common standards and metrics can facilitate effective tracking and evaluation, enabling a comprehensive understanding of industry performance on a global scale. Addressing this challenge is vital to foster international collaboration and ensure a coordinated response to industrial transformation.

7.3 Initiative Towards Zero Emission—India and Global Perspective

7.3.1 India's Zero-Emission Initiatives:

India committed to achieving net zero by 2070 at the COP26 Climate Summit 2021. By 2030, it will increase its non-fossil energy capacity to 500 gigawatts, fulfil 50% of its energy needs from renewable sources, and cut the total projected carbon emissions by one billion tonnes. Additionally, by 2030, India will lower the carbon intensity of its economy by more than 45%. However, the path is not swift, and the pledge has huge implications for a country of nearly 1.4 billion people, which is one-sixth of the global population. To achieve the 2070 target, India needs to achieve five main decarbonisation goals:

1. *Phase out coal in the Industrial Sector by 2065*: The industrial sector relies heavily on coal. India has to phase out coal in this sector to achieve net zero by 2065. This involves:
   a. Transitioning to cleaner energy sources such as natural gas, renewable energy and hydrogen;
   b. Implementing energy-efficient technologies;
   c. Encouraging Industrial innovation and adoption of low-carbon processes.
2. *Phase out all other use of coal by 2060:* In addition to the industrial sector, coal is used for electricity generation and other applications. To meet the net-zero target, India must phase out all other use of coal by 2060; this requires:
   a. Replacing coal-fired power plants with renewable energy plants;
   b. supporting workers and communities dependent on the coal industry through just transition strategy;
   c. investing in alternative energy sources and infrastructure to ensure a stable energy supply.
3. *Achieve 557 GW of wind-based generation capacity by 2050*: Wind energy is another important part of India's renewable energy mix. India needs to focus on:
   a. Expanding onshore and offshore wind farms;
   b. developing policies and incentives to attract investment in wind energy projects;
   c. strengthening grid connectivity to facilitate the transmission of wind power;
   d. encouraging research and development in wind turbine technology to increase efficiency and reduce cost.
4. *Expand to 68 GW of nuclear-powered generation capacity by 2050:* Nuclear energy provides a reliable and low-carbon energy source. To reach 68 GW of nuclear capacity by 2050, India must:
   a. Construct new power plants and upgrade existing ones;

b. address regulatory and safety concerns associated with nuclear energy;
c. foster international collaboration and partnerships for nuclear technology and expertise;
d. develop a robust nuclear waste management strategy.

These steps will mitigate climate change, drive economic growth, create jobs and improve energy security. The journey towards a sustainable future is challenging, but with determination and innovation, India will lead the global fight against climate change.

A significant transformation is also required in India's transportation sector to meet the 2070 target. The CEEW report projects that by 2070:

- Electric cars are projected to dominate, constituting 84% of all car sales;
- electric trucks are expected to make up 79% of total truck sales, with the remaining 21% being powered by hydrogen fuel;
- biofuel blends are anticipated to play a significant role, providing 84% of the petrol needs across cars, trucks and even the aviation sector.

To drive this transition, the Indian government has implemented the Faster Adoption and Manufacturing of Hybrid and Electric Vehicles (FAME) scheme (Government of India ministry of information and broadcasting 2022). The scheme has two phases, Phase I (2015–2019) which focuses on promoting electric and hybrid vehicles, deploying electric buses and setting up charging infrastructure, and Phase II (2019–2024) which focuses on supporting the electrification of public and shared transportation, demand incentives for electric vehicles and creating charging infrastructure. Though the sale of electric cars has increased and 6000 charging stations were established by 2021, it is still a small fraction of the overall mandate that is to be achieved (Deshpande 2023). India is expected to invest $10 trillion in green technologies and energy efficiency initiatives.

India has also undertaken significant collaborations and international partnerships to further its zero-emission agenda. A notable example is the partnership between IIT Kanpur and the Kuppam Area Development Authority (KADA), which aims to transform Kuppam, Andhra Pradesh, into the nation's first net-zero emissions constituency through renewable energy and sustainable practises. Moreover, India actively participates in international climate agreements, advocating for technology transfer and financial support from developed countries to accelerate its low-carbon transition. The road to net zero is steep, but with sustained determination, innovation and global cooperation, India is poised to lead the global fight against climate change.

7.3.2 *Global Zero-Emission Initiatives*

Several countries have undertaken notable initiatives to support global zero-emission goal. The European countries has committed to achieve climate neutral by 2050 with policies promoting renewable energy, energy efficiency and sustainable transport.

Many EU countries have set targets for phasing out internal combustion engine vehicles by 2035 (Sen et al. 2025). Similarly, United States have introduced regulations to reduce air pollution, including promoting zero-emission vehicles. Federal Tax credits and rebates are available for purchasing electric vehicles (Global EV Outlook 2023). In Asia, Japan has taken a proactive approach by encouraging the adoption of hydrogen fuel cell vehicles through subsidies and infrastructure development. Whereas, Bhutan stands out as a global leader in sustainability, with long-standing policies focused on sustainable forestry management and the protection of national parks, which cover nearly 40% of the country. These conservation efforts, coupled with Bhutan's reliance on hydropower for clean energy, have enabled the nation to achieve net-zero emissions, serving as an exemplary model for balancing development and environmental stewardship (Wood 2022).

7.4 Conclusion and Suggestions

To achieve zero emissions across all sectors, energy and emission-intensive industries must no longer rely on other sectors to bear the burden of climate change mitigation. The development of green industrial policies is essential, particularly for industries that are slow to change and deeply entrenched in society. As a result, industrial transformation is expected to become a central focus in future international climate policy and negotiations.

Although carbon pricing has been a key tool in climate policy, its impact on reducing emissions in energy and emissions-intensive industries has been limited. To address this, we propose a comprehensive industrial policy framework focused on achieving zero emissions within these industries. Our framework, drawing on engineering, economics, policy, governance and innovation studies, comprises six key pillars: setting clear policy direction, promoting knowledge creation and innovation, shaping markets, building governance capacity, ensuring international policy coherence and addressing socio-economic consequences. Effective policies should include a variety of mitigation options, such as managing materials demand, promoting electrification and implementing carbon capture and storage (CCS) technologies.

Governments need to take proactive and assertive measures to meet the challenges of achieving zero emissions in industry. This involves establishing clear roadmaps and policies and reshaping markets to support the transition to zero-emission solutions. Governments should expand research and innovation policies to focus on industry-specific zero-emission technologies and develop their capacity and knowledge regarding industrial transformation. Additionally, they must elevate this issue in international climate and trade negotiations to ensure it remains prominent on the global agenda.

The strategy must also be adaptable and responsive to evolving technologies, systems and socio-cultural norms. It is essential to tailor industrial policy strategies to specific sub-sectors and consider national and regional contexts, such as access to cleaner feedstocks and energy sources. The design and implementation of strategies, interventions and instruments should be sensitive to each context's unique characteristics, including economic development, governance traditions, resource availability and industrial landscape. Given that significant process changes in heavily impacted sectors may require one or two investment cycles, it is crucial to reevaluate industrial development now, as 2050 approaches.

AI Discloser No AI tool has been used for writing the manuscript.

References

Additional profits of sectors and firms from the EU ETS. 2008–2019 - CE Delft - EN (2021) Retrieved June 22, 2023, from https://cedelft.eu/publications/additional-profits-of-sectors-and-firms-from-the-eu-ets/

Aiginger K (2015) Industrial policy for a sustainable growth path. In: New perspectives on industrial policy for a modern Britain, pp 364–394. https://doi.org/10.1093/ACPROF:OSO/9780198706205.003.0019

Aiginger K, Rodrik D (2020) Rebirth of industrial policy and an agenda for the twenty-first century. J Ind Compet Trade 20(2):189–207. https://doi.org/10.1007/S10842-019-00322-3/TABLES/2

Andersson FN, Karpestam P (2014) Carbon Management The Australian carbon tax: a step in the right direction but not enough. https://doi.org/10.4155/cmt.12.18

Arthur WB (1994) Increasing returns and path dependence in the economy. University of Michigan Press. https://doi.org/10.3998/mpub.10029

Ball J (2018) Hot air won't fly: the new climate consensus that carbon pricing isn't cutting It. Joule 2:2491–2494. https://doi.org/10.1016/j.joule.2018.11.019

Bataille C, Åhman M, Neuhoff K, Nilsson LJ, Fischedick M, Lechtenböhmer S, Solano-Rodriquez B, Denis-Ryan A, Stiebert S, Waisman H, Sartor O, Rahbar S (2018) A review of technology and policy deep decarbonization pathway options for making energy-intensive industry production consistent with the Paris Agreement. J Clean Prod 187:960–973. https://doi.org/10.1016/J.JCLEPRO.2018.03.107

Bataille CGF (2020) Physical and policy pathways to net-zero emissions industry. Wiley Interdisciplin Rev: Clim Change 11(2):e633. https://doi.org/10.1002/WCC.633

Bauer F, Hansen T, Hellsmark H (2018) Innovation in the bioeconomy-dynamics of biorefinery innovation networks. Technol Anal Strateg Manag 30:935–947. https://doi.org/10.1080/09537325.2018.1425386

Busch J, Foxon TJ, Taylor PG (2018) Designing industrial strategy for a low carbon transformation. https://doi.org/10.1016/j.eist.2018.07.005

Cai X, Zhu B, Zhang H, Li L, Xie M (2020) Can direct environmental regulation promote green technology innovation in heavily polluting industries? evidence from Chinese listed companies. Sci Total Environ 746. https://doi.org/10.1016/J.SCITOTENV.2020.140810

Cass N, Schwanen T, Shove E (2018) Infrastructures, intersections and societal transformations. Technol Forecast Soc Chang 137:160–167. https://doi.org/10.1016/J.TECHFORE.2018.07.039

CO_2 Emissions in 2022–Analysis - IEA. (02 March 2023) Retrieved June 17, 2023, from https://www.iea.org/reports/co2-emissions-in-2022

Deshpande M (2023) Electrifying progress: analysis of India's FAME Scheme For Electric Vehicles-IMPRI Impact And Policy Research Institute. IMPRI. https://www.impriindia.com/insights/analysis-of-indias-fame-scheme/

Dietz S, Fruitiere C, Garcia-Manas C, Irwin W, Rauis B, Sullivan R (2018) An assessment of climate action by high-carbon global corporations Article (Accepted version) (Refereed) Original citation

Directive 2008/98/EC of the European Parliament and of the C... - EUR-Lex. (22.11.2008) https://eur-lex.europa.eu/legal-content/EN/LSU/?uri=CELEX:32008L0098. Retrieved June 22, 2023

Edelenbosch OY, Kermeli K, Crijns-Graus W, Worrell E, Bibas R, Fais B, Fujimori S, Kyle P, Sano F, van Vuuren DP (2017) Comparing projections of industrial energy demand and greenhouse gas emissions in long-term energy models. Energy (Oxford) 122:701–710. https://doi.org/10.1016/J.ENERGY.2017.01.017

Edmondson D, Kern F, Rogge K (2018) The co-evolution of policy mixes and socio-technical systems: towards a conceptual framework of policy mix feedback in sustainability transitions. Res Policy 48. https://doi.org/10.1016/j.respol.2018.03.010

Edquist C (1997) Systems of innovation: technologies, institutions and organizations

Energy Agency I (2022) Skills development and inclusivity for clean energy transitions.

Energy Technology Perspectives 2020–Analysis - IEA. (10 September 2020) https://www.iea.org/reports/energy-technology-perspectives-2020. Retrieved June 18, 2023

Eskander SMSU, Fankhauser S (2020) Reduction in greenhouse gas emissions by national climate legislation

Ghisellini P, Cialani C, Ulgiati S (2016) A review on circular economy: the expected transition to a balanced interplay of environmental and economic systems. J Clean Prod 114:11–32. https://doi.org/10.1016/j.jclepro.2015.09.007

Global EV Outlook 2023 (2023) Policy developments–Global EV Outlook 2023–Analysis-IEA

Godin B (2009) National innovation system: the system approach in historical perspective. Sci Technol Human Values 34(4):476–501. https://doi.org/10.1177/0162243908329187

Government of India ministry of information and broadcasting (2022) FAME India Faster Adoption and Manufacturing of (Hybrid &) Electric Vehicles in India

Grabas C, Nützenadel A (2014) Industrial policy in Europe after 1945: Wealth, power and economic development in the cold war. In: Industrial policy in Europe after 1945: wealth, power and economic development in the cold war, 1–388. https://doi.org/10.1057/9781137329905/COVER

Green industry development in different types of regions, 27 2163 (2019) (testimony of Markus Grillitsch and Teis Hansen). https://doi.org/10.1080/09654313.2019.1648385

IAMC 1.5°C Scenario Explorer and Data hosted by IIASA (release 1.x). (2018) https://data.ene.iiasa.ac.at/DOI/SR15/08-2018.15429/. Retrieved June 17, 2023

Johansson B, Bauer F, Nilsson LJ (2020) Assessing low carbon transitions a conceptual model

Just transition: targeted training of workers can turn challenges into opportunities | European Economic and Social Committee (7 December 2021) https://www.eesc.europa.eu/en/news-media/news/just-transition-targeted-training-workers-can-turn-challenges-opportunities. Retrieved June 22, 2023

Karltorp K, Sandén BA (2012) Explaining regime destabilisation in the pulp and paper industry. Environ Innov Soc Trans 2:66–81. https://doi.org/10.1016/J.EIST.2011.12.001

Kivimaa P, Kern F (2016) Creative destruction or mere niche support? innovation policy mixes for sustainability transitions. Res Policy 45(1):205–217. https://doi.org/10.1016/J.RESPOL.2015.09.008

Lechtenböhmer S, Nilsson LJ, Åhman M, Schneider C (2015) Decarbonising the energy intensive basic materials industry through electrification - implications for future EU electricity demand. SDEWES, pp 1–16

Lechtenböhmer S, Schneider C, Roche Y, Höller S (2015) Re-industrialisation and low-carbon economy—can they go together? results from stakeholder-based scenarios for energy-intensive industries in the German State of

Masson-Delmotte V, Zhai P, Pörtner HO, Roberts D, Skea J, Shukla PR, Pirani A, Moufouma-Okia W, Péan C, Pidcock R, Connors S, Matthews JBR, Chen Y, Zhou X, Gomis MI, Lonnoy E, Maycock T, Tignor M, Waterfield T (2021) Global warming of 1.5°C An IPCC Special Report on the impacts of global warming of 1.5°C above pre-industrial levels and related global greenhouse gas emission pathways, in the context of strengthening the global response to the threat of climate change, sustainable development, and efforts to eradicate poverty Summary

for Policymakers Edited by Science Officer Science Assistant Graphics Officer Working Group I Technical Support Unit
Mathews JA (2013) The renewable energies technology surge: A new techno-economic paradigm in the making? Futures 46:10–22. https://doi.org/10.1016/J.FUTURES.2012.12.001
Meckling J, Allan BB (2020) The evolution of ideas in global climate policy. Nat Climate Change 10(5):434–438. https://doi.org/10.1038/s41558-020-0739-7
Meckling J, Sterner T, Wagner G (2017) Policy sequencing toward decarbonization. Nat Energy 2(12):918–922. https://doi.org/10.1038/s41560-017-0025-8
Mickwitz P (2003) A framework for evaluating environmental policy instruments: context and key concepts. Evaluation 9(4):415–436. https://doi.org/10.1177/135638900300900404
Mowery DC, Nelson RR, Martin BR (2010) Technology policy and global warming: why new policy models are needed (or why putting new wine in old bottles won't work). Res Policy 39(8):1011–1023. https://doi.org/10.1016/J.RESPOL.2010.05.008
Nahm J (2021) Collaborative advantage: forging green industries in the new global economy. Oxford University Press. https://doi.org/10.1093/oso/9780197555361.001.0001
Nemet GF (2009) Demand-pull, technology-push, and government-led incentives for non-incremental technical change. Res Policy 38(5):700–709. https://doi.org/10.1016/j.respol.2009.01.004
Nielsen TD, Hasselbalch J, Holmberg K, Stripple J (2020) Politics and the plastic crisis: a review throughout the plastic life cycle. Wiley Interdisciplinary Rev Energy Environ 9(1):1–18. https://doi.org/10.1002/wene.360
Organisation for Economic Co-operation and Development (2009) Eco-innovation in industry : enabling green growth, 276
Policy Briefing: Forging a carbon-neutral heavy industry by 2050: How Europe can seize the opportunity I Corporate Leaders Groups. (3 October 2019)https://www.corporateleadersgroup.com/reports-evidence-and-insights/forging-a-carbon-neutral-heavy-industry-by-2050-how-europe-can-seize-the-opportunity. Retrieved June 19, 2023
Rissman J, Bataille C, Masanet E, Aden N, Morrow WR, Zhou N, Elliott N, Dell R, Heeren N, Huckestein B, Cresko J, Miller SA, Roy J, Fennell P, Cremmins B, Koch Blank T, Hone D, Williams ED, de la Rue du Can S, … Helseth J (2020) Technologies and policies to decarbonize global industry: Review and assessment of mitigation drivers through 2070. Appl Energy 266:114848. https://doi.org/10.1016/J.APENERGY.2020.114848
Rodrik D (2014) Green industrial policy. Oxf Rev Econ Policy 30(3):469–491. https://doi.org/10.1093/OXREP/GRU025
Rogge KS, Johnstone P (2017) Exploring the role of phaseout policies for lowcarbon energy transitions: the case of the German Energiewende Exploring the role of phase-out policies for low-carbon energy transitions: The case of the German Energiewende. https://doi.org/10.1016/j.erss.2017.10.004
Rosenbloom D, Markard J, Geels FW, Fuenfschilling L (2020) Opinion: why carbon pricing is not sufficient to mitigate climate change-and how "sustainability transition policy" can help. Proc Natl Acad Sci USA 117(16):8664–8668. https://doi.org/10.1073/PNAS.2004093117
Schot J, Steinmueller WE (2018) Three frames for innovation policy: R&D, systems of innovation and transformative change. Res Policy 47(9):1554–1567. https://doi.org/10.1016/J.RESPOL.2018.08.011
Sci-Hub |(2017) The transition of energy intensive processing industries towards deep decarbonization: characteristics and implications for future research. Renew Sustain Energy Rev 79:1303–1313. https://doi.org/10.1016/j.rser.2017.05.156
Sen A, Teter J, Miller J (2025) Vision 2050: Update on the global zero-emission vehicle transition in 2024. Int Council Clean Transport
Seto KC, Davis SJ, Mitchell RB, Stokes EC, Unruh G, Ürge-Vorsatz D (2016) Carbon lock-in: types, causes, and policy implications. 41:425–452. https://doi.org/10.1146/ANNUREV-ENVIRON-110615-085934

Smith P, Davis SJ, Creutzig F, Fuss S, Minx J, Gabrielle B, Kato E, Jackson RB, Cowie A, Kriegler E, Van Vuuren DP, Rogelj J, Ciais P, Milne J, Canadell JG, McCollum D, Peters G, Andrew R, Krey V, … Yongsung C (2016) Biophysical and economic limits to negative CO_2 emissions. Nat Clim Change 6(1):42–50. https://doi.org/10.1038/NCLIMATE2870

Sood A, Tellis GJ (2005) Technological evolution and radical innovation. J Mark 69(3):152–168

Springer Science+business Media, #, and Springer L (2020) Rebirth of industrial policy and an agenda for the twenty-first century. J Industr Compet Trade 20:189–207. https://doi.org/10.1007/s10842-019-00322-3

Sun L, Miao C, Yang L (2017) Ecological-economic efficiency evaluation of green technology innovation in strategic emerging industries based on entropy weighted TOPSIS method. Ecol Ind 73:554–558

Upham P, Kivimaa P, Mickwitz Å (2014) Climate policy innovation: a sociotechnical transitions perspective. Environ Polit 23. https://doi.org/10.1080/09644016.2014.923632

Waisman H, Bataille C, Winkler H, Jotzo F, Shukla P, Colombier M, Buira D, Criqui P, Fischedick M, Kainuma M, La Rovere E, Pye S, Safonov G, Siagian U, Teng F, Virdis MR, Williams J, Young S, Anandarajah G, … Trollip H (2019) A pathway design framework for national low greenhouse gas emission development strategies. Nat Clim Change 9(4):261–268. https://doi.org/10.1038/S41558-019-0442-8

Warwick K (2013) Beyond Industrial Policy. 2. https://doi.org/10.1787/5k4869clw0xp-en

Weaver S (2024) Overview of national climate change advisory councils

Weber KM, Rohracher H (2012) Legitimizing research, technology and innovation policies for transformative change. Res Policy 41(6):1037–1047

Wesseling JH, Lechtenböhmer S, Åhman M, Nilsson LJ, Worrell E, Coenen L (2017) The transition of energy intensive processing industries towards deep decarbonization: characteristics and implications for future research. Renew Sustain Energy Rev 79(January):1303–1313. https://doi.org/10.1016/j.rser.2017.05.156

Wood J (2022) Which countries have already reached net zero emissions? | World Economic Forum

World Energy Employment–Analysis-IEA (2024) https://www.iea.org/reports/world-energy-employment. Retrieved June 22, 2023

Yang M, Chen L, Wang J, Msigwa G, Osman AI, Fawzy S, Rooney DW, Yap PS (2023) Circular economy strategies for combating climate change and other environmental issues. Environ Chem Lett 21(1):55–80. https://doi.org/10.1007/s10311-022-01499-6

Chapter 8
Role of Participatory Approach and Local Governance in Achieving Climate Justice for Cities

Payal Thaorey and Abid Mustafa Khan

Abstract Climate justice is a multidimensional approach that recognizes the disproportionate climate effects in a broader sense on marginalized communities and strives to protect them. As an epicentre of both emissions and vulnerability, urban cities must be at the forefront of participatory governance strategies that advance equitable and sustainable development results. This paper focuses on the context and background for investigating the relationship between participatory approaches and local governance within the framework of climate justice in cities. Addressing national and international frameworks, legal mechanisms and also the Sustainable Development Goals (SDGs), the manuscript argues for the rights-based, bottom-up paradigm that necessitates the active involvement of citizens on the streets, district, city, region, and country levels in urban climate planning. Local governments get opportunities as well as challenges to align climate action with social justice principles within their cities. Empirical and critical analysis are assembled reflecting on how systematic integration of participatory urban planning into local governmental policies can yield more resilient, adaptive or just outcomes. This paper uses examples from India, such as Pune's SWaCH cooperative, and references to global models from developed nations such as Denmark and the USA to provide a comparative perspective on best practices. This research will make an attempt to propose a "Public Participatory Process (PPP)" model to institutionalize civic engagement and uphold the democratic legitimacy of climate governance. In conclusion, given recent developments in the fields of adaptation and climate governance, the paper aims both to emphasize the immediate requirement for structured and inclusive context-sensitive climate change frameworks to enable citizens and local governing bodies to collaborate and contribute more effectively to climate justice and change.

Keywords Climate justice · Participatory governance · Local government · Urban sustainability · Public participation · Sustainable development goals

P. Thaorey (✉) · A. M. Khan
Post Graduate Department, Dr. Babasaheb Ambedkar School of Law, Rashtrasant Tukadoji Maharaj Nagpur University, Nagpur, India
e-mail: payal.mundafale@gmail.com

K. Mukherjee et al. (eds.), *Urban Climate*, The Urban Book Series,
https://doi.org/10.1007/978-3-032-17895-4_8

8.1 Introduction

Climate change presents vast societal challenges, particularly impacting urban areas due to the compounding effects of environmental issues and socioeconomic disparities. However, cities have the potential to lead climate change response through participatory local governance that endorses climate justice. This chapter defends Local communities' participation in decision-making, supported by local governments, can provide equitable responses to climate impacts. A strictly top-down approach may neglect local diversity and worsen existing inequalities. A more inclusive strategy engages a variety of urban stakeholders, fostering climate policies that reflect community needs, strengths and inclusive democratic values. The collaboration between local government and community groups in creating the Sustainable City Action Plan is an example of these synergies at work (Sahasranaman and Kapur 2014). With workshops, dialogues and surveys, community inputs shaped sustainability initiatives, demonstrating how participatory governance drives energy, emission and waste reduction.

This paper promotes the pairing of participatory processes and empowered local government to achieve climate justice in cities. By enabling diverse communities to engage in decision-making and backing their initiatives, local authorities can tailor impactful and fair climate solutions. These synergies can lead to transformative change, a critical need as climate impacts intensify.

8.1.1 Climate Change and Cities

Cities and climate change are two deeply intertwined issues affecting our world. As urbanization increases globally, more people inhabit cities which are major contributors to greenhouse gas emissions driving planetary warming. Meanwhile, cities are especially vulnerable to climate change, which includes the rise in sea levels, extreme heat and natural disasters. This interdependence demands cross-disciplinary research and cooperation to build the sustainable and more equitable urban futures for the realization of climate justice.

Cities are a primary engine of the climate change because of human activities concentrated within them. The majority of emissions stem from energy and transportation systems powering urban infrastructure and movement. As rural populations continue migrating to cities, total emissions will keep rising without transitions to low-carbon systems (While and Whitehead 2013). Coastal megacities are especially at risk, with many facing potential inundation or uninhabitability if sea levels continue rising according to current projections. Beyond sea level impacts, cities worldwide will struggle with more frequent and intense heat waves, flooding from intense rainfall and storm events, as well as connected issues like energy and food insecurity without adapting infrastructure, policies and social services. Moreover, climate pressures may force large-scale migration and place new demands on cities to accept climate refugees fleeing impacts in other regions.

However, cities also offer opportunities due to their density, resources and capacity for innovation. When constructed and managed sustainably, cities can reduce individual environmental footprints through shared transportation, efficient energy and water systems, and concentration of services.

Cities serve as the driving force behind capitalist growth. As individuals strive for an improved quality of life, they often relocate from rural to urban regions. In urban areas, people benefit from higher incomes, granting them the means to acquire a wider range of consumer goods, including private transportation and larger residences. The move towards urbanization increases the need for electricity in both residential and commercial sectors. While urban dwellers generate more emissions per capita compared to those in rural areas, their household sizes are usually smaller. Moreover, urbanization fosters the exploration and dissemination of new ideas and innovations (Kahn 2009).

Climate change and urbanization create a feedback loop, as each impacts and transforms the other in complex ways. Climate shifts like prolonged drought or more intense storms reshape development possibilities and infrastructure needs in cities. Adaptation challenges like flood defence or energy resilience become long-term priorities integrated into all planning and budgeting. Meanwhile, cities shape regional and global climate through their collective emissions profiles and energy/land-use decisions amplified at a vast scale. As climate concerns take hold, scholars argue we have entered a new era of "climate-conditioned urbanization" requiring cross-sectoral integration between climate science, urban planning, policymaking and community organizing.

Cities' identity is often related to its population size. Natural increase as well as migration both contributes in population size. From the year 2020 to 2023, an average of 5% population growth is visible in Indian metro cities like Mumbai. Since migration costs are zero in India and this freedom is granted to every citizen under Article 19 of the Constitution of India, people do migrate from villages and towns to cities in search of livelihood, education opportunities and healthcare. Both the natural increase and migration add to the existing population of the cities and thereby demand new infrastructure and sharing of resources available for city population like water and land with the newly added population. Further, there is no denial of the fact that such an increase in city population does give rise to new forms of challenges/problems (like more infrastructure and less green space, pollution) which may contribute to climate change as well.

When cities face such inevitable challenges, the best way to deal with these challenges is to have well-organized urban planning and design. Historical evidence shows that town planning in earlier eras considered climate and its challenges and accordingly designed the structures. For example, the ventilation system in Mughal forts in India. Similarly, ancient Indian practices like organic farming, traditional irrigation systems, community-based water management systems, etc., provide significant methods of sustainable city planning (Biswas 2024). However, the increase in city population generates a demand for more adaptive and sustainable urban design that can accommodate this growing population. According to Ahluwalia et al. (2014), between 1951 and 2011, India's urban population grew by 3.5 times, from 62 million to over 377 million people. This rapid urbanization is set to continue, as the proportion of the urban population in India is expected to increase to 40% by 2030. Moreover, the number of megacities in India has also risen dramatically, with cities such as Delhi, Mumbai and Kolkata contributing to more than 30% of the total urban population. This rapid urbanization highlights the need for sustainable development solutions that integrate climate resilience with infrastructure planning. It provides an understanding of the progression of urban design as cities must now adapt to the pressures of increasing population while addressing environmental and social challenges.

It becomes apparent that the climate has been a significant factor in urban development. This has been a consistent factor, shaping the urban landscape over the years. Therefore, it's safe to say that climate considerations have long been integral to urban planning. This is a trend that continues to persist in contemporary urban development initiatives. Overall, scholarly understanding of this relationship and available policy options continues deepening in response to the urgency and complexity of jointly addressing urban and climatic challenges. Solutions must recognize practical limitations while still pursuing emissions reductions and resilient development.

8.1.2 Cities and Customized Climate Change Planning

One-size-fits-all approach will not suffice given cities' diverse capacities and starting points. National climate policies can enable local action but must consider resource distribution and governance realities on the ground. Communities everywhere are testing innovative solutions through pilot projects and sharing lessons learned. However, transformative change ultimately requires cooperation between local, regional and international partners along with participation from private sector, researchers and civil society. The scale of these intersecting crises demands united, justice-oriented and evidence-based action from all levels of governance and sectors of society. Therefore, based upon the unique requirements of the city, the climate planning process is required to be conducted and accordingly the stakeholders' participation must be claimed. Earth's natural climatic variability shows that the climate continues to fluctuate regardless of human activity. The geographic and demographic features continuously contribute to changing the climate of cities and, therefore, rather than searching for unattainable fixes to end climate change permanently, policies and projects must establish resilient, adaptable societies and systems able to withstand inevitable climate shifts over generations to come.

The city's innovative Cloudburst Management Plan, which involves implementing climate-friendly streets, has given Copenhagen a reputation of being a leader in climate adaptation, especially after a brief but extreme flood crisis in 2011. This plan was made as a direct response to the severe flooding brought about by powerful rainstorms that uncovered the city's vulnerability to extreme weather. Through public participation and collaborative planning, Copenhagen's planning authorities acted proactively in incorporating the community. The Cloudburst Management Plan in Denmark, for instance, redesigned urban spaces to do double-duty, both mitigating floods and providing recreational areas. This included co-design workshops in which local citizens actively participated in co-creating projects such as Enghaveparken, a park meant to hold excess rainwater, and climate-sensitive cycling infrastructure meaning both some kind of flood protection and a means for a sustainable urban mobility (City of Copenhagen 2014; Sörensen and Torfing 2019).

Copenhagen's devotion to inclusivity in urban design has been critical to its long-term resilience. By soliciting citizen input for this decision, the city not only tackled flooding here and now, but also deepened civic trust in climate adaptation over time. As this example shows, bottom-up inclusive governance can make climate adaptation measures more effective and legitimate by involving those affected at the local level to provide regionally relevant and accepted measures. Copenhagen has established itself as a planetary leader in climate-speedy urban planning through this multi-actor strategy (City of Copenhagen 2014; Sörensen and Torfing 2019).

The perspective of urban climate change as an unavoidable but manageable fact over and beyond its framing as a problem results in more practical, tailored and sustainable solutions. The problems of both urbanization and planetary warming are inextricably tangled, indeed requiring their solution to be approached collaboratively, going across academic disciplines and other societal sectors, to include particularly the local governments. Uncontrolled foes like climate change cannot be vanquished through grandiose pledges realised by technocratically complex solutions, but through progressive efforts to limit dangers within a context of common understanding and resources. Grounded, however, in long-term, evidence-based thinking alongside justice and equity, policies can create resilient communities that will be more robust to fluctuations in global climate for generations to come.

8.2 Methodology

Climate justice is not only about environmental issues, and the role of the local government is the key to achieving them. It then realizes that the participatory mindset and the role of stakeholders are important. The city life has been growing over the years through urbanization, which has raised the question of its sustainability. The SDG 11 emphasizes the importance of not only providing sustainable city solutions through technocratic but also integrating with local governments and public participation. The vision is to acknowledge the role of urbanization in cities contributing to climate change and showcasing that city, depending on its governance, customizes the action plan towards achieving SDG 11.

The paper uses a qualitative, analytical approach framed through a thematic analysis of participatory governance and the role of the local government in achieving the urban climate justice. Bringing together insights from normative interdisciplinary literature (including legal and policy discourses and global examples), this study takes a critical perspective on the intersection of the participatory frameworks and the climate adaptation strategies, specifically in terms of the urban governance, climate resilience, human rights and participatory development. The article does not deliver a systematic review but a conceptual and illustrative analysis of participatory urbanism in India (for instance, the experience of Pune's KKPKP) and the developed world (for instance, Portland, Copenhagen, Freiburg). Based on an analysis of normative principles, existing legislation and relevant examples, the article offers an integrated view for local governance on institutionalising inclusive decision-making processes for environmental justice in cities.

8.3 Climate Justice

Climate change effects are deeply entangled across borders and generations, from rising seas to worsening natural resources. No country alone can solve the crisis yet collective action remains elusive due to short-term thinking and geopolitical games allowing major polluters to continue unabated exploitation with impunity. To properly address these intersecting injustices, new multilevel and participatory global governance arrangements are required that empower public oversight while protecting common assets for future populations. Climate justice (Robinson 2018) is an approach to climate action that focuses on the unequal impacts of climate change on marginalized or otherwise vulnerable populations. The objective of climate justice is to ensure a fair distribution of the impacts of climate change and the actions taken to reduce it. Climate justice is a type of environmental justice (Schlosberg and Collins 2014). As the world increasingly urbanizes, local governments serve an essential part in providing climate justice.

8.3.1 Urbanization and Cities

Globally, more people now live in the urban areas than the rural areas, with millions migrating to cities each year. According to the UN, urban populations will nearly double to over 6 billion by 2050. Asia and Africa will account for 90% of this rapid urban growth (UN 2010). In India, the extent of urban transition matches global trends. Years of accelerated rural–urban migration have transformed India into a predominantly urban society, bringing opportunities but also tremendous challenges. Urbanization is one of the most significant global changes taking place in the twenty-first century. By 2050, over 90% of the world's population is expected to live in cities (Fig. 8.1), with most growth in Asia and Africa where poverty is concentrated today (Ahluwalia et al. 2014). How this rapid increase in urbanization is managed over the coming decades will shape whether nations can reduce poverty through sustainable development. Yet, challenges from insufficient infrastructure, pollution and inequality are also substantial if cities develop unsustainably.

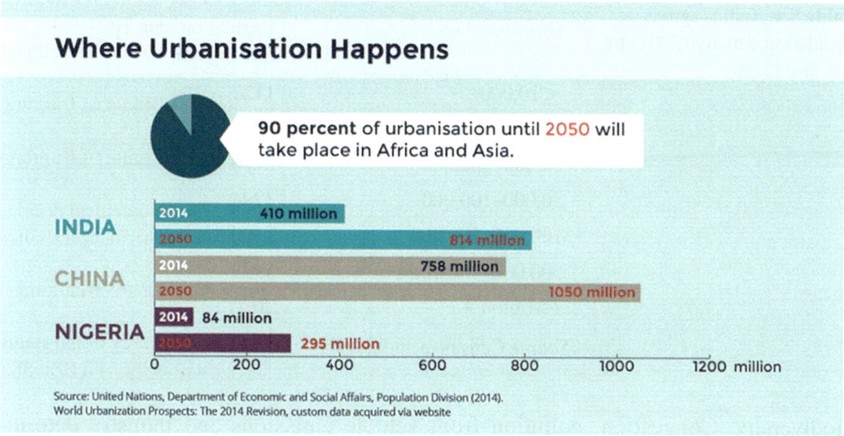

Fig. 8.1 World urbanization. *Source* UN development of economic and social affairs, population division (2014)

There are both push and pull factors driving sustained rural–urban population shifts worldwide. In developing countries like India, rural push factors stem from limits on arable land and scarce livelihood options for growing rural populations. Agricultural mechanization also displaces rural labourers. Meanwhile, cities represent centres of economic activity and sources of hope for better lives. Their promise of employment, amenities, education and healthcare facilities pulls migrants from the countryside.

India's booming economy further incentivizes urban migration. Fast-growing industries concentrate in cities, generating white-collar and blue-collar jobs. New infrastructure investments make proximity to cities more achievable, enabling suburban sprawl. Even changing social mores encourage independence through urban lifestyles over agrarian village life. Ultimately, limited capacity of rural systems to sufficiently support populations fuels push–pull dynamics propelling urbanization.

Post-independence, accelerated economic development and free movement policies rapidly transformed India's demographic makeup. Between 1951 and 2011, India's urban population grew a massive 3.5 times from 62 million to over 377 million. Its share of total population rose from 17 to 31%, one of the highest urban transition rates globally. Four metro cities—Delhi, Mumbai, Kolkata and Chennai alone absorb over 10% of urban residents. 35 megacities exceeding 1 million people account for 38% of urban population on just 2.5% of land (Table 8.1). But their phenomenal growth far outstrips development of economic bases and provision of infrastructure. Sprawling slums illustrate shortfalls in affordable housing supply, while overwhelmed transit systems and stretched water/power grids symbolize deeper issues.

Rising populations concentrate environmental pressures on land, energy, water, food systems through ballooning consumption patterns. Loss of green spaces and conversion of agricultural/forest lands to paved urban fabric impacts habitats and

Table 8.1 India—gross population density (2001) by city size class

City size class	Population density (per sq. km.)
< 10,000	1127
10,000–20,000	1462
20,000–50,000	2100
50,000–100,000	3294
100,000–500,000	5300
500,000–1 Million	5929
1 Million +	20,713

Source Census of India 2001

biodiversity. Congestion, pollution from vehicle emissions and industry deteriorate urban living standards and public health. Cultural assimilation challenges rural migrants adapting to city life too. Socio-economic inequalities also run disparity between urban poor living in slums lacking basic services versus affluent groups. Marginalized groups face exploitation and higher disease risks. According to the 2011 Census, about 17.4% of India's urban population live in the slums, which translates to over 65 million people. These areas are often characterized by the scarcity of basic infrastructure such as the reliable water supply, sanitation and the electricity. More than 70% of slum dwellers live in informal settlements with no secure tenure and the experience heightened vulnerability to environmental hazards such as flooding, pollution and the climate change. Natural disasters increasingly imperil burgeoning coastal and seismic zone city populations unprepared for impacts of climate change as well.

To maximize benefits of urbanization while mitigating pressures, the paradigm shift in the urban planning and governance is needed. Growing decentralization helps distribute burdens but accessible public transportation is paramount to link dispersed communities. Mixing land uses and increasing housing densities leverages existing infrastructure more efficiently. Transitioning to renewable energy sources and improving resource efficiency can reduce environmental footprints. Green building codes that mandate rooftop gardens, greywater recycling etc. create more sustainable urban forms. Public–private sector collaborations leverage collective skills/finance for basic service delivery including waste management. In the Union Budget for 2025–26, an amount of INR 96,777 crore (nearly 1000 billion) is allocated for urban development. There is a 17 per cent hike over the previous year's budgeted amount of INR 82,576 crore (or INR 825.7 billion) (Fig. 8.2).

Most crucially, long-term strategic planning through participatory processes ensures all communities' needs and aspirations shape future development visions sustainably and equitably. By balancing interconnected economic, social and environmental progress, cities represent opportunities for economic prosperity, social inclusion and enhanced resilience in the era ahead. With collaborative effort, policy innovation and political will, urban sustainability goals can be achieved.

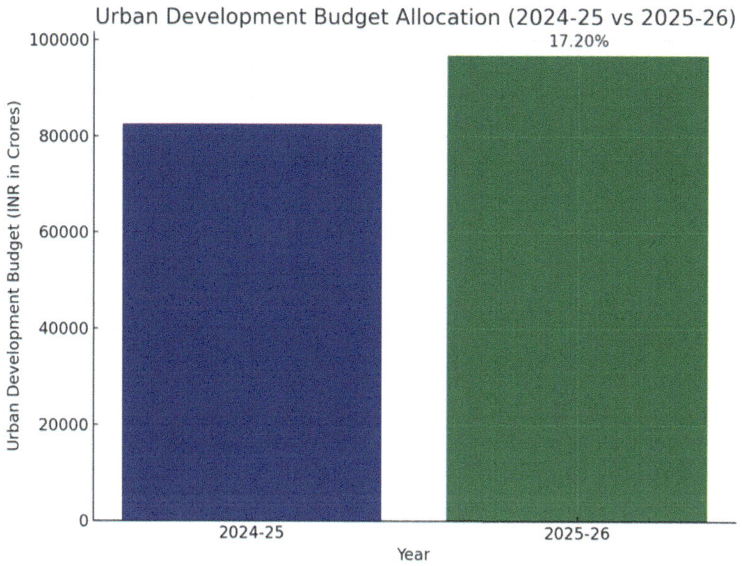

Fig. 8.2 India—urban development budget allocation. *Source* Union Budget of India 2025–26

8.3.2 *Sustainable Cities and Local Government*

In September 2015, world leaders came together at the United Nations and adopted the 2030 Agenda for Sustainable Development (SDSN 2013). The Sustainable Development Goals (SDGs) address the major challenges, i.e. economic, social and environmental in the next fifteen years. Critically, the SDGs recognize that sustainability cannot be achieved without tackling issues related to urbanization and cities. This is because cities have become the predominant form of human habitation universally that will serve a crucial part in determining whether the world achieves the sustainability vision.

The SDGs which build upon the successes of the Millennium Development Goals but pursue a broader and more integrated vision of sustainability relevant to all countries. In addition to reducing poverty, the SDGs address issues like inequality, climate change, production and use that is sustainable, as well as equity and tranquillity. When crafting the post-2015 agenda to succeed the Millennium Development Goals, world leaders must carefully consider urbanization. While the MDGs successfully highlighted dimensions of extreme poverty, the SDGs must similarly educate on critical urban challenges and opportunities between now and 2030 relating to slums, infrastructure, ecosystems, resilience and planning.

Each goal has specific, measurable targets to achieve the overall objectives. Goal 11 says "Make cities and human settlements inclusive, safe, resilient and sustainable" through targets like upgrading slums, improving transport access, reducing disasters and pollution. Collectively, the SDGs provide a comprehensive blueprint for economies, societies and the environment (Kanuri et al. 2016). Most importantly, a goal centred on cities represents perhaps the surest way to realize opportunities from urbanization and ensure prosperity, equity and environmental protection for both current and future generations worldwide.

(a) Localization of SDGs: Localizing the SDGs refers to adapting them to local contexts and mobilizing stakeholders for implementation. The local government is closest to citizens; these authorities will play a leading role in achieving SDG targets through service delivery around issues like housing, transport, basic services and sanitation. An estimated 65% of targets may remain unmet without local engagement.

By 2050, approximately 75% of the global population will be settled in cities, compared to the current 50% global urbanization rate. Cities occupy just 3% land but account for over 80% of global GDP. This will also affect significant resource use and emissions—cities currently generate over 60% of carbon emissions while occupying just 2% of the land on Earth. However, cities are also hubs of economic activity, business and innovation. If developed sustainably, they offer opportunities to boost living standards while reducing humanity's environmental impact.

Many targets under Goal 11 on sustainable cities depend on action at the urban scale. Cities face both development challenges like poverty and pollution as well as opportunities to provide efficient infrastructure, mobility networks and other services. Therefore, it is important for the local governments to implement these goals in such a way that it will be beneficial to the people as well as the city. Pursuing an integrated, city-focused approach to achieving the SDGs presents a viable roadmap.

Localization involves defining locally relevant goals and collecting disaggregated data to ensure no one is left behind. Empirical evidence has demonstrated that government interventions have the potential to enable cities to build resilience and safeguard themselves against the challenges posed by the climate change. According to the Maharashtra Government's SDG report for 2022-23 shows the analysis of initiatives undertaken by the Maharashtra Government through its local governments in order to achieve the SDG 11 (Fig. 8.3).

According to the above report, Maharashtra government through its municipal corporations of cities has received 68.2% in 2021 and 84% proceeds on its waste generated. The door-to-door waste collection is 90% for both years and this initiative is helping the cities in keeping clean and hygienic. For around 50% of Maharashtra cities, the smart city plan is ready, and around 70% of the population in cities have access to the public transport. All these initiatives help the governments to achieve SDGs11.

However, the government can't regulate the weather and nature, but certainly priority-based designing of policies, risk assessment and allocation of climate finance for making the cities sustainable will certainly justify the importance of local government in dealing with climate change. Cities vulnerable to the climate change can employ a combination of public policies and market incentives to mitigate risks and minimize the financial burden associated with adaptation measures.

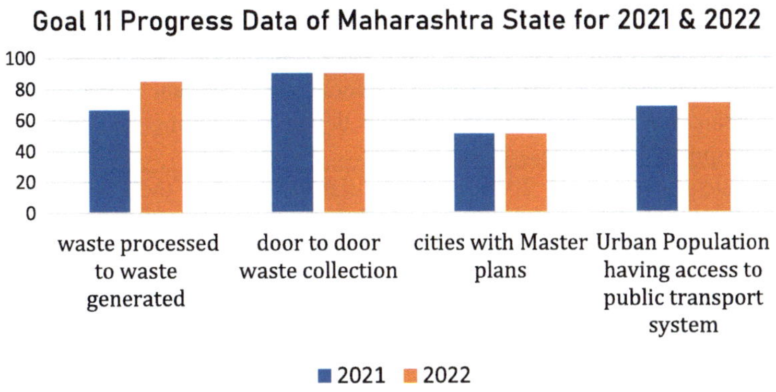

Fig. 8.3 India—Goal 11 progress of Maharashtra State, India. *Source* Maharashtra Government's SDG report 2022–23

(b) Top-down or Bottom Up: Both top-down national policy and bottom-up local leadership initiatives can enable localization. Cities are also supporting this process by networking through commitments to initiatives like the Global Covenant of Mayors for Climate and Energy, which provides a platform to access finance and partnerships and gain guidance on sustainable development. Through this covenant, cities will take accountability for decreasing emissions, increasing resilience to climate change and committing to sustainable urban growth. It also encourages collaboration among cities by sharing knowledge, resources and best practices, which can help align local actions with global climate goals, including those enshrined in the Paris Agreement. Governance at all layers of government must coordinate to ensure that equitable outcomes are achieved in all sectors.

Working through their leadership role on sustainability, cities can mobilize system-wide SDG progress through a number of channels. They can help raise public awareness on development challenges, catalyse multi-sector partnerships, address local poverty and infrastructure deficits, promote innovative solutions to local issues, and enhance urban resilience. It is expected that towards upcoming New Urban Agenda efforts on the ground would crystallize defining pathways for an integrated sustainable urbanization either through top-down or bottom-up approach depending on the form of governance a particular nation has mayors and local governments can also shape sustainability outcomes in powerful ways, in addition to direct management responsibilities. First, they can mobilize political will by building public awareness of the social, economic and environmental challenges that communities face. Second, they offer a space to catalyse concerted action between a varieties of urban stakeholders via collaborative solution-seeking in relation to sustainability objectives.

(c) Development Gaps: Cities also act as a major agent to address place-based development deficits. Alleviating poverty, expanding access to fundamental services, and building resilient infrastructure that meet local needs and utilize local assets require integrated, multi-sector solutions that cross traditional boundaries and are best delivered by local governments. Local leaders also advocate for new methods of service delivery, planning and infrastructure development to fill these gaps.

(d) Climate resilience: In 2016, global initiatives such as the Global Covenant of Mayors for Climate and Energy emerged as significant efforts. Cities have become key indicators of sustainability and pivotal participants in global initiatives addressing resilience and the challenges posed by climate change. It united a coalition of mayors and city leaders committed to climate action and sustainable development in cities. Urban authorities govern initiatives to help reduce emissions, increase adaptation efforts, mitigate disaster risk and aid recovery interventions. Good resilience strategies, if integrated with how a city is run, can also strengthen the ability of communities to go in pace with the environmental and economic shocks.

(e) Slums Within Cities: The current population residing in urban slums is continuously increasing across world, and this number is expected to increase significantly unless effective strategies are implemented to improve living conditions. Unplanned urban sprawl risks exacerbating environmental degradation and trapping communities in long-term cycles of poverty. For example, the report of the slum statistics of Maharashtra Government shows a progressive trend in the slum population in cities during the period 2011–2017 (Fig. 8.4).

However, well-guided compact cities have tremendous potential to power economic growth, generate employment, enhance livelihoods and reduce humanity's

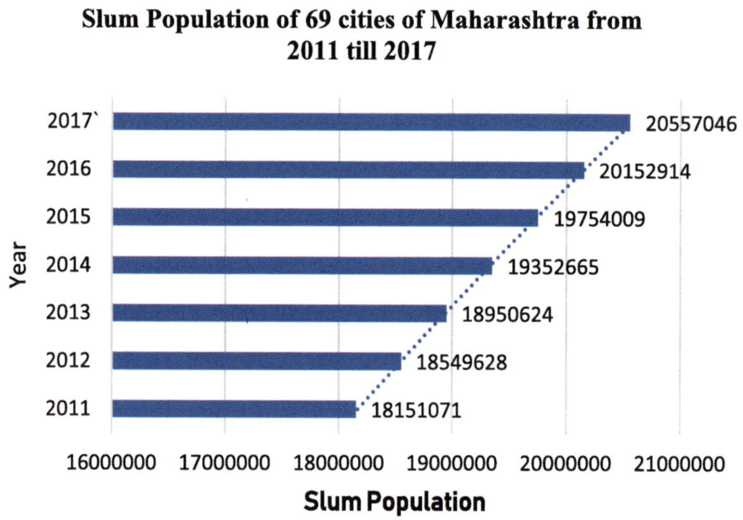

Fig. 8.4 Slum population of Maharashtra state, India. Source: Maharashtra Government's slum statistics 2011–17

ecological footprint. Addressing interconnected urban issues demands practical, locally tailored solutions bringing together diverse stakeholders from government, non-profit sector and commercial community. A distinct goal focused on cities could perform a significant part in this by providing a framework for collaboration, coordination and integrated approaches across sectors. It would also give visibility and support to local governments already advancing sustainability priorities through innovative policies and projects.

A standalone urban SDG offers several implementation advantages compared to merely mainstreaming urban challenges within broader goals. It would incentivize coherent policy development, planning, investment and private sector engagement at national and sub-national levels. Civil society would gain opportunities to participate successfully in the local decision-making and demand accountability.

Any urban framework must be carefully scoped and limited to ensure applicability across diverse local contexts globally, while also complementing universal goals on health, education and other issues. With continued leadership from local governments, U.N. agencies, and urban networks, an SDG tailored to realities on the ground could help accelerate progress on sustainability if mobilized through an explicit commitment to building inclusive, safe, resilient and the sustainable cities for all. By adapting targets to local contexts, cities can pursue priorities through collaborative partnerships across territories. If supported by clear frameworks like the upcoming New Urban Agenda, locally led efforts offer the best approach to realizing development benefits from urbanization while addressing inequity and environmental impacts.

8.3.3 Local Government Climate Justice Process

Local governments are closest to their residents and are therefore best positioned to understand local needs and challenges. This proximity allows them to tailor climate action to specific vulnerabilities and circumstances within their communities. Localization of human rights is possible only with strategic involvement of stakeholders in decision-making. This creates an obligation on the local governments to make better policies and more effective implementation of climate justice initiatives. Also, their political will and commitment can accelerate and sustain these efforts. The following are the steps to be followed by the local governments to achieve with climate justice (Fig. 8.5).

(1) Step 1—Urban Planning: Through integrated urban planning (Crispi et al. 2022), local governments can contribute in climate justice by promoting climate-resilient development with low-carbon outcomes. Climate justice begins with planning sustainable frameworks to establish the foundational infrastructure of a city by determining land uses, transportation networks, building densities and access to services. When designed with sustainability and climate change mitigation in mind, planning processes can support more compact, walkable

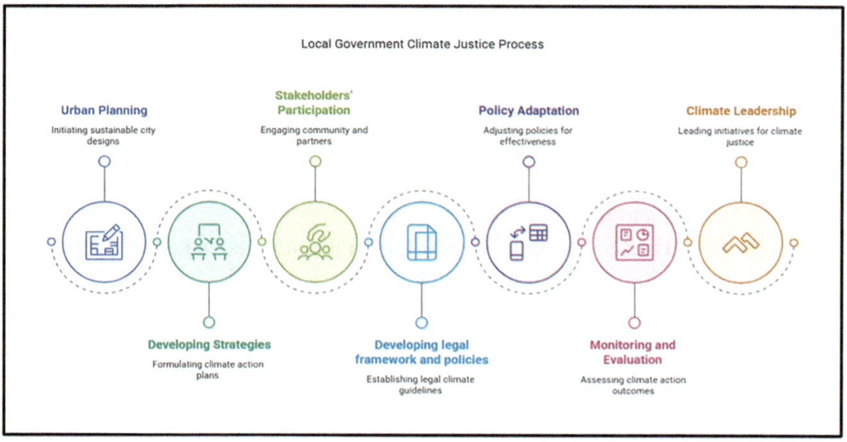

Fig. 8.5 Steps of local government climate justice process. *Source* UN-Habitat

urban forms that generate fewer emissions. By prioritizing public transport links, bike lanes, and mixed-use zoning that collocates residences, workplaces and amenities, urban planning frameworks can help shift trip-making away from high-polluting private vehicles towards low-carbon mobility options. Planning also influences the built environment via development standards, which provide rules for energy usage of buildings. Planning imparts the critical role in the climate change adaptation and mitigation by steering development away from vulnerable coastal and flood-prone areas via regulations like setbacks. Incorporating projected climate hazards such as the sea level rise and increased flooding into siting of infrastructure and land use designations protects against future climate impacts.

(2) Step 2—Developing Strategies: Designing appropriate climate strategies demands a thorough understanding of a city's existing vulnerabilities and environmental challenges. Comprehensive greenhouse gas inventories and climate risk assessments provide the foundation for evidence-based policymaking. Quantified emissions tracking sources from transportation, buildings, waste and industry enables cities to identify where to focus mitigation efforts. Meanwhile, detailed climate impact mapping that considers site-specific exposures and the differentiated vulnerabilities of populations informs well-targeted adaptation planning. Combining emissions and risk data presents a holistic picture of a city's relationship with the climate crisis and forms an important baseline against which to measure progress. Gaining public input early and often enhances the legitimacy, feasibility and justice of climate policies. Interactive community workshops, focus groups and online surveys bring diverse local perspectives into the planning process. This helps ensure strategies address priority concerns and incorporate traditional knowledge. Associated emissions and resilience outcomes are analysed to identify preferred pathways. The local

governments can have strategy for designing building through a building code strengthened through requirements for on-site renewable energy systems, electric vehicle charging and energy efficiency retrofits. Meanwhile, nature-based resilience solutions like green infrastructure, urban forestry and coastal buffers can be incorporated into landscape ordinances and coastal zone management plans.

(3) Step 3—Stakeholders' Participation: As it is a process, public participation does not refer to a single event. It requires public involvement consisting of a series of activities and actions. Public participation is a collaborative process that benefits both the agency and the public, allowing for mutual learning and advantages. By enabling effective public participation, the values of the public can be identified and incorporated into decisions that have a direct impact on them. The consultation process as part of public participation is expected to provide sufficient time for effective and meaningful exchange of information between parties (Voss 2014). Political exclusions stemming from poverty can limit participation. Effective public participation must be based on reciprocal regard for local sovereignty, loyalty, trust and structured discussion among stakeholders. A detailed discussion on the participatory approach will be done in the next section.

(4) Step 4—Developing legal framework and policies: Strong legal frameworks are also needed to underpin climate action. Multiple policy scenarios can then be modelled that vary land uses, densities, infrastructure investments and green space provision under different policy packages. Due to uncertainties, climate policies must incorporate flexibility through adaptive management approaches. The consequences of climate change enhance underlying societal disparities, threatening the most marginalized urban residents. It is therefore imperative that climate planning and policy address issues of environmental justice.

(5) Step 5—Policy Adaptation: Adaptation (Brooks 2013) processes must ensure representation and inclusion of vulnerable groups early on through community-led organizations. Impact assessments evaluating distribution of benefits and burdens across demographic lines will help in making the policies adaptable. Superficial, unreasonable and stagnant policies will not contribute to achieving climate justice. Similarly, policies which are difficult to implement and are coercive are certainly not adaptable. Achieving climate justice demands both evaluative criteria and deeper understanding of contextual factors influencing vulnerability and adaptive capacities. With commitment to inclusive, evidence-based planning, cities can help deliver climate solutions that leave no one behind.

(6) Step 6—Monitoring and Evaluation: This step helps in the achievement of climate justice by monitoring and evaluating of tasks, policies implemented by the government for achieving climate justice. Climate policy's outcomes and advancements have to be monitored in adherence with climate justice principles. Monitoring and evaluation which involve the collection, monitoring and analysis of data during the implementation of a strategy to assess the progress and impact. While determining the indicators it is necessary for monitoring and measuring progress towards the desired goals; it is important to strike the

balance between the convenience and the comprehensiveness. Hence, it should be discouraged to have complex mechanism of M&E which creates a lot of hassle (Raoul Wallenberg Institute 2022)

(7) Step 7—Climate Leadership: Cities have both a moral imperative and strategic opportunity to demonstrate ambitious climate leadership through coordinated local action. A good leadership can enforce leveraging planning frameworks, regulatory tools, and processes of evaluation and engagement within the communities so as to have flourishing, equitable and climate-resilient futures. With cities playing such a critical role, responsive leadership with a vision to build a sustainable, low-carbon urbanization city may represent our best chance at averting a climate crisis.

8.4 Local Government and Participatory Approach for Sustainable Cities—A Synergy

Public participation (Berry et al. 2019) offers an avenue to help address tensions arising from these competing priorities by empowering communities to have a voice in decision-making processes that impact them. When the public is engaged in dialogue and given opportunities to potentially shape pollution-related decisions, society gains a mechanism to articulate concerns over environmental degradation affecting their quality of life.

As specified by the UNHRC (2018) local governments are meant to respect, protect and fulfil human rights standards in accordance with their legal framework and the limits of their individual states. Also, connecting human rights grounded strategy with local governance provide an evident mechanism to accelerate movement towards inclusive and sustainable development for everyone. The human rights-based approach (HRBA) is "a conceptual framework for the process of human development that is normatively based on international human rights standards and operationally directed to promoting and protecting human rights." The UN formed a Unified Consensus on HRBA for Development Partnerships which highlighted "Participation and inclusion" as a principle of human rights. However, quoting UN as "Every human being has the right to active, free and meaningful participation in, contribution to and enjoyment of civil, economic, social, cultural and political development in which human rights and fundamental freedoms to be realized".

Explanation of Participation: Participation is the cornerstone of the Human Rights Based Approach (HRBA) (Fig. 8.6). It's entitling all to participate in the political and public life. Endeavours this will empowers the people or groups and helps eradicate the inequality and marginalization. The right to participate is also linked with the other fundamental human rights, such as the right to freedom of peaceful assembly and association, the right to know, the freedom of opinion. When local governments want to address local climate change issues, challenges and priorities, they must ensure full participation of the stakeholders.

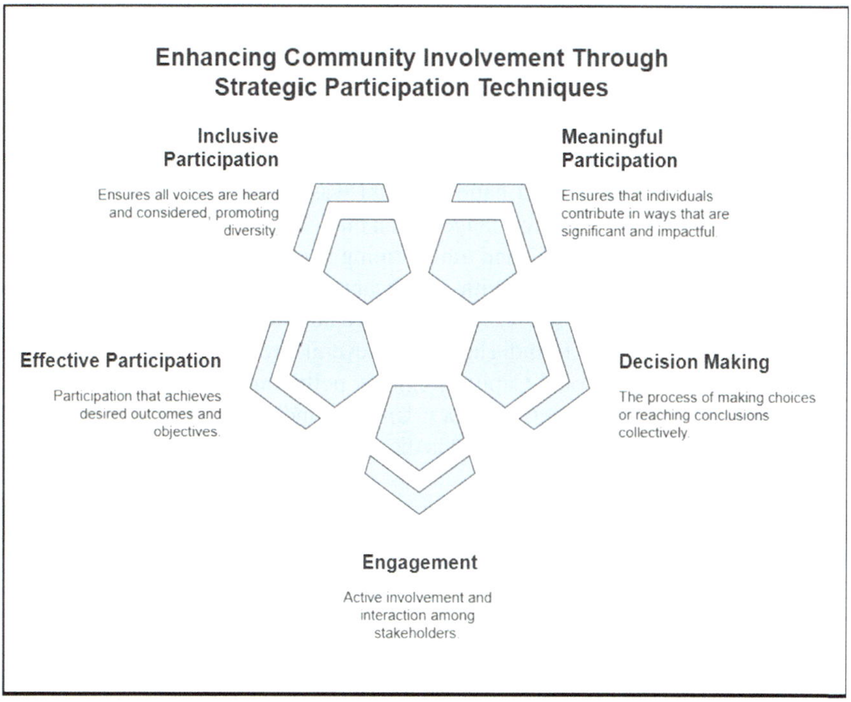

Fig. 8.6 Strategic participation techniques. *Source* UN-Habitat

(a) Meaningful Participation: Expressive and meaningful participation strengthens social cohesion by promoting inclusion, empowerment and a sense that stakeholders have ownership over projects due to their involvement. When individuals and communities express their concerns to authorities then it will certainly benefit the local government in policy making and building trust between groups. Organizations also benefit from showcasing their understanding and adherence to social conditions and applicable laws and regulations. Consultation among stakeholders contributes to ongoing learning and improvement rather than simple exploitation of local circumstances. There are varying levels of participation that differentially impact social quality and inclusion.

(b) Decision-Making: Providing true decision-making involvement represents the participatory ideal, while consultation seeks input without necessarily incorporating perspectives shared. Information sharing about plans without soliciting feedback represents a more limited form of engagement. Each mode confers different degrees of social empowerment and inclusiveness within communities. For issues as interconnected as pollution and climate change, participatory governance helps reconcile diverse stakeholder priorities and enables collaborative problem-solving to address unintended consequences. Building shared understanding can alleviate misconceptions that fuel opposition to policies or projects.

The city of Delhi in India serves as an excellent example for applying climate justice criteria (Hughes 2013). The city's relevance stems from the fact that it is part of a larger pattern of action towards climate change in developing and newly industrialized nations. Delhi was among the first major global cities to launch a climate change action plan. The city continues to revise its methods for confronting and transforming to climate change. Despite being a rapidly expanding city with significant service shortcomings, it reflects government's willpower in climate justice decision-making.

The Supreme Court of India has issued several directives regarding measures to reduce the National Capital Region's pollution. One significant recommendation from the M.C. Mehta v. Union of India (vehicular pollution) case (Rosencranz and Jackson 2003) is the conversion of all diesel buses in Delhi by December 2002.

(c) Engagement: Effective participation requires proactive, transparent and sincere engagement embedded systematically from project planning through community relationship management. Participation encompasses a spectrum of engagement models from information sharing to delegation of decision powers. Experience shows that restricting participation to superficial checkpoints after key decisions risks conflict by neglecting cooperation. Earlier involvement creates opportunities to better align priorities through leveraging local expertise together with technical knowledge. Continuous stakeholder engagement builds trust and organizational legitimacy. Productive engagement demonstrates consistency, accountability and accessibility of participatory processes.

(d) Effective Participation: Issues like climate change compound governance challenges due to widespread social, economic, and ecological impacts unfolding across spatial and temporal scales. Participation itself includes multiple dimensions that must also be navigated to ensure ethical and effective adaptation decision-making. Consideration of shifting stakeholder salience over time and place matters for long-term social quality impacts. Following the logic of identifying paradoxes in wicked problems, several contradictions emerge specific to participation design for climate adaptation governance. Copenhagen's adaptation to extreme climate events—particularly the 2011 flood crisis—led to its globally recognized Cloudburst Management Plan. The city's planning authorities embraced public participation in redesigning urban spaces, turning parks and streets into dual-purpose flood mitigation zones and recreational areas. Through co-design workshops, local citizens helped shape projects like Enghaveparken and climate-sensitive cycling infrastructure (City of Copenhagen 2014; Sörensen and Torfing 2019). Copenhagen's approach exemplifies how inclusive governance can drive long-term resilience and deepen civic trust in climate adaptation efforts.

(e) Inclusive Participation: Participation must be inclusive of all segments within the society. It must not be a "tokenistic participation" rather efforts must be taken by the local government to include each and every group within its society so as to no one must be left alone. In India, disable-friendly environmental justice is a least prioritised and therefore lacks inclusivity. However, determining who participates and to what extent poses dilemmas. While aiming for more decentralized solutions, power imbalances still shape participatory processes and outcomes if not carefully considered from the start. Lessons from development studies demonstrate how participation can inadvertently reinforce marginalization of certain groups without proper safeguarding of inclusiveness and representation.

8.5 Comparative Case Studies

For detailed understanding, three examples from different parts of the globe will be discussed here, in order to understand the success factors and challenges faced by local government in implementing public participation for achievement of SDG 11 (Table 8.2).

(A) Pune, India: Participatory Approach and Empowerment

The Kagad Kach Patra Kashtakari Panchayat (KKPKP) initiative in Pune City, India is one of the prominent examples which has paved the way in emphasising synergy between "Participatory Approach" and local government in order to have sustainable development. The Pune Municipal Corporation (PMC) signed an agreement with the KKPKP to form a Solid Waste Collection and Handling (SWaCH) Cooperative Society which focused on the sustainable solid waste management while preserving right to work with dignity of the waste workers. Through this collaboration the PMC saved INR 16M per annum on the recyclable scrap and the informal scrap workers organisation earn around INR 3,50,000 PER DAY. Further, The SWaCH Cooperative expanded its efforts to include services like composting, responsible disposal of e-waste and organized activities to clean up the city's water bodies in collaboration with the citizens of Pune city. (Chikarmane 2012).

(B) Copenhagen, Denmark: Urban Water Adaptation

Copenhagen's adaptation to extreme climate events—particularly the 2011 flood crisis—led to its globally recognized Cloudburst Management Plan. The city's

Table 8.2 Comparative case studies

	City	Approach	Success factors	Challenges
(A)	Pune (India)	Waste management via cooperative (KKPKP)	Local employment, policy inclusion	Limited scaling
(B)	Copenhagen	Urban water adaptation	High community trust, infrastructure finance	High cost
(C)	Portland	Participatory budgeting	Inclusive process	Political resistance

planning authorities embraced public participation in redesigning urban spaces, turning parks and streets into dual-purpose flood mitigation zones and recreational areas. Through co-design workshops, local citizens helped shape projects like Enghaveparken and climate-sensitive cycling infrastructure (City of Copenhagen 2014; Sörensen and Torfing 2019). Copenhagen's approach exemplifies how inclusive governance can drive long-term resilience and deepen civic trust in climate adaptation efforts.

(C) Portland, USA: Participatory Budgeting and Green Infrastructure Planning

Portland, Oregon stands out for its civic innovation in integrating participatory budgeting and community-driven environmental planning. The city's "Green Streets" programme combines stormwater management with public space enhancement and was developed through interactive community engagement sessions. These included public mapping, neighbourhood design charrettes, and iterative consultations with underrepresented groups. Furthermore, Portland established a Climate Equity Steering Committee, composed of community representatives from frontline populations, to ensure that climate policy implementation is not only environmentally sound but socially equitable (City of Portland 2016, 2019).

(D) Freiburg, Germany: Energy Cooperatives and Eco-Districts

In Germany, Freiburg has emerged as a model of bottom-up energy transition through its community-led energy cooperatives. These initiatives empower citizens to co-own and manage renewable energy infrastructure, such as solar panel installations and local microgrids (Rohracher and Späth 2014). Freiburg's Vauban eco-district, which limits car use, prioritizes pedestrian and bike-friendly spaces and includes energy-efficient housing, was co-developed with extensive public involvement (Scheurer and Newman 2009). Residents participated through participatory design forums and community boards, creating a neighbourhood that embodies sustainable, inclusive urban living.

Comparative Lessons for Urban Governance: These global examples offer critical lessons for Indian cities and others in the developing world seeking to embed participation in climate governance. Key takeaways include (Table 8.3):

- Equity Integration: Embedding social equity within climate action frameworks through advisory councils or steering committees.
- Decentralized Power: Granting decision-making authority to citizen groups or cooperatives to foster ownership.
- Systems Thinking: Adopting integrated urban approaches, linking mobility, housing, water and energy with **participatory processes.**

These cities exemplify that how participatory approaches can transcend political boundaries and economic stages of development. The synergy between the empowered local governments and active citizenry can also yield climate solutions that are both technically effective and socially transformative.

Table 8.3 Global case comparisons

City	Key practice	Participation mechanism	Climate justice outcome
Portland, USA	Green infrastructure and budgeting	Community workshops, equity steering committee	Equitable resource allocation, flood resilience
Copenhagen, DK	Climate-resilient urban design	Co-design workshops with residents	Inclusive adaptation planning, safer public spaces
Freiburg, GER	Renewable energy cooperatives, eco-district planning	Citizen-owned energy systems, participatory design boards	Energy democracy, low-carbon and inclusive urban lifestyle

8.6 Challenges and Paradoxes in Participatory Approach

Examining participation as a wicked problem in itself, some dimensions challenging determination of appropriate levels and methods (Sprain 2016) include:

1. Problem definition, managing dissent, temporal perspectives and scaling across jurisdictional levels.
2. Expert framing of issues may differ from public values yet allowing fully unconstrained definition risks deprioritizing adaptation needs. Achieving legitimacy requires respectfully accounting for disagreement even as consensus is often pursued to enable coordination.
3. Sustained participation proves difficult long-term without risk of politicization, yet one-off engagements fail to support dynamic governance needs.
4. Multi-level coordination balances local bottom-up knowledge with holistic multi-scale assessments, though scale mismatches can mute certain voices.

Considering the huge population, climate dynamics, top-down governance, concentration of powers, limited budgets and conservative willingness of people often create paradox which affect participation. Following are the reasons which affect effective participation.

I. Democratic Legitimacy Paradox: Invitation of all perspectives risks gridlock, yet exclusion compromises inclusiveness.
II. Representation Paradox: Direct engagement of all stakeholders proves impractical due to scale, yet representative structures may marginalize some.
III. Technical-Political Paradox: Adaptation portrays as technical, yet participation inherently political in determining views shaping responses.
IV. Governance Paradox: Inviting non-state actors expands participation but lacks democratic norms risking powerful interest co-optation without safeguards.

To effectively address these paradoxes, it is imperative to innovate approaches to participation. It remains insufficient to merely dismiss concerns regarding power dynamics, representation, or marginalisation through superficial technical solutions.

Concepts such as establishing workable agreements that balance dissent with coordination, and recognising participation as an inherent democratic right of citizenship rather than a privilege contingent upon specific viewpoints, contribute to reinforcing procedural justice and equity. At the same time, governance frameworks must embed democratic theory rather than exist as mere post-political vessels. Potential lies in flexible, learning-oriented processes which respect social-structural dimensions of participation including power and conflicts for democratically good for inclusive and equitable climate adaptation in the long run. Scrutiny of dynamics such as marginalization and dissensus remains a prerequisite for participatory climate governance to provoke vigorous but legitimate public deliberation that generates just, permanent solutions.

Environmental issues such as pollution and the climate change exacerbate pre-existing sustainability challenges through interconnected societal, economic and ecological consequences that span scales. Public participation provides mechanisms through which to address the tension between the demands of growth, protection and inclusion through a collaborative, learning-based model of governance. But even the wisdom of how much participation, and what sort, is optimal is not without contradictions that need common deliberately iterative re-development of, and protection of, procedural justice from being yoked to accidental co-optation or marginalisation of views. Framing participatory processes as an activation of rights to citizenship has potential implications for climate adaptation governance by incorporating/involving knowledge and perspectives shaping robust, publicly endorsed responses to time-bound sustainability challenges in an inclusive, dynamic and equitable manner.

8.7 Solutions to Enhance Participatory Approach for Sustainable Climate Justice

Based on the above discussion, it can be analysed that cities must design a "Public Participatory Process—PPP" which can effectively address the climate change issues. This process can be designed considering the following drivers-

a. **Enabling Adaptability and Learning**—Based on the liking/expertise of the citizens who wish to voluntarily work with the municipal corporations, the Local Government must consult with these people. This consultation will help the LGs to explore the sustainable ways for dealing with the climate issues. Once it is done, further sub-committees can be formed (within the volunteers) for identifying the targeted geographical area and stakeholders. The democratic decision-making structure and bottom-up delivery model will empower citizens to customize their approach to climate justice in a flexible and adaptable way. This will enable them to experiment and learn as they progress.
b. **Designing for Inclusion**—From the very beginning, the participatory processes should be carefully designed to prioritize inclusion and justice. The marginalised stakeholders from all fields (irrespective of making any discrimination like

gender, class, caste, race or religion) within the ambit of municipal corporation must be brought together for achieving the sustainable climate justice. Factors like geographic/demographic representation can be considered for enhancing public engagement. Further, depending upon the priority requirement the inclusion factor can be customised by the LGs. The local bodies can make efforts to engage and assist the families of marginalized stakeholders in accessing education and vocational training, enabling them to secure a stable income. While the initiative aims to uphold the principles of inclusion and justice, it is important to be mindful of the internal governance model where representative leaders speak for the marginalized stakeholders. Without conscious facilitation, this model may inadvertently lead to majoritarianism, favouring more vocal spokespersons and potentially undermining broader inclusion goals within the community.

c. **Embracing Complexity**—The primary objective of every local government must be to make its city sustainable by managing the green demands and supplies of its residents. The local government must manage to build all interlinkages and interdependencies between the resources and facilities available within the city for achieving the sustainability. An organized plan that incorporates citizen-led participation and strategic outreach efforts will definitely help the cities in dealing with the challenges which it is facing due to climate change.

8.8 Climate Justice and Local Government—Future Prospects

Intermediary bodies could help monitor resource use and emissions across borders. Joint management councils could oversee critical shared resources like water and fisheries. An international climate court could help hold major polluters accountable through binding enforcement while financing mechanisms support communities facing loss and damage. Global democratic representation also needs strengthening to remedy imbalances that currently underrepresent climate-vulnerable regions within influential economic institutions. A second UN parliamentary chamber open to universal membership could help ensure all have a say on pressing global concerns.

Climate justice movements rightly identify climate change as a human rights crisis requiring participatory governance where citizens meaningfully influence decisions directly impacting their lives and communities (Skillington 2017). New deliberative forums are needed to channel public input into international decision-making and policy formation, building mutual understanding between stakeholders. Inclusiveness is crucial in debates defining subjects of rights and justice across generations—future populations deserve a voice in decisions determining their planetary inheritance. Intergenerational representation through entities like ombudsperson offices could help place sustainability at the forefront of policymaking.

With political will and public pressure, the international community has windows of opportunity to build on principles that are already enshrined in foundational UN documents that promote equality, participation, and shared responsibility. Through new statutes, charters and procedures the norm of democratic experimentalism could characterize institutional "cosmopolitanization" and enable continuous improvement. Only by creating systems of mutual self-determination where every community enjoys rights and responsibilities with respect for each other can humanity create a just, sustainable and peaceful common future in the age of deep global integration and shared vulnerability to environmental threats in limited and vulnerable planetary systems. The nature of the risks necessitates a frank and principled response that establishes the authority of citizens across borders to protect life-creating systems as common heritage. But with imagination and coordination, arrangements suitable for our globalized world can guide development down a road of justice for all the lived and yet-to-be-lived lives affected.

While green cities can be an attractive solution for urbanization (Broto 2018), achieving such a model requires addressing social, environmental and economic justice. Visions gracing cities green are easy to come by, but authenticity can only come from identifying the city as a unique living context expressed within the community's environment, economic practices, settlement patterns, and cultural practices: not a model imposed. Justice and sustainability are intertwined goals that emerge from bottom-up engagement and localized solutions, not top-down dictation.

Combating the existential threat of climate change demands global urgency and equity in cooperation across every sector of society. This requires new multilevel governance arrangements that promote public participation and guardrails for oversight; treat the social cohesion shared living spaces as a common heritage; and remediate power imbalances that prioritize polluting interests over vulnerable communities. A just transition to renewable energy and green jobs must work for fossil fuel-dependent workers and regions in this country. May offer venues to implement systems of reciprocal self-determination where humanity can create sustainable prosperity within planetary boundaries while respecting the rights of affected peoples and future generations. Within cities sustainability is a grassroots endeavour in which cooperation recognizes cities as spatialized and dynamic social–ecological systems shaped and reshaped by social actors struggling within real-life existing unrealities. Democratic experimentalism, combined with creativity and a commitment to justice, provides humanity with the means to respond to the climate challenge through new institutional arrangements suitable for our globalized era.

8.9 Conclusion

Cities are a huge part of the solution on tackling the existential threat of climate change, as leaders to help on sustainability by enforcing climate-friendly regulations, policies, and initiatives. Urban centres are critical foci of both global prosperity and greenhouse gas emissions, and must pursue low-carbon transitions based on inclusive, justice-focused modalities that lift up diverse communities. Climate action is intricately tied to the principles of climate governance; to do so, global multilateralism needs to be matched by local empowerment, and to ensure this, we need robust participation and coordination at all levels, layer by layer, from global, national and local actors.

This paper has shown that participatory processes can be integrated into decision-making processes, allowing cities to develop tailored solutions that meet local needs in an equitable way while enhancing social cohesion. Local governments can deliver climate solutions that leave no one behind by planning for sustainable urban forms, designing evidence-based strategies in conjunction with communities, and linking adaptation policies with frameworks of legal rights. There is potential in exploring synergies between participatory approaches and contexts of empowered local governance to design impactful responses to climate in socio-culturally relevant ways.

While there are daunting challenges in the present, opportunities too are unlocked through projects like the SDGs and movements pushing for climate justice, demanding global arrangements that respond better to vulnerable populations. As scientific understanding deepens about potential risks of the new technology, new governance experiments are necessary to redress power imbalances and protect common resources for all those affected, both present and future. While sustainability certainly poses the question of the lifestyle choices made at individual and institutional levels, it is more fundamentally a moral project, one which places humanity in a position to solve many of the environmental, social and economic crises on overlapping fronts, if humanity chooses to cooperate. A just transition that allows for the flourishing of lives within planetary boundaries and accessible to all is still possible this century with political will supplemented by ongoing grassroots mobilization.

Funding Not applicable. This research did not receive any specific grant from any funding agencies.

Availability of Data and Materials Not applicable. Data sharing does not apply to this article as no datasets were generated or analysed during the current study.

Declarations

Ethical Approval Not applicable. This study did not involve human participants or animals, and no ethical approval was required.

Clinical Trial Number Not applicable. This study did not involve any clinical trials or related human interventions requiring clinical trial registration.

Consent to Publish Declaration Not applicable. The study does not include any content or data that requires individual consent for publication.

Consent to Participate Declaration Not applicable. As the study did not involve human participants, no consent to participate was required.

References

Ahluwalia IJ, Kanbur R, Mohanty PK (2014) Urbanisation in India: challenges, opportunities and the way forward. SAGE

Berry LH, Koski J, Verkuijl C, Strambo C, Piggot G (2019) Making space: how public participation shapes environmental decision-making. SEI Discussion Brief, Stockholm Environment Institute. Retrieved from: https://www.sei.org/publications/making-space-how-public-participation-shapes-environmental-decision-making/

Biswas D (2024) Sustainable development through ancient Indian practices and knowledge system: lessons from history. Soc Sci Rev Multi J 2(2):106–123

Brooks T (2013) The real challenge of climate change. PS: Politic Sci Politic 46(1):34–36. https://doi.org/10.1017/S1049096512001412

Broto VC (2018) Green city promises and "just sustainabilities." RCC Perspect 2018(1):55–64. https://doi.org/10.5282/rcc/8219

Chikarmane P (2012) Integrating waste pickers into municipal solid waste management in Pune. WIEGO Policy Brief (Urban Policies), India

City of Copenhagen (2014) Cloudburst management plan. City of Copenhagen

Crispi G, Sietchiping R, Ojoleck B, Ricca V, Amin A, Njuguna S, Charles A, Cárcamo AMBdM, Quinlan V, Gatsinzi JR (2022) Urban planning law for climate smart cities: urban law module. UN-Habitat. Retrieved from: https://unhabitat.org/urban-planning-law-for-climate-smart-cities-urban-law-module

Hughes S (2013) Justice in urban climate change adaptation: criteria and application to Delhi. Ecol Soc 18(4):48. https://doi.org/10.5751/ES-05929-180448

Kahn ME (2009) Urban growth and climate change. Ann Rev Resour Econ 1:333–349. http://www.jstor.org/stable/43202497

Kanuri C, Revi A, Espey J, Kuhle H (2016) Cities and a territorial approach to the SDGs. In: Sustainable development solutions network (ed) Getting started with the SDGs in cities: a guide for stakeholders. Sustainable Development Solutions Network, pp 7–19. https://doi.org/10.1787/5-km959d3

Raoul Wallenberg Institute (2022) Localising human rights in the context of SDGs: a Handbook for Cities. Raoul Wallenberg Institute of Human Rights and Humanitarian Law, Lund, Sweden

Robinson M (2018) Climate justice: hope, resilience, and the fight for a sustainable future. Bloomsbury Publishing

Rohracher H, Späth P (2014) The interplay of urban energy policy and socio-technical transitions: the eco-cities of Graz and Freiburg in retrospect. Urban Stud 51(7):1415–1431. https://doi.org/10.1177/0042098013500360

Rosencranz A, Jackson M (2003) The Delhi Pollution Case: The Supreme Court of India and the limits of Judicial Power. Columbia J Environ Law 223

Sahasranaman A, Kapur V (2014) The practice of PPP in urban infrastructure. In: Ahluwalia IJ, Kanbur R, Mohanty PK (eds) Urbanisation in India: challenges, opportunities and the way forward. SAGE, pp 141–168

Scheurer J, Newman P (2009, January) Vauban: a European model bridging the green and brown agendas (Case study prepared for Revisiting Urban Planning: Global Report on Human Settlements 2009). United Nations Human Settlements Programme (UN-Habitat). Retrieved from https://www.researchgate.net/publication/228970605_Vauban_A_European_Model_Bridging_the_Green_and_Brown_Agendas

Schlosberg D, Collins LB (2014) From environmental to climate justice: climate change and the discourse of environmental justice. WIREs Clim Change 5(3):359–374

Skillington T (2017) Climate justice and human rights. Palgrave Macmillan. https://doi.org/10.1057/978-1-137-02281-3

Sörensen E, Torfing J (2019) Democratic innovations and the role of participation in climate adaptation planning: The case of Copenhagen's Cloudburst Management Plan. Urban Stud 56(7):1371–1389. https://doi.org/10.1177/0042098018796250

Sprain L (2016) Paradoxes of public participation in climate change governance. Good Soc 25(1):62–80. https://doi.org/10.5325/goodsociety.25.1.0062

Sustainable Development Solutions Network (SDSN) (2013) Why the world needs an urban sustainable development goal. Sustainable Development Solutions Network. http://www.jstor.com/stable/resrep16098

UN (United Nations) (2010) World Urbanization Prospects, the 2009 Revision: highlights. United Nations, New York

United Nations High Commissioner for Human Rights (UNHRC) (2018, April 18) Summary of the intersessional panel discussion on the role of local government in the promotion and protection of human rights (A/HRC/38/22). United Nations. https://daccess-ods.un.org/access.nsf/Get?DS=A/HRC/38/22&Lang=E

Voss H (2014) Environmental public participation in the UK. Int J Soc Qual 4(1):26–40. https://doi.org/10.3167/IJSQ.2014.040103

While A, Whitehead M (2013) Cities, urbanisation and climate change. Urban Stud 50(7):1325–1331

Chapter 9
Cities in Transition: A Case Study-Based Geospatial Investigation for Impact of Urban Expansion on Ecology in Kolkata Municipal Corporation and Its Surroundings

Sreya Mukherjee and Sayani Mukhopadhyay

Abstract Urbanization is a multifaceted process entwined with rural–urban migration and the provision of infrastructural development; it is radically reconstituting the environment of cities throughout the world, but especially in the Global South. This study pertains to land use/land cover (LULC) changes and ecological effects of urbanization in Kolkata, India, one of the most populous and fastest-growing cities in the country. The study analyzed LULC changes through supervised classification of satellite images, Landsat 5 TM for 2001 and Landsat 8 OLI/TIRS for 2021. Land surface temperature (LST) and Soil-Adjusted Vegetation Index (SAVI) have also been computed to determine ecological impacts. Results show that over the last 20 years, the built-up area has shown its maximum increase at the cost of agricultural land and vegetation cover. Most of the expansion has been observed in the peri-urban fringe on the outskirts of the city core. The study also revealed differences in LST, where some areas were registering increases in LST due to the conversion of natural land cover to a built-up environment. The SAVI decline within the core city indicates a deteriorating vegetation health and land quality subjected to pressure from urbanization. These findings bear testimony to the ecological consequences of quick urbanization and stress the dependency on sustainable urban planning and policy intervention to counteract the impacts and allow the city to withstand them in the future.

Keywords Urban growth · Land-use change · Land surface temperature · Soil-adjusted vegetation index · Future dynamics

---

S. Mukherjee (✉) · S. Mukhopadhyay
Department of Geography, Asutosh College, Kolkata, West Bengal, India
e-mail: sreya.mukherjee@asutoshcollege.in

S. Mukhopadhyay
e-mail: sayani.mukhopadhyay@asutoshcollege.in

K. Mukherjee et al. (eds.), *Urban Climate*, The Urban Book Series,
https://doi.org/10.1007/978-3-032-17895-4_9

9.1 Introduction

Urbanization has a complex meaning that includes population movements from rural areas in search of better economic opportunities. Not only does this process lead to the physical expansion of certain urban areas, but it also requires the establishment of base infrastructure: roadways and transport systems, water supply and sewerage, along with housing and basic amenities such as healthcare, education, and recreation (Shao et al. 2021). Urbanization is on the rise all across the globe, with the pace being much higher in the Global South, where cities are experiencing an unprecedented influx of migrants, thereby creating a huge sprawl. According to the United Nations (2018), about 68% of the world population is expected to live in urban areas by 2050, against 55% in 2018. This acceleration in urban growth creates severe problems for the cities due to the concomitant pressure put on land resources and infrastructure, which are already stretched beyond imagination, especially in the developing world (Seto et al. 2012).

In the case of India, the urban population rose from about 25.85 million in 1901, representing 10.84% of the total population, to 377 million in 2011, constituting 31.2% of the population (Sarkar 2020). This peculiar acceleration of urbanization, particularly the period from 27.8% in 2001 to 31.2% in 2011, which is further expected to be the increasing trend, has been backed by economic growth and rural-to-urban migration (Ahluwalia 2016). West Bengal, an important state in India, is showing the same trend with an increase in urban population from 27.5% in 1991 to 31.9% in 2011, thus surpassing the national mean. Kolkata, being the capital of the state, has developed into an area thoroughly urbanized and classified as 100% urban population, thus revealing the direct impact of urbanization on its economic and social texture.

As a consequence of these physical transformations of space from the late seventies onward, the fate of Kolkata and other major urban centers has been sealed by the unplanned extension of urban settlements into the countryside, mostly in the absence of any coordinated land-use planning framework. The rapid growth of urban settlements will continue worldwide, especially in Asia and Africa, where poverty and socioeconomic factors present challenges to migrants (UN Habitat 2020). The environmental consequences of rapid urban expansion are deep and widespread, from contributing to deforestation and degrading landscapes and green spaces to fueling climate change (James 2023). While the literature provides evidence in these respects, it also discusses how urbanization tends to worsen pollution, reduce access to crucial resources, heighten the levels of poverty, and raise the occurrence of extreme weather conditions and, as such, pose threats to urban resilience (IPCC 2014).

Even while extensive studies regarding urbanization impacts and the necessity for sustainable development practices abound (Parnell 2016), a conspicuous knowledge gap remains regarding concrete adaptive strategies for local contexts in the rapidly urbanizing regions of the Global South. In contrast, existing literature has tended to focus on urbanization trends in general and has often neglected localized approaches to adaptive planning that would limit the negative effects of urban sprawl. Recent studies express the urgent need for urban planners and policymakers to adopt inclusive, resilient, and adaptive approaches to urban population centers addressing the dilemma of resource scarcity, climate change, and social inequality towards sustainable urban futures (Liu et al. 2014).

Due to the interlocking complexities of sustainable development and the unpredictability of future growth, consideration must be given to resilient urban planning above all else. Cities that fail to engage in these challenges might as well be obsolete, having little capacity to deal with the other pressing issues of resource scarcity, climate change, and urban poverty. Development of the research agenda along these lines will greatly aid in the processes of sustainable urbanization and improvement of the lives of millions of urban sector dwellers of this world.

Therefore, the major objectives of this study would be as follows:

1. To detect the change in land use and land cover over the last three decades.
2. To identify the ecological impact of urbanization.
3. To evaluate the efficacy of analyzing the future dynamics of urban change as a policy intervention.

9.2 Study Area

Kolkata is one of the most populous and one of the rapidly growing cities in India. Kolkata is an unplanned city that has extended its area gradually just to support the influx of immigrants from different parts of the state or country (Mitra et al. 2012), which results in continuous expansion of the urban areas in the outskirts and comparatively less flourished surrounding areas. Kolkata, one of India's biggest metropolises, is accompanied by a fast-accelerating and unplanned process of urbanization. This is fundamentally an outcome of continuous immigration from the rest of India and the world, absorbing the local zones by the periphery–for example, extending up into nearby villages. This research focused on the districts of Kolkata and even the periphery regions; it brings much valuable information about LCLU changes during this spell of rapid urbanization (Gupta and Bhagat 2023) Kolkata is a city whose history can be traced back over 300 years, which makes it much older than the majority of its urban institutions in India. However, the city came into pre-eminence during the British colonial rule and attracted an assorted crowd, thus making it cosmopolitan. It has added people from different religions, communities, and cultures to the already diverse demography of the city and left behind the trends of the city.

Considering KMC as the core, concentric buffers of 1 km distance up to 10 km have been created (Fig. 9.1). The buffers are not even on all the parts as only the places situated under the Kolkata Urban Agglomeration (KUA) area have been considered. With the help of this methodology, different distances from the city center can be evaluated for the details of LCLU changes, thereby providing a better insight regarding the dynamism of urbanization in Kolkata.

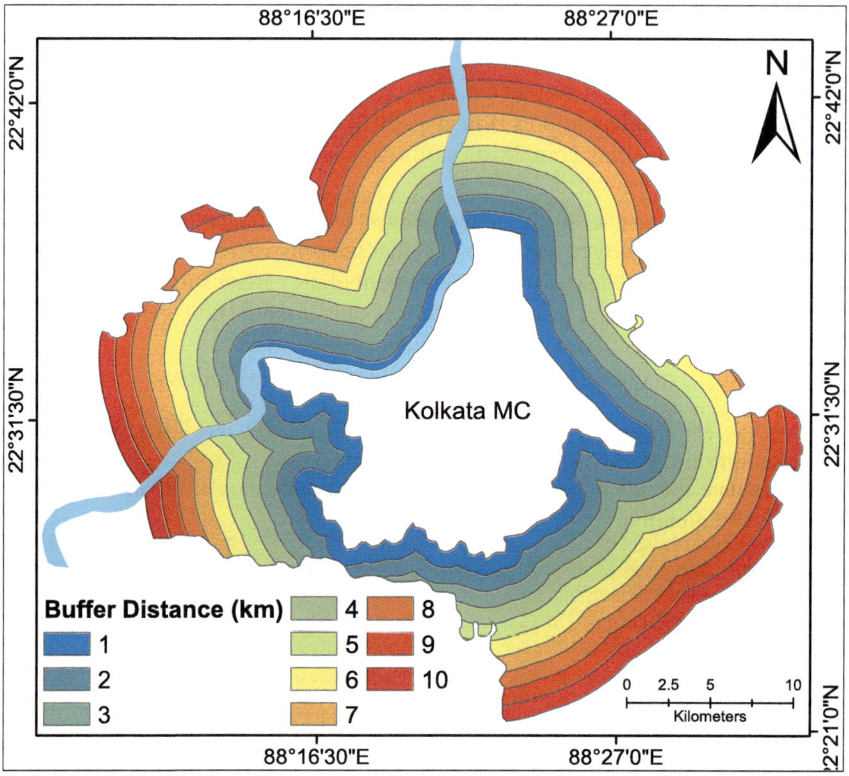

Fig. 9.1 Location of the Kolkata Municipal Corporation along with corresponding 10 buffer regions at 1 km. interval

9.3 Methodology

9.3.1 Data Consideration

This research employed satellite-based multispectral images obtained from the USGS Earth Explorer archive for 2001 and 2021 to examine changes in land use over time and their regional distribution. The photos were acquired in January to ensure low cloud cover and to mitigate seasonal fluctuations in land use patterns. The analysis utilized cloud-free imagery obtained from Landsat 5 TM (2001) and Landsat 8 OLI/TIRS (2021). Each satellite provided numerous optical bands, including 6 bands for TM and 8 bands for OLI/TIRS. These bands captured the reflectance of the land surface across different electromagnetic spectrums. This enables the differentiation of various land surface characteristics. Both sensors offer periodic surveillance and a spatial resolution of around 30 m, allowing for meticulous examination of land utilization patterns across extensive regions at frequent intervals. After capturing the image, we conducted radiometric calibration and atmospheric correction using the

DOS1 module in the QGIS 3.10 environment. This was done with the help of the Semi-Automatic Classification (SCP) plugin. This correction reduces the impact of variations in the roughness of the terrain and disturbances in the atmosphere on the amount of radiant radiation.

9.3.2 Extraction of Land Use and Land Cover Using Supervised Classification Method

Following preprocessing, the calibrated pictures were cropped to the designated project region and employed for classification in the ArcGIS 10.2.2 environment using the Supervised Classification method. This approach entails the iterative prediction of classes from training datasets provided by the author, utilizing variations in reflectance across different bands and prior knowledge. The classification included five land use categories: vegetative land, agricultural land, waterbodies, built-up places, and fallow land. Training samples were selected for each class based on their spectral signatures, following the methods provided by Lillesand et al. (1979), and then classified using the Maximum Likelihood Classifier (MLC) algorithm. The Kappa Coefficient, which measures the statistical significance of the classification matrix, remained above 0.85 for each classified image, suggesting a generally acceptable outcome.

9.3.3 Estimation of Land Surface Temperature (LST) and Soil-Adjusted Vegetation Index (SAVI)

To compute LST, initially, the digital number (DN) readings from the pixels in the Thermal Band (Band 6 for Landsat 5 TM and Band 10 for Landsat 8 OLI/TIRS) were translated into spectral radiance at the sensor (L_λ):

$$L_\lambda = \frac{L_{\max} - L_{\min}}{\mathrm{QCAL}_{\max} - \mathrm{QCAL}_{\min}} \times \mathrm{QCAL} + L_{\min} \tag{9.1}$$

where L_λ = spectral radiance sensed at the sensor; $L_{\max}$ = maximum radiance; $L_{\min}$ = minimum radiance; $\mathrm{QCAL}_{\max}$ = maximum DN value; $\mathrm{QCAL}_{\min}$ = minimum DN; QCAL = DN value of a particular pixel. Thereafter, spectral radiance sensed at the sensor was changed into brightness temperature using Eq. (9.2):

$$\text{Brightness Temperature} = \frac{K_2}{\ln\left(\frac{K_1}{L_\lambda} + 1\right)} \tag{9.2}$$

where K_2, K_1 = Thermal Constant for Thermal Bands of Landsat images provided in metadata file of the image.

Next, the Proportion of Vegetation (PV) was calculated using Eq. (9.3) which was further used to extract land surface emissivity (LSE) by Eq. (9.4) for the Landsat-5 TM image and Eq. (9.5) for the Landsat-8 OLI/TIRS image, which measures the transfer of heat energy.

$$P_V = \frac{\text{NDVI} - \text{NDVI}_{\text{min}}}{\text{NDVI}_{\text{max}} - \text{NDVI}_{\text{min}}} \tag{9.3}$$

$$\text{LSE} = 0.004\,\text{Pv} + 0.986 \tag{9.4}$$

or,

$$\text{LSE} = 0.02644\,\text{Pv} + 0.96356 \tag{9.5}$$

Finally, the Land Surface Temperature (LST) has been extracted using Eq. (9.6):

$$\text{LST} = \frac{\text{BT}}{1 + \left(\lambda * \frac{\text{BT}}{\rho}\right) \times \ln(\text{LSE})} - 273.15 \tag{9.6}$$

where LST = Land Surface Temperature; λ = Average wavelength of Thermal Band; ρ = h*(c/σ) in which; "h" is Planck's Constant (6.626 × 10–34); "c" is the speed of light (3 × 108 m/s); "σ" is Boltzmann's Constant (1.38 × 10–23 J/K).

The Soil-Adjusted Vegetation Index (SAVI) improves upon NDVI in that it minimizes the effect of soil background, which is responsible for erroneous vegetation index calculation, especially in sparse vegetation areas. Due to the influence of soil brightness variation, NDVI may not give reliable results, while SAVI incorporates a correction factor for soil brightness (L) and is, therefore, a more reliable tool for analyzing vegetation in arid and semi-arid areas where soil exposure is more. This adjustment would ensure better capacity for vegetation monitoring under changing soil conditions. The SAVI has been calculated using Eq. (9.7):

$$\text{SAVI} = \frac{(\text{NIR} - \text{Red}) * (1 + X_{\text{SAVI}})}{\text{NIR} + \text{Red} + X_{\text{SAVI}}} \tag{9.7}$$

where NIR and Red correspond to near infra-red and red multispectral band; XSAVI represents soil adjusted factor for SAVI.

Finally, Soil Moisture Index (SMI) has been calculated as it provides the basis for drought monitoring, irrigation water management, flood prediction, soil health assessments, and climate studies. It will know how much water is available for crops, over or under-water them, run-off risk assessments, and the long-term moisture trend of assessment change for climate.

$$SMI = \frac{LST_{max} - LST}{LST_{max} - LST_{min}} \tag{9.8}$$

A detailed methodological framework of the present work has been provided in Fig. 9.2.

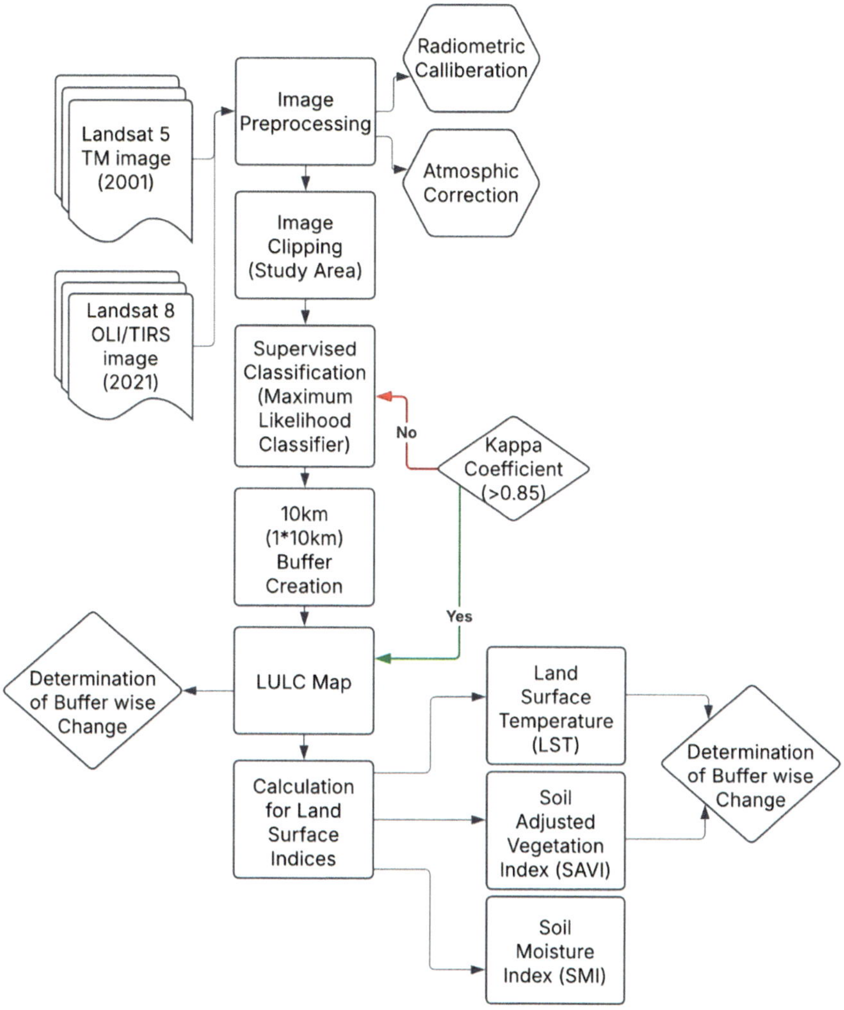

Fig. 9.2 Detailed methodological framework of the study

9.4 Result and Discussion

9.4.1 Change in Land Use and Land Cover

A huge change in land use and land cover has been observed in and around Kolkata over the last two decades. As the population grows, so does Kolkata's urban sprawl. Agricultural land, wetlands, and natural habitats are turned into residential, commercial, and industrial sites during this enlargement. Satellites and suburbs were developed for the congestion relief of the city center residents. Urbanization has led to a depletion of green spaces and vegetation coverage. Buildings, roads, or some other types of construction are taking over the parks and gardens that used to be covered with vegetation.

In the core KMC, there was limited land available for transformation, yet in the eastern and some southern areas, agricultural land has been converted to built-up areas and some water bodies have been filled for construction. In buffer zone 1, the entire southern part, previously covered in vegetation, has been converted to built-up areas due to high demand attributed to population growth. Similar conditions are observed from buffer 2 to buffer 5, where almost all areas, except for some water bodies and small patches of vegetation, have been converted to built-up areas. This zone is predominantly inhabited by the lower middle class who cannot afford to live in the main city but need to commute regularly for work, hence they prefer closer proximity to the city. In the last four buffers, water bodies, vegetation, and agricultural land are still visible in the southern, southwestern, and southeastern parts, but the remaining areas have been transformed.

The industrial section of Kolkata has undergone an expansion causing the transformation of open land into industrial zones. These zones consist of new industrial estates, warehouses, and factories. The rise in industrial activities has been responsible for pollution of air and water in some places. From Fig. 9.4, it can be concluded that the built-up area is capturing the land that was previously covered by vegetation and cultivable land. The highest change in built-up can be observed in buffer 1 which includes Bidhannagar, Laketown, Howrah, and some parts of Behala as this area is less distant from the core city. But as we move outward from the core city, an increasing rate of built-up is observed. The higher availability of land suitable for development in outer buffers supports this occurrence. Degradation of agricultural land is higher in outer buffers as agricultural land used to be higher in the outskirts, due to continuous population and urban growth natural agricultural land is getting converted into artificial concrete land to support the increasing population. Vegetation coverage is getting lost due to this phase of transformation. Maximum change in vegetation cover has been observed in buffer 1 due to proximity to the core city. From Fig. 9.3, it is evident that all the buffers are showing loss in vegetative and agricultural land (Fig. 9.4).

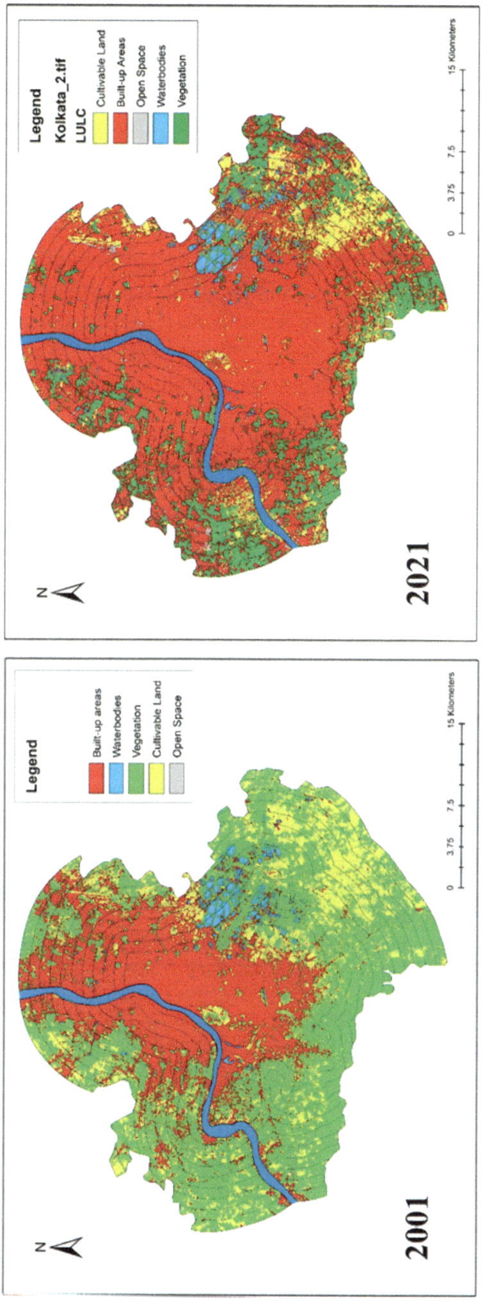

Fig. 9.3 Spatio-temporal distribution of land use and land cover in KMC and its surroundings for 2001 and 2021

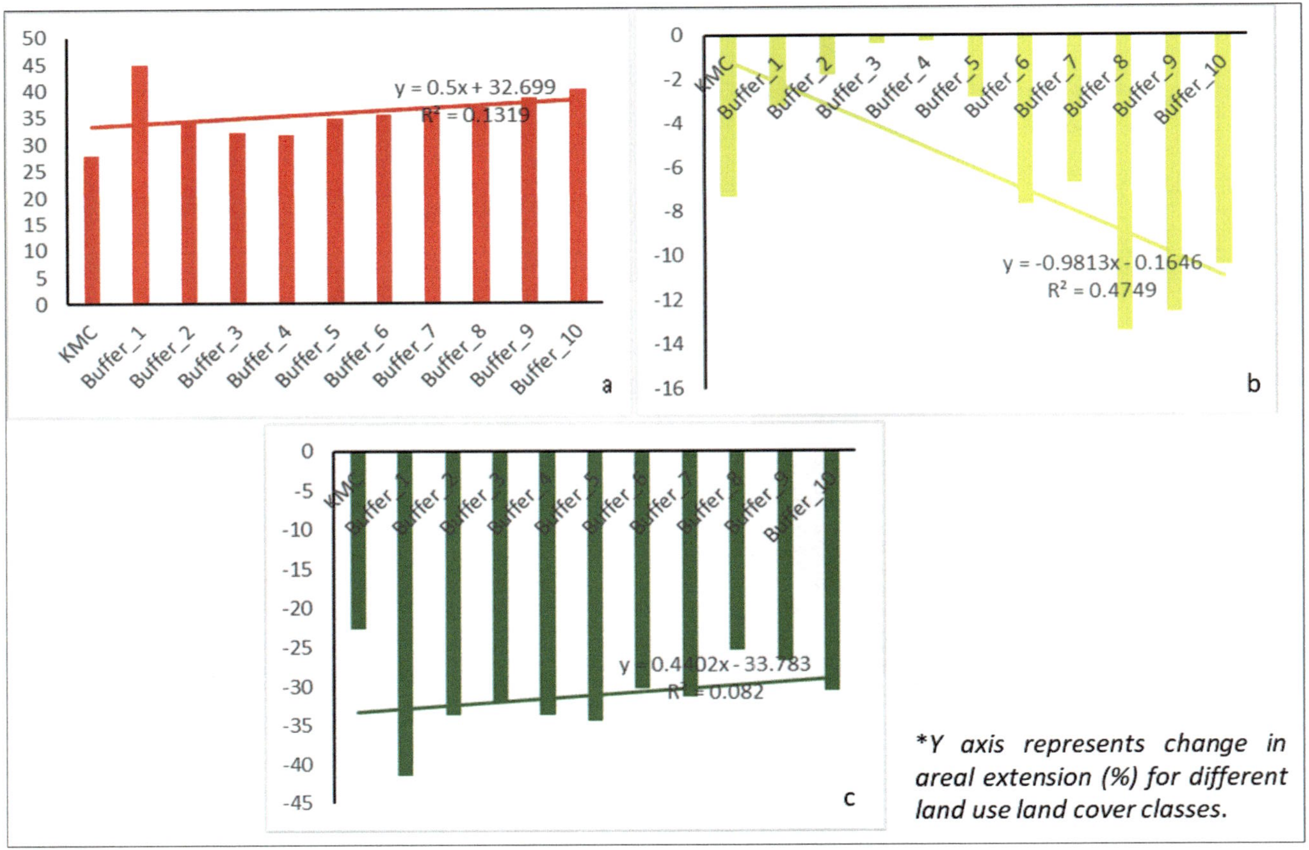

Fig. 9.4 Buffer-wise land use and land cover change **a** Built-up area **b** Agricultural land **c** Vegetation cover

9.4.2 Ecological Impact of Urbanization

Urbanization profoundly affects ecology, leading to significant alterations in natural land cover and ecosystems. As cities expand, they encroach upon and fragment natural coverings, resulting in loss of biodiversity and disruption of ecological processes. Deforestation and habitat destruction to make way for infrastructure and buildings reduce the available habitat for native species, leading to local extinctions and population declines. Urban areas also introduce pollutants such as chemicals, noise, and light, which can degrade air and water quality, disrupt wildlife behavior, and hence impact human health. Changes in land cover, such as the transformation of vegetation with impervious surfaces like concrete, alter water flow patterns, leading to increased runoff and decreased groundwater recharge. Urbanization also affects temperature, producing urban heat islands that can alter the local climate and exacerbate heat-related stress.

The reduction in land surface temperature (LST) observed in the core city, particularly the KMC area, during 2021 can be attributed to the impacts of the COVID-19 pandemic. With reduced human activity and industrial operations due to lockdowns and movement restrictions, there was less heat generated from sources like vehicles and factories, leading to a decreasing trend in urban temperatures. Despite a reduction in the general land surface temperature (LST), it is important to point out that areas to the western side of buffer zones 5–8 have recorded an increase in LST. Similarly, the north-western extremes of buffer zones 9 and 10 show a rise in LST.

From Fig. 9.5, it can be seen that in the core area, i.e. KMC, mean LST has decreased sharply. The most important causes for the decline in LST include the effects of the COVID-19 pandemic and the severe lockdowns during 2020–2021. Some recessed intense activities in industry, decreased vehicular emissions, and diminished operations of business establishments contributed significantly to the reduction in urban heat generation. The period under lockdown has induced lesser human-induced thermal activities, thus resulting in a much lower urban heat island (UHI) effect at the city center. It could also be that tree cover and other urban green activities, including rooftop gardening and water bodies' restoration, support cooling in the core. Great changes in building materials, such as reflective rooftops and energy-efficient structures, also might have contributed to heat accumulation reduction. In the buffers, the mean LST is less than the core as the concentration of the built-up area is comparatively less. However, continuous urbanization has saturated the core KMC area. So, increasing the rate of concretization is observed in the buffers, except for the last buffer (9–10 km) each buffer shows increasing LST. Outer buffers are depicting a lesser increasing rate compared to the immediate buffer to the core KMC, as urban expansion always starts keeping the city in the center and spreads outside (Fig. 9.6).

This change can be attributed to the ongoing transformation from natural land cover to man-made land use, such as urbanization and infrastructure development. This change exacerbates heat stress, disrupts ecosystems, hampers water supply, contributes to the formation of urban heat islands, worsens air quality, and decreases

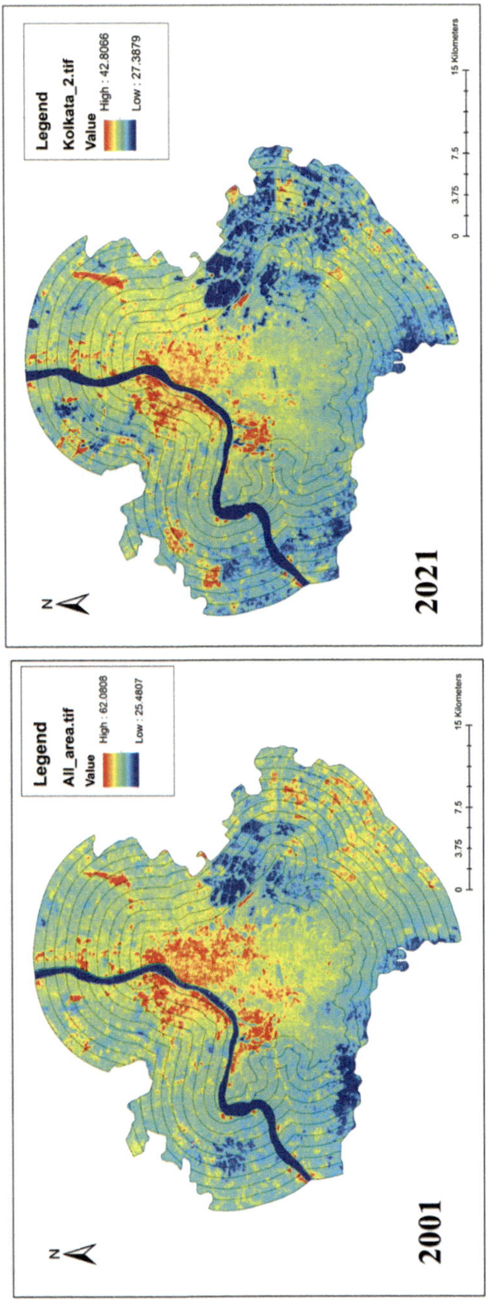

Fig. 9.5 Buffer-wise land surface temperature of 2001 and 2021

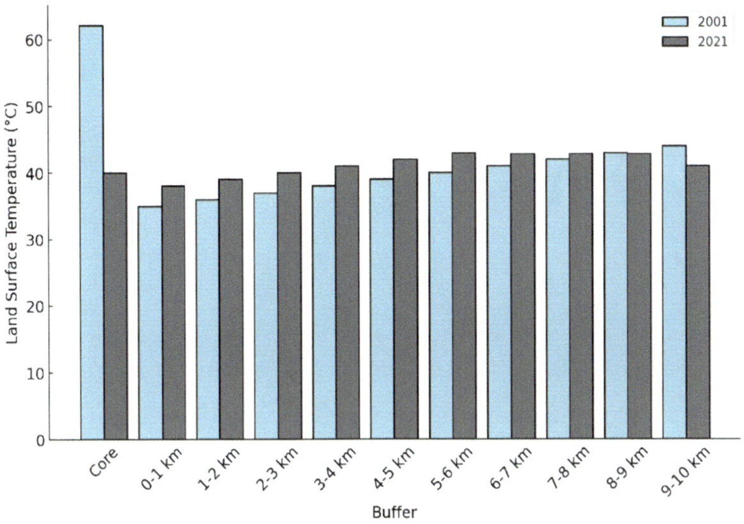

Fig. 9.6 Buffer-wise mean LST distribution of 2001 and 2021

agricultural productivity. Overall, the contrast in LST trends highlights the complex interplay between natural processes and human activities, emphasizing the importance of sustainable urban planning and environmental management to mitigate the loss of vegetation coverage.

In the KMC area, the Soil-Adjusted Vegetation Index (SAVI) has shown a decrease over the last two decades, indicating a decline in vegetation and overall land quality. During 2001–2021, the SAVI value was reduced by 19.04% in the study area, demarcating a decrease in vegetation density or health in the region. This decline in vegetation can be attributed to several factors associated with urbanization and human activities. Firstly, rapid urbanization often leads to converting green spaces into built-up areas, resulting in the loss of vegetation cover. Additionally, increased pollution, including air and water pollution from industrial and vehicular sources, can negatively impact plant health and growth. The urban heat island effect, caused by replacing natural vegetation with impervious concrete surfaces, can also contribute to vegetation. Furthermore, changes in land use patterns, such as deforestation, agricultural expansion, or infrastructure development, can further exacerbate the decline in vegetation and land quality (Fig. 9.7).

It is evident from Fig. 9.8 that mean SAVI values have progressed from the core to outer buffers, having the minimum core values indicating poorer vegetation conditions and increasingly high values in the surrounding area. Such a trend reflects the very urbanized and limited extent of green cover of the city center: here dense infrastructure, pollution, and heavy human activity render the vegetation health very low. In contrast, the outer buffers would generally have more greenery, including farmland or natural cover, and hence have a comparatively greater SAVI index they also signify having a better plant cover.

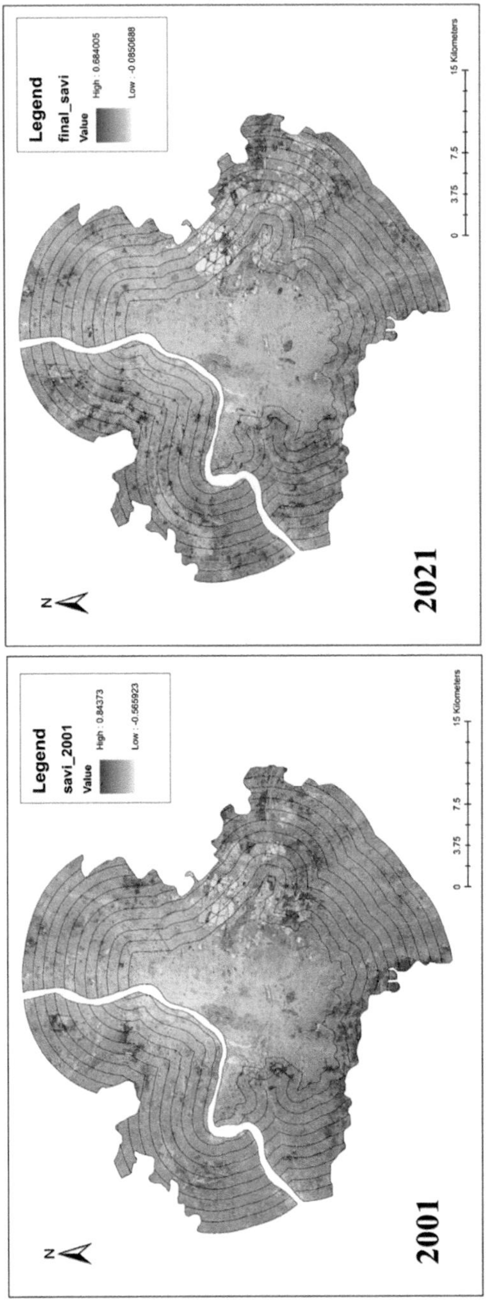

Fig. 9.7 Buffer-wise soil-adjusted vegetation index for 2001 and 2021

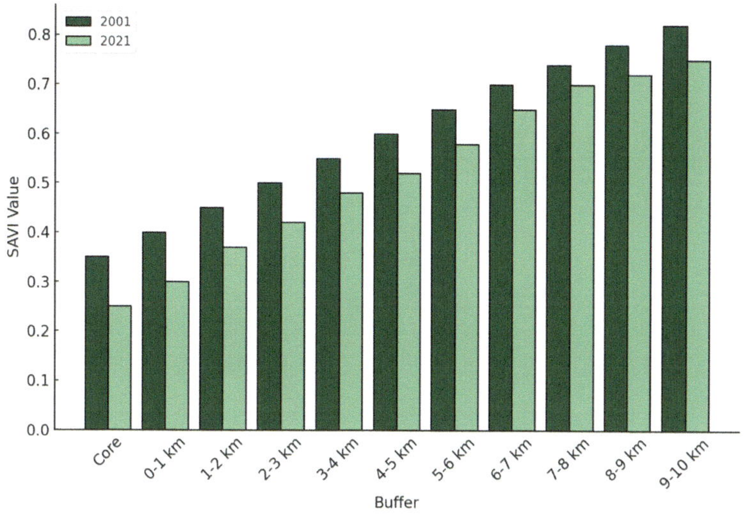

Fig. 9.8 Buffer-wise comparative trend of SAVI (2001 vs 2021)

Over some time, urban expansion and population growth will bring about land-use changes affecting the condition of the vegetation in the core as new developments replace the already existing green patches. These green patches may also be affected by some environmental initiatives such as reforestation, the creation of suburban parks, and stricter land-use regulations. Such changes may also occur due to changes in climatic trends, rainfall distribution, as well as changes in soil management practice as they all have been affecting the SAVI values between 2001 and 2021. Overall showing that lowering the SAVI in the core emphasizes the need for targeted greening measures and sustainable urban planning to reduce the loss of vegetation due to such built environments. Overall, the decrease in SAVI in the KMC area reflects the ongoing challenges associated with urbanization and highlights the need for sustainable land management practices to mitigate further deterioration of vegetation and land quality (Fig. 9.9).

The comparative analysis of the Soil Moisture Index (SMI) from 2001 and 2021 showed that a marked decline in soil moisture occurred in Northern, Eastern and Western parts of the buffers covering Baranagar, Howrah, Saltlake, and Rajarhat New Town. This change can be attributed to the increasing urbanization and sudden change in LULC in and around the mentioned places. This also implies persistent dry conditions in the area, which is affected by varying factors like increasing temperature, land-use changes, and deforestation. Inversely correlated, SMI stands concerning land surface temperature (LST), where an increase in LST leads to a decrease in soil moisture due to increased evaporation. Increasing trends in LST from the past two decades could be pointed out as a reason for drying in the SMI maps. In a positive correlation, as SMI goes high, the amount of vegetation cover keeps soil moisture high. Therefore, a decrease in vegetation over time with a corresponding low SAVI

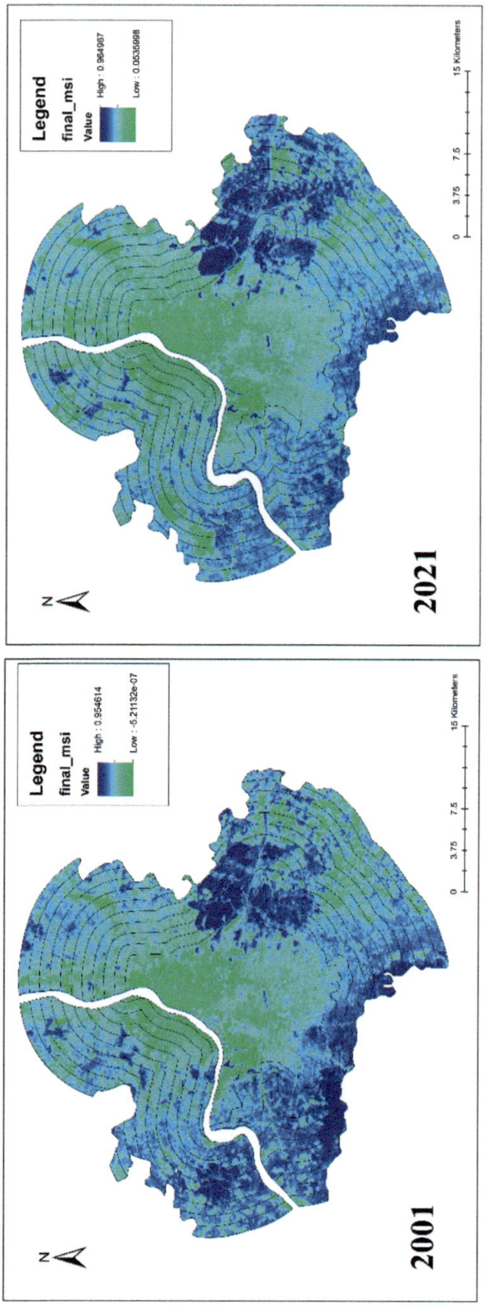

Fig. 9.9 Buffer-wise soil moisture index for 2001 and 2021

could contribute to reduced soil moisture and land degradation. The trend indicates that soil moisture distribution is being affected by climate change and anthropogenic activities and consequently pushes for the adoption of sustainable land and water management practices.

9.4.3 *Prognostication of Future Dynamics of Urban Change as a Policy Intervention*

The future dynamics of urban change in Kolkata and its surroundings are expected to be characterized by rapid urbanization, driven by population growth and economic development. This will lead to continued expansion of built-up areas at the expense of agricultural land and green spaces, exacerbating existing challenges like traffic congestion and pollution. Government policies, urban planning strategies, and technological advancements will play crucial roles in shaping the urban landscape, with a focus on promoting sustainable development and improving infrastructure. According to estimates by the United Nations, Kolkata's population is projected to reach over 17 million by 2030, an increase of around 2.4 million from 2015. This population growth, coupled with economic expansion, will result in the continued expansion of built-up areas. Data from the Kolkata Municipal Corporation (KMC) shows that between 2001 and 2011, the urban population of Kolkata increased by 11.1%, with a corresponding decrease in rural population. This trend is expected to continue, leading to the conversion of agricultural land and green spaces into residential and commercial zones. For instance, satellite imagery analysis indicates that between 2000 and 2020, Kolkata's built-up area expanded by approximately 57%.

Figure 9.10 shows a negative correlation between LST and the SAVI throughout various buffer zones in the KMC for the year 2001. The strongest correlation was observed in the buffer zones between 1 and 2 km ($r = -0.45$), while the weakest correlation was in the zone between 8 and 9 km ($r = -0.24$). This inverse relationship suggests that areas with higher vegetation cover (higher SAVI) tend to experience lower temperatures due to evapotranspiration cooling off the surface and shading that limits heat absorption. On the other hand, areas of higher urbanization with lower vegetation cover act like urban heat islands (UHIs) where the heat is retained by building materials such as concrete and asphalt, which have very high thermal inertia. The correlation weakens in the outer buffer zones because of more vegetation cover and less urbanization.

However, this urban expansion comes with challenges, including traffic congestion and pollution. Kolkata is already one of the most congested cities in India, with an average traffic speed of only 20 km/h (Chakrabartty and Gupta 2014). Air pollution levels are also concerning, with Kolkata consistently ranking among the cities with the highest levels of PM2.5 pollutants. Continued urbanization is likely to exacerbate these issues, placing additional strain on infrastructure and public services.

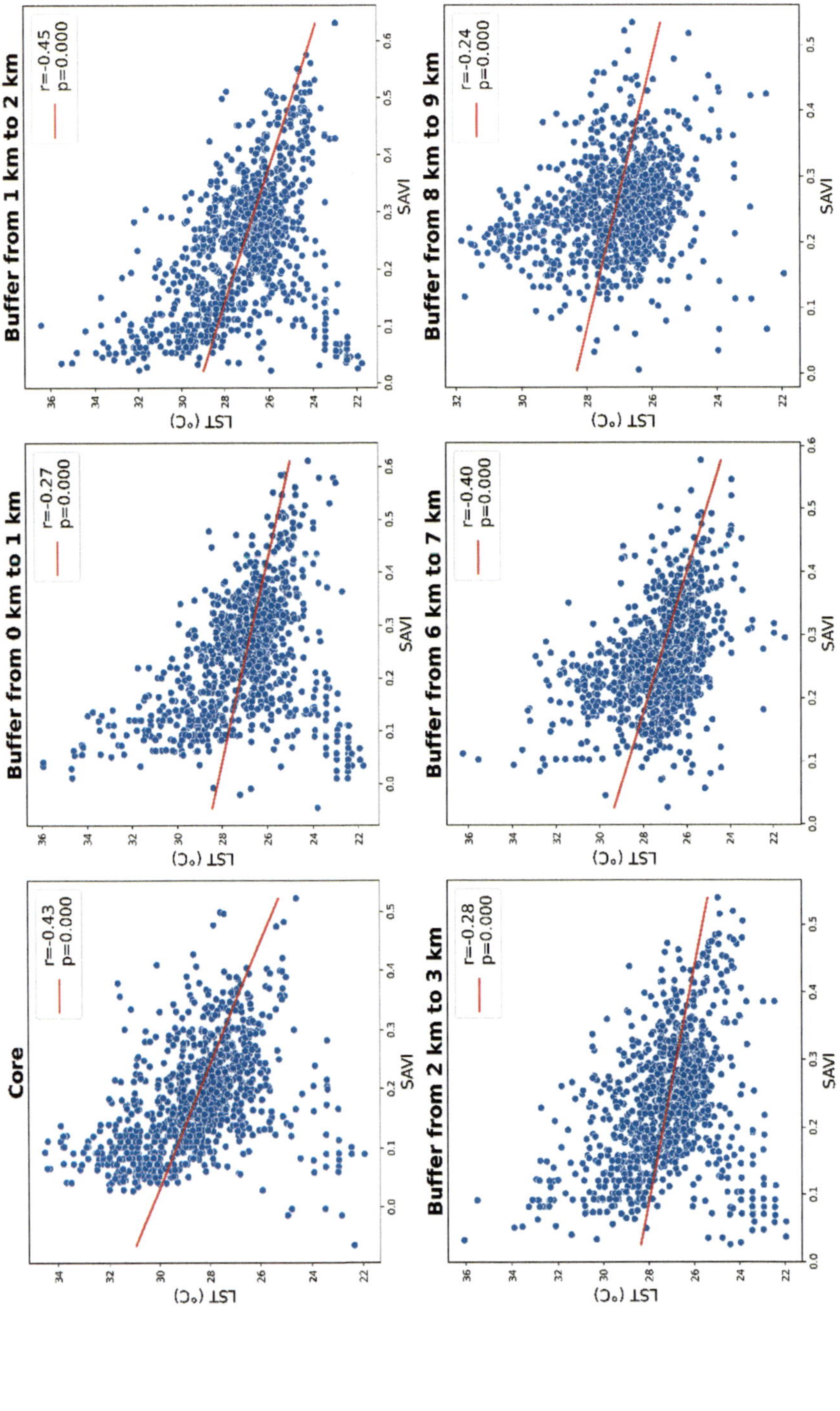

Fig. 9.10 Significant correlation ($\alpha = 0.05$) between LST and SAVI in core KMC and its different buffer zones in 2001

Figure 9.11 shows that in the year 2021, across multiple buffer zones of KMC, there was a negative correlation between LST and SAVI with a stronger correlation in the 9 km to 10-km buffer ($r = -0.41$) and a weaker correlation in the 0 km to 1 km buffer (r = -0.07). Compared to 2001, the correlation has especially weakened in the core and inner buffer zones, suggesting less vegetation influence on surface temperature. Declining vegetation influence can be attributed to urban expansion, development of built-up space, and diminution of green cover to cool the land surface by evapotranspiration and shading. However, in the outer buffer zones, it denotes that vegetation has continued to act as an important temperature regulator in places where the urbanization process is less vigorous. But due to continuous urban expansion, these areas have started getting affected, which will show its impact soon if proper measures are not taken.

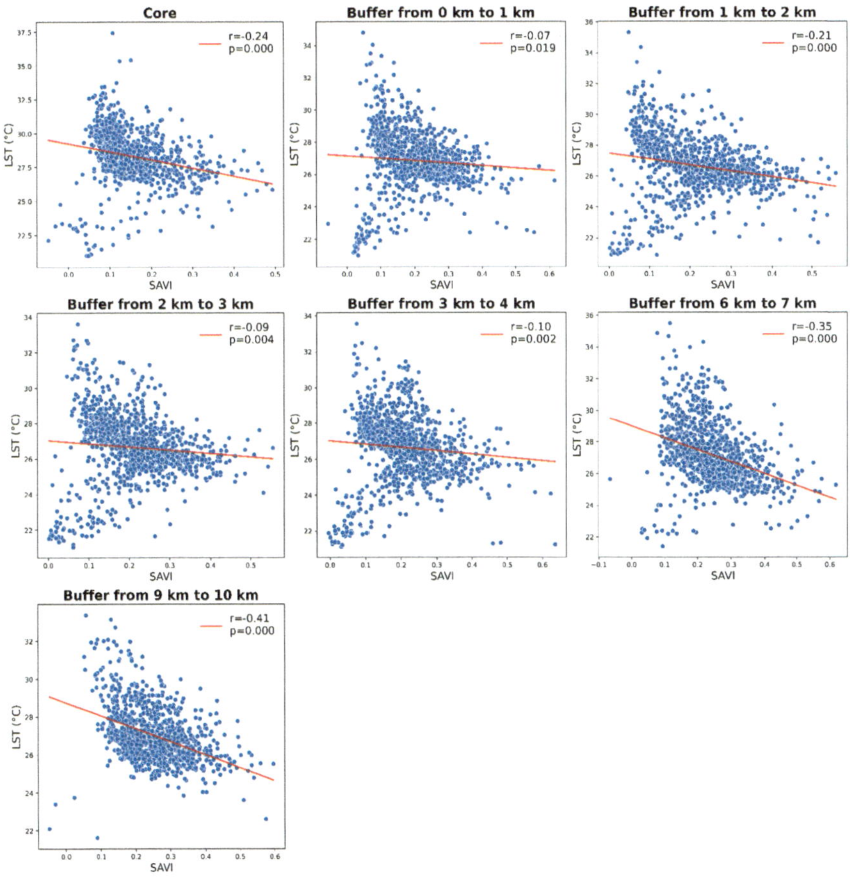

Fig. 9.11 Significant correlation ($\alpha = 0.05$) between LST and SAVI in core KMC and its different buffer zones in 2021

To address these challenges, government policies and urban planning strategies will be crucial. The Kolkata Metropolitan Development Authority (KMDA) has outlined plans for sustainable urban development, focusing on improving transportation infrastructure, creating green spaces, and enhancing public services. For instance, the East–West Metro corridor aims to improve connectivity and reduce traffic congestion. Additionally, initiatives like the Green Cities Mission, aiming to increase green cover and promote sustainable development, are being implemented.

However, the region will also face risks associated with climate change. Kolkata is particularly vulnerable to flooding due to its low-lying geography and proximity to the Bay of Bengal. Data from the Indian National Centre for Ocean Information Services (INCOIS) indicates that sea levels along the Kolkata coast have risen by around 8 cm over the past two decades (Mandal et al. 2018). Extreme weather events, such as cyclones and heavy rainfall, are becoming more frequent and severe. To mitigate these risks, adaptation measures such as improving drainage systems, constructing flood barriers, and implementing sustainable water management practices are essential. Addressing the future dynamics of urban change in Kolkata and its surroundings requires a holistic approach that balances economic growth with environmental sustainability. Collaboration among stakeholders, including government agencies, urban planners, and the public, is crucial to ensure the long-term livability and resilience of the region.

9.5 Conclusion

The drastic change in land use patterns, including built-up areas, vegetation coverage, and cultivable land, is evident as we move away from the main city center, especially in areas where land transformation is more feasible. This transformation is characterized by a substantial increase in built-up areas in outer buffers, accompanied by a decline in vegetation cover and cultivable land. These changes are not uniformly distributed across all buffers, indicating uneven urban growth. Such alterations in land use profoundly affect various ecological parameters, including land surface temperature, vegetation quality, and water quality, all of which are adversely impacted by urban expansion. Over the past two decades, the surrounding areas of Kolkata have experienced significant negative effects due to this urban growth.

Several factors contribute to urban growth in specific locations, including building density, population concentration, distance from the river, distance from the main town, and distance from the transport network. These factors have been identified as significant drivers of land use change and have a notable impact on ecological parameters. For instance, areas with higher building density and population concentration tend to experience more rapid urban growth, while proximity to rivers and transportation networks also influence development patterns. Understanding these factors and their impact on ecological parameters is crucial for effective urban planning and management to ensure sustainable development and mitigate the adverse effects of urban expansion on the environment and human well-being.

The urban harmony is very essential for sustainable urban development, which emphasizes social security with a demand for accelerated funding of social programs to further enhance the social security system and urban infrastructure (Liang et al. 2016). Labor efficiency refers to equal ecological and economic performance since it puts forward the measurement of the balance between economic growth and its adverse impacts on ecology. Thus, sustainable urban development could be facilitated by incorporating green spaces and blue spaces, as well as promoting urban agriculture and garden farming. To increase sustainability, ecological structures with energy-efficient solutions and good spatial configuration must be pursued. Transportation in urban areas is a major contributor to the degradation of the environment, so e-transport modes are adopted to reduce the negative impact. In addition, promoting various biodegradable products for both home and industrial use will help combat pollution and make urban centers more environmentally friendly and resilient.

AI Disclosure

The authors confirm that none of the text, figures, or data in this manuscript were produced using generative AI tools. Nevertheless, during the editing stage, ChatGPT and other AI-based tools were only utilized for grammar correction and language improvement. The authors used standard geospatial and statistical software to manually make all analytical decisions, process data, interpret results, and create visualizations.

References

Ahluwalia IJ (2016) Challenges of urbanisation in India. Contemp Issues Dev Econ, pp 163–177. https://doi.org/10.1057/9781137529749_10

Chakrabartty A, Gupta S (2014) Traffic congestion in the metropolitan city of Kolkata. J Infrastruct Dev 6(1):43–59. https://doi.org/10.1177/0974930614543046

Gupta D, Bhagat RB (2023) Dynamics of urbanisation and its impact on land cover and land use changes in Kolkata, India (1971 to 2011), pp 1–32

UN habitat (2020) World cities report 2020, the value of sustainable urbanization

IPCC (2014) Weather and climate extreme events in a changing climate coordinating. https://doi.org/10.1017/9781009157896.013.1514

James N (2023) Urbanization and its impact on environmental sustainability. J Appl Geogr Stud 3(1):54–66

Liang X, Zhang W, Chen L, Deng F (2016) Sustainable urban development capacity measure—A case study in Jiangsu Province, China. Sustainability (Switzerland) 8(3). https://doi.org/10.3390/su8030270

Lillesand TM, Kiefer RW, Chipman JW (1979) Remote sensing and image interpretation

Liu H, Zhou G, Wennersten R, Frostell B (2014) Analysis of sustainable urban development approaches in China. Habitat Int 41:24–32. https://doi.org/10.1016/j.habitatint.2013.06.005

Mandal UK, Nayak DB, Samui A, Jana AK, Mullick S, Lama TD (2018) Trend of Sea-level-rise in West Bengal Coast. J Indian Soc Coast Agric Res 36(2):64–73

Mitra C, Shepherd M, Jordan T (2012) Assessment and dynamics of urban growth in the City of Kolkata. In: Dutt AK, Wadhwa V, Thakur B, Costa FJ (eds) Facets of social geography international and Indian perspectives (Issue January), pp 541–555. https://doi.org/10.1017/UPO9788175969360.031

Parnell S (2016) Defining a global urban development agenda. World Dev 78:529–540. https://doi.org/10.1016/j.worlddev.2015.10.028

Sarkar R (2020) Association of urbanisation with demographic dynamics in India. Geo J 85(3):779–803. https://doi.org/10.1007/s10708-019-09988-y

Seto KC, Reenberg A, Boone CG, Fragkias M, Haase D, Langanke T, Marcotullio P, Munroe DK, Olah B, Simon D (2012) Urban land teleconnections and sustainability. Sustain Sci 109(20):7687–7692. https://doi.org/10.1073/pnas.1117622109

Shao Z, Sumari NS, Portnov A, Ujoh F, Mandela PJ (2021) Geo-spatial information science urban sprawl and its impact on sustainable urban development: a combination of remote sensing and social media data. Geo-Spat Inf Sci 24(2):241–255. https://doi.org/10.1080/10095020.2020.1787800

Chapter 10
Legal Implications of Green Building Initiatives in the Context of Climate Change and Sustainable Development Goals

Udai Shankar, Bhupal Bhattacharya, and Ripon Bhattacharjee

Abstract The future of the planet is clouded by climate change, which puts the environmental resources in jeopardy. Surprisingly, buildings are important contributors to this looming issue, consuming energy and emitting greenhouse gases. However, there is a ray of hope by way of implementing construction activities sustainably. This study digs into the legal implications of implementing these potent solutions, paving the road for a greener future in line with the SDGs. As sustainable construction gains traction around the world, it is critical to examine the legal frameworks that govern it. This study dissects both national and international legislation that influence the certification and enforcement of construction works. It dissects the rich fabric of zoning regulations and environmental requirements that shape their legal landscape. However, legal integration is more than just laws and regulations. This pertains to the management of potential hazards and obligations encountered by individuals engaged in construction activities. To ensure the success of these critical endeavors, this research addresses liability concerns and investigates issues concerning dispute resolution processes. It seeks to facilitate smooth integration with current frameworks for sustainable construction by demystifying the legal concerns of green building. This paper aims to make it easier for green building projects to fit into current legal systems involved in sustainable construction by critically analyzing the available literature.

Keywords Sustainable construction · Green building initiatives · Climate change · Legal implications · Sustainable development goals · Environmental regulations

U. Shankar
RGSOIPL, IIT Kharagpur, Kharagpur, West Bengal, India
e-mail: uday@iitkgp.ac.in

B. Bhattacharya
Department of Law, Raiganj University, A State Aided University of Govt. of West Bengal, Raiganj, West Bengal, India

R. Bhattacharjee (✉)
National Law University, Tripura, Agartala, Tripura, India
e-mail: ripon.bhattacharjee69@gmail.com

K. Mukherjee et al. (eds.), *Urban Climate*, The Urban Book Series,
https://doi.org/10.1007/978-3-032-17895-4_10

10.1 Introduction

The ongoing impacts of climate change and global warming are posing a hazard to the environmental resources and the very future of humanity (Mora et al. 2018). Concrete homes are the primary participants in this disaster and largely responsible for international power intake and greenhouse fuel emissions (Shahsavari and Akbari 2018). However, with the adoption of the idea of sustainable improvement, a shining desire emerges in the shape of sustainable construction practices.

The idea of sustainable development, in balancing both environmental and developmental activities, has for the first time received considerable recognition from the year 1987 onwards after the Brundtland commission Report (Brundtland 1987). This commission has emphasized for the developmental activities to be carried out but not at the cost of the environmental resources (Borowy 2021). As the urgency of addressing climate change has become increasingly undeniable, construction companies have begun to progressively shift from conventional practices toward adopting sustainable methods (Nisbet 2014). Green building initiatives, which promote energy performance, renewable energy use, and environmentally responsible materials, offer a promising path toward a more sustainable future.

The legal framework behind green initiatives, which encompass a range of sustainable construction projects and technologies, has a vital role to play in encouraging sustainable development as well as facing challenges. The laws and regulations of green projects, incentives, and certifications across different countries differ because they are mostly based on the various economic, political, and environmental contexts of each country. While green building practices serve as a powerful tool to combat climate change and achieve sustainable development goals, the legal framework often struggles to keep pace with this evolving landscape (Pinkse 2012). Existing regulations may not provide adequate incentives or even allow for the implementation of these crucial practices.

Despite the initial cost, research shows that green building practices offer significant long-term benefits (Zhang et al. 2018). Green homes demonstrably reduce energy consumption and greenhouse gas emissions, thereby mitigating climate change and its environmental consequences (Platjouw 2009). Additionally, these practices directly contribute to achieving the United Nations Sustainable Development Goals, particularly those focused on responsible consumption and production, such as SDG 12, and climate action, SDG 13. Green buildings often provide long-term cost savings through reduced energy and water use, ultimately leading to job creation and economic opportunities.

With nearly 40% of all energy consumption and 30% of carbon emissions (Khan et al. 2014) generated from greenhouse gases from the construction sector, the impacts of climate change on both the environment and the survival of humans are increasing (Li et al. 2019). However, sustainable construction practices do present a path to lessen the impact of these challenges. This study aims to:

1. Find out about existing national and international legal frameworks on the process of green building initiatives.

2. Identify the gaps in the current regulations in ensuring strict enforcement liability and standardization;
3. Decide upon actionable issues associated with the successful implementation of legal frameworks with the Sustainable Development Goals, concentrating on SDG 11 (Sustainable Cities) and SDG 13 (Climate Action).

10.2 Legal Framework in the Clean Integration of Building Initiatives

The legal framework closely depends on various specific contexts, (Reeder 2010) including tax subsidies for green building projects. Developed nations with larger tax bases would possibly desire better stages of green certification. Developing economies may rely more on upfront offers, like grants, to offset the initial cost increase of green materials (Onuoha et al. 2018). During recessions, direct subsidies might be more attractive than tax breaks due to immediate cash flow benefits. Additionally, incentives can be tailored to various kinds of projects. Residential projects might benefit from tax breaks on green construction, while commercial buildings could see advantages from property tax discounts for achieving green standards. Careful analysis is often lacking in determining the optimal mix of tax breaks and subsidies for a particular scenario (Serres et al. 2010). Monitoring and adjustments are crucial to ensure these incentives achieve their intended goals of increasing green building adoption.

10.3 Legal Incentives for Wider Adoption

An incentive or certification is one of the crucial provisions of this legal framework in cases of green projects (Circo 2007). Tax breaks, subsidies and grants, although they provide significant increases in investment in sustainable projects (Taghizadeh-Hesary 2020), can be implemented differently in different regions. To illustrate, Germany has a strong DGNB Gold Green Building Standards, which is connected with significant tax incentives (Raab 2015). The various financial benefits attract the investors by depicting a commitment towards the sustainability process.

However, India usually relies mostly on grants for projects related to low-cost housing (Gopalan 2015) in complying with the conditions for achieving sustainability. These grants aim to decrease the cost of green projects but don't always strive for the same level of rigorous standards as tax incentives do. In many developing countries, nonenforcement of the green policies is considered to be a serious violation of the policies (Waked 2010). Developing nations such as India thus consequently struggle with the absence of standardized enforcement mechanisms in establishing trust and predeterminability of green projects (Koulopoulos 2006).

Established green building certification programs like Leadership in Energy and Environmental Design offer a valuable framework for sustainable construction (Vierra 2016a). However, participation in these programs remains limited (Zhang et al. 2019). The legal incentives can be utilized to strengthen these programs and achieve wider adoption of sustainable building practices by way of linking green building certification levels to the permitting process (Olubunmi et al. 2016). Projects achieving higher certification levels (e.g., LEED Gold or Platinum) require faster permitting timelines or simplified regulatory approvals. This incentivizes surpassing minimum standards and pursuing more ambitious sustainability goals.

Public agencies and municipalities, as major consumers of construction services, can prioritize green building practices in their projects (Darko et al. 2017). This could involve mandating green building certification (e.g., LEED) for a certain percentage of public projects. This would increase demand for certified buildings, encourage private developers to participate in these programs, and establish minimum green building standards for public construction, even if full certification isn't required. This ensures that a baseline level of sustainability is achieved. Furthermore, strengthening existing tax breaks or subsidies by tying them to achieving specific green building certification levels can incentivize the wider adoption of these practices. Projects achieving higher certification levels would receive greater financial benefits, making certification a more attractive and financially rewarding option. Moreover, establishing public recognition programs or awards for developers and designers who design and construct certified green buildings can ignite a spirit of competition within the industry, supplementing accelerative innovation in sustainable design and construction. Media coverage of these awards programs can amplify the recognition companies receive, boosting their reputation for sustainability and attracting environmentally conscious tenants, buyers, and investors.

10.4 Public Procurement and Sustainable Construction

Public procurement guidelines can be a powerful tool to stimulate demand for green building technologies and materials (Burja 2009). The larger a government's budget for a construction project is, the greater the potential impact on the demand for green technologies and materials (Hwang 2012). Countries with significant public infrastructure spending can significantly influence the market. The type of projects undertaken by the government also plays a decisive role. Large-scale public buildings like schools, hospitals, and government offices can showcase the benefits of green technology and materials, encouraging wider adoption within the private sector.

Policies can mandate the usage of innovative construction technologies and substances for all public projects, or they could encourage their use through incentives and preferential treatment (Olubunmi et al. 2016). Mandatory rules have a stronger impact on demand but may additionally face challenges in terms of cost and feasibility. Policies may be specific, outlining a set list of accredited innovative technology and materials, or they may be more flexible, focusing on reaching particular environmental performance goals. Specificity guarantees adherence to standards; however, flexibility permits for innovation and adaptation to new technology. The success of the policies pivots on the availability of green technologies and materials at competitive prices (Hafezalkotob 2017).

10.5 Regulatory Challenges in Implementation

Government projects offer a guaranteed market for green technologies and materials, encouraging manufacturers to invest in research and development and increase production capacity (Dutż and Sharma 2012a). This can lead to economies of scale and lower prices for everyone. Government projects serve as a model for the private sector. The successful implementation of green features in public buildings can demonstrate the feasibility and benefits of these technologies, encouraging private developers to follow suit.

Public procurement policies may face challenges such as higher initial costs for green projects and limited availability of green materials in certain regions. In addition, balancing stringent environmental standards with cost-effectiveness requires careful planning and policy design. Despite these challenges, public procurement policies can be a powerful tool for stimulating demand for innovative construction technologies and materials. Their effectiveness depends on the level of public spending, the design and implementation of the guidelines, and prevailing market conditions.

Effective eco-friendly building regulations require clear goals for energy savings, water conservation, and air pollution reduction (Chandel et al. 2016). These science-based goals should be regularly updated to reflect technological advancements and evolving environmental needs. The regulations, if they allow flexibility in achieving these goals, may foster innovation in green building technologies without dictating specific choices. Standardized green building rating systems are crucial. Everyone involved in the construction process, from builders and designers to permitting authorities, needs a clear understanding of these benchmarks. Independent verification can ensure adherence to these standards. Stricter regulations for larger projects that pose a greater environmental impact are necessary. Incentivising highly innovative eco-friendly projects through faster approvals, tax breaks, and public recognition is crucial. Educational programs are essential to equip all stakeholders with the knowledge to implement these regulations, measure green building performance, and stay informed about new technologies. Regular monitoring and evaluation are essential to adapt the regulations to incorporate new technologies in addressing evolving environmental needs and learn from existing green buildings. Benchmarking green buildings allows for continuous improvement and development of more effective regulations in the future.

10.6 Challenges and Possibilities in Standardizing Building Codes

Implementing universal green building regulations across vast regions or the entire globe presents significant challenges that require solutions for effective and adaptable application. A critical issue is the immense variation in climates and resources across locations. Strategies successful in one region might be ineffective elsewhere. For instance, prioritizing energy performance for heating is crucial in cold climates, while water conservation becomes paramount in arid regions. Therefore, regulations need to be flexible and adaptable to address local needs.

Economic disparity presents another hurdle. Developed nations may possess the resources to enforce stricter green building standards, whereas developing countries might prioritize affordability and basic infrastructure needs (Golubchikov and Badyina 2012b). Overly stringent requirements for developing countries could hinder construction activity and restrict access to affordable housing. Striking a balance between environmental protection and affordability is essential. Construction traditions and readily available materials also differ significantly between regions. A standardized code might not account for these variations, potentially hindering innovation in local building practices or creating challenges for compliance due to limited access to specific materials. Enforcement capacity is another concern. Standardization necessitates robust enforcement mechanisms. Developing countries might lack the resources or infrastructure to enforce standardized codes across large areas. Standardized codes could limit the ability of local governments to tailor regulations to their specific environmental and economic contexts. Local expertise and knowledge could be disregarded in favor of a universal standard.

10.7 Opportunities for Standardizing Green Building Codes

Implementing standardized green building regulations globally offers significant advantages despite the challenges involved. Firstly, a level playing field is established for green buildings everywhere. This fosters increased investment in eco-friendly construction and collaboration on innovative technologies. This not only sparks new ideas but also ensures that the most effective practices are adopted worldwide.

Secondly, consistent regulations streamline the planning and permitting process for green buildings, particularly for multinational corporations. This reduces costs and expedites project timelines, making green construction more achievable. Furthermore, standardized rules enable the benchmarking of the performance of green buildings across different regions. This facilitates knowledge sharing and the development of even more sustainable building practices. Additionally, as public awareness of the environmental and economic benefits of green buildings grows, their demand increases. This incentivizes builders to prioritize environmentally responsible construction practices. Ultimately, standardized green building rules contribute to a sustainable future for the planet. This includes combating climate change, conserving resources, and promoting responsible production and consumption practices.

10.8 Finding the Right Balance

A foundational plan outlining key sustainability goals is crucial. This plan should establish common objectives while allowing regions to adapt based on factors like climate, access to resources, and economic realities. Implementing regulations with different levels of stringency based on building type, size, and location is a useful approach. This ensures that environmental benefits are achieved while catering to the specific needs and circumstances of different localities. Focusing on achieving specific green outcomes rather than mandating specific technologies or materials is key. This approach encourages regions to creatively utilize available resources and fosters innovation in developing local solutions for sustainability. Addressing the said challenges and opportunities of combining standardized and adaptable rules, green building regulations can become a powerful tool for driving sustainability in the construction industry. This balanced approach fosters clear guidelines while simultaneously encouraging new ideas and adaptations, ultimately propelling the building industry toward an environmentally responsible and resilient future.

10.9 Liability and Dispute Resolution

Uncertainties regarding liability for novel green building technologies pose significant barriers to industry innovation. The uncertainty about the liability in the legal framework of green projects poses another major problem (Nambiar 2010). When it comes to adopting new technologies on an experimental basis, it poses a problem for the proper determination of accountability (Lerner 1999). The problem is that if a green technology doesn't work because of a flaw in design, there consequently arise various environmental concerns resulting in attracting enormous liabilities (Shedroff 2009). There may be disputes and delays in resolving some differences among developers, contractors, and certifiers as to which roles and responsibilities each has in ensuring the success of this technology (Ferme and Zuo 2018).

Green building codes and regulations are not uniform, and this complicates the legal ground for sustainable development (Vierra 2016b). The European Union's Energy Performance of Buildings Directive has helped much in standardizing energy efficiency across member states; however, some regions, especially those in the tropics, are faced with dissimilar issues (Liu et al. 2010). For instance, in

Indonesia, the building codes need to be followed considering unique local conditions, prioritizing cooling methods adopted, which are more typically dealt with in temperate climates (Stepani 2022). Many Western-centric standards are taken as a common formula, mostly ignoring the ground reality (Obeng-Odoom 2016). Tropical regions with high humidity require commonly practiced ventilation systems and other innovative architectural alternatives (Mirrahimi et al. 2016).

To overcome these hurdles and incentivize the adoption of innovative solutions, establishing clear legal frameworks is crucial. Several strategies can promote a legal environment conducive to green construction innovation, such as:

i. **Standardization and Transparency**: Defining clear performance standards and implementing transparent certification processes for green building technologies will establish benchmarks for assessing liability and build confidence among stakeholders, including investors, developers, and contractors.
ii. **Shared Liability and Warranties**: Adopting a shared liability model can allocate responsibility appropriately, while comprehensive warranty provisions for new technologies ensure accountability and recourse if problems arise. Additionally, considering statutory limits on liability during the initial adoption phase can incentivize innovation by protecting early adopters.
iii. **Dispute Resolution Mechanisms**: Promoting alternative dispute resolution mechanisms and specialized green-building tribunals can expedite the resolution of legal disputes and ensure fair outcomes for all parties involved.
iv. **Knowledge Sharing and Education**: Facilitating knowledge sharing and education through initiatives like information dissemination programs can enhance understanding and compliance among legal professionals and construction industry stakeholders.

10.10 Conclusion

This paper explores the significant challenges posed by unclear legal frameworks surrounding novel green building technologies. These uncertainties can impede progress and discourage the adoption of environmentally sustainable construction methods. However, a closer examination of legal systems and potential solutions reveals several key steps to address these issues and promote innovation in green building.

10.10.1 Strategies for Advancing Sustainability

i. **Standardization and Transparency**: Establishing clear performance standards and implementing transparent certification processes for green building technologies are crucial. This approach fosters a framework for assessing liability while ensuring adherence to safety and environmental regulations.

Increased clarity reduces uncertainties for all stakeholders, including investors, developers, and contractors.

ii. **Shared Liability and Warranties**: A shared liability model can effectively allocate responsibility, while comprehensive warranty provisions for new technologies provide accountability and recourse in case of unforeseen issues. Furthermore, considering statutory limits on liability during the initial adoption phase incentivizes innovation by mitigating risk for early adopters.

iii. **Efficient Dispute Resolution**: Promoting alternative dispute resolution mechanisms and specialized green building tribunals can expedite the resolution of legal disputes and ensure fair outcomes for all parties involved.

iv. **Knowledge Sharing and Education**: Facilitating knowledge sharing and education initiatives is essential. By disseminating information and best practices, legal professionals and construction industry stakeholders can gain a deeper understanding of the legal landscape, fostering better compliance and encouraging the adoption of sustainable building practices.

v. **Pre-approval Programme**: Establishing a system for pre-approving commonly used innovative building technologies and materials. This involves rigorous testing and assessment to ensure they meet environmental and safety standards. Projects utilizing pre-approved materials can benefit from faster permitting as their safety and effectiveness are already established.

vi. **Performance-Based Standards**: Moving away from prescriptive rules that dictate specific materials or technologies to sustainable technologies may yield a better result. Instead, implementing performance-based guidelines that focus on achieving targeted environmental goals is the need of the hour. This allows for flexibility in innovative building design and fosters the use of new technologies, as long as projects demonstrably meet established environmental and safety performance goals.

vii. **Specialized Permitting Teams**: Creating dedicated teams within permitting agencies that focus solely on green building projects. These teams can develop expertise in innovative building practices and technologies, leading to faster and informed evaluation methods compared to traditional permitting teams.

viii. **Streamlined Online Systems**: Developed and utilised online systems for filing permit applications, tracking progress, and receiving approvals. This enhances transparency, improves communication efficiency, and allows for faster feedback between applicants and the permitting agency.

ix. **Risk-Based Review**: Implementing a risk-based review system that prioritizes projects with potentially higher environmental or safety risks for more rigorous scrutiny. Lower-risk green building initiatives, especially those using pre-approved technologies, can benefit from a faster and more streamlined permitting process.

x. **Post-construction Verification**: Implement a system for post-construction verification to ensure that innovative building features are installed and operating as designed. This verification process can be balanced with streamlined permitting by focusing on a risk-based approach, concentrating on high-risk projects for more intensive verification processes.

xi. **Standardized Codes**: Developing nationwide or local standardized green building codes that are clear, consistent, and easily accessible to all stakeholders involved in construction projects. This reduces confusion and simplifies the permitting process for projects adhering to these standardized codes.
xii. **Public–Private Partnerships**: Encouraging public–private partnerships to develop and implement green permitting strategies. This can leverage the knowledge of the private sector in streamlining approaches while ensuring adherence to public safety and environmental policies.

If legal uncertainties are addressed and innovative solutions are properly and effectively explored, the legal system can play a pivotal role in advancing efforts to combat climate change and achieve sustainable development goals in the construction sector. Through collaborative efforts and a commitment to overcoming existing challenges, a better future characterized by more sustainable and innovative green building practices can be ensured.

10.11 Scope and Implications for Future Study

With the increasing significance of sustainable construction in the battle against global warming and parallel with promoters of Sustainable Development Goals, future research in this area is bound to be fruitful. In the future, it is possible to explore better-evolving approaches to legal frameworks in different countries, especially emerging economies, to see how they manage the specific difficulties related to the specification of the construction sector in the context of sustainability. Moreover, it is recommended for future research to evaluate the effectiveness of existing dispute resolution processes and liability models for sustainable construction. A further area for exploring the legal integration of sustainable construction into existing frameworks is yet to be accomplished. With the development of sustainable construction worldwide, there is an emerging need to assess the impact of such legal frameworks on the economy to determine whether they are indeed effective in enhancing the incentives or the absence thereof for investment in green building projects.

References

Borowy I (2021) The social dimension of sustainable development at the UN: from Brundtland to the SDGs. In: The struggle for social sustainability, pp 89–108. Policy Press
Brundtland GH (1987) What is sustainable development. Our Common Future 8(9)
Burja A (2009) Using green public procurement (GPP) for sustainable consumption and production. J Eur Environ Plan Law 6(3):319–338
Chandel SS, Sharma A, Marwaha BM (2016) Review of energy efficiency initiatives and regulations for residential buildings in India. Renew Sustain Energy Rev 54(2016):1443–1458

Circo CJ (2007) Using mandates and incentives to promote sustainable construction and green building projects in the private sector: a call for more state land use policy initiatives. Penn St l Rev 112(2007):731

Darko A, Chan APC, Ameyaw EE, He B-J, Olanipekun AO (2017) Examining issues influencing green building technologies adoption: The United States green building experts' perspectives. Energy Build 144(2017):320–332

Dutz MA, Sharma S (2012) Green growth, technology and innovation. World Bank Policy Res Work Pap 5932

Ferme L, Zuo J, Rameezdeen R (2018) Improving collaboration among stakeholders in green building projects: role of early contractor involvement. J Leg Aff Disput Resolut Eng Constr 10(4):04518020

Golubchikov O, Badyina A (2012) Sustainable housing for sustainable cities: a policy framework for developing countries. Nairobi, Kenya: UN-HABITAT

Gopalan K, Venkataraman M (2015) Affordable housing: policy and practice in India. IIMB Manag Rev 27(2):129–140

Hafezalkotob A (2017) Competition, cooperation, and coopetition of green supply chains under regulations on energy saving levels. Transp Res Part E: Logist Transp Rev 97(2017):228–250

Hwang B-G, Tan JS (2012) Green building project management: obstacles and solutions for sustainable development. Sustain Dev 20(5):335–349

Khan MA, Khan MZ, Zaman K, Naz L (2014) Global estimates of energy consumption and greenhouse gas emissions. Renew Sustain Energy Rev 29(2014):336–344

Koulopoulos TM, Roloff T (2006) Smart sourcing: driving innovation and growth through outsourcing. Simon and Schuster

Lerner JS, Tetlock PE (1999) Accounting for the effects of accountability. Psychol Bull 125(2):255

Li YL, Han MY, Liu SY, Chen GQ (2019) Energy consumption and greenhouse gas emissions by buildings: a multi-scale perspective. Build Environ 151(2019):240–250

Liu F, Meyer AS, Hogan JF (2010) Mainstreaming building energy efficiency codes in developing countries: global experiences and lessons from early adopters, vol 204. World Bank Publications

Mirrahimi S, Mohamed MF, Haw LC, NlN I, Yusoff WFM, Aflaki A (2016) The effect of building envelope on the thermal comfort and energy saving for high-rise buildings in hot–humid climate. Renew Sustain Energy Rev 53(2016):1508–1519

Mora C, Spirandelli D, Franklin EC, Lynham J, Kantar MB, Miles W, Smith CZ et al (2018) Broad threat to humanity from cumulative climate hazards intensified by greenhouse gas emissions. Nat Clim Chang 8(12):1062–1071

Nambiar S (2010) Paradigm of 'Green' adjudication: developing principles for indian environmental decision making in disputes involving scientific uncertainity. ILI Law Review 1(1):1

Nisbet MC (2014) Disruptive ideas: public intellectuals and their arguments for action on climate change. Wiley Interdiscip Rev: Clim Chang 5(6):809–823

Obeng-Odoom F (2016) Reconstructing urban economics: towards a political economy of the built environment. Bloomsbury Publishing

Olubunmi OA, Xia PB, Skitmore M (2016) Green building incentives: a review. Renew Sustain Energy Rev 59(2016):1611–1621

Onuoha IJ, Aliagha GU, Rahman MSA (2018) Modelling the effects of green building incentives and green building skills on supply factors affecting green commercial property investment. Renew Sustain Energy Rev 90(2018):814–823

Pinkse J, Kolk A (2012) Addressing the climate change—sustainable development nexus: The role of multi stakeholder partnerships. Bus Society 51(1):176–210

Platjouw FM (2009) Reducing greenhouse gas emissions at home or abroad? The implications of Kyoto's supplementarity requirement for the present and future climate change regime. Rev Eur Community Int Environ Law 18(3):244–256

Raab S (2015) Key diversity in existing green buildings standards: GBEL, DGNB, LEED and ÖGNB-focus on new construction and commercial buildings. PhD diss, Wien, p 2015

Reeder L (2010) Guide to green building rating systems: understanding LEED, Green Globes, Energy Star, the National Green Building Standard, and more, vol 12. John Wiley & Sons
De Serres A, Murtin F, Nicoletti G (2010) A framework for assessing green growth policies
Shahsavari A, Akbari M (2018) Potential of solar energy in developing countries for reducing energy-related emissions. Renew Sustain Energy Rev 90(2018):275–291
Shedroff N (2009) Design is the problem: the future of design must be sustainable. Rosenfeld Media
Stepani HMN, Emmanuel R (2022) How much green is really "cool"? Target setting for thermal comfort enhancement in a warm, humid city (Jakarta, Indonesia). Atmosphere 13(2):184
Taghizadeh-Hesary F, Yoshino N (2020) Sustainable solutions for green financing and investment in renewable energy projects. Energies 13(4):788
Vierra S (2016) Green building standards and certification systems. National Institute of Building Sciences, Washington, DC
Vierra S (2016) Green building standards and certification systems. National Institute of Building Sciences, Washington, DC 62
Waked DI (2010) Antitrust enforcement in developing countries: reasons for enforcement & non-enforcement using resource-based evidence. In: 5th annual conference on empirical legal studies paper
Zhang L, Wu J, Liu H (2018) Turning green into gold: A review on the economics of green buildings. J Clean Prod 172(2018):2234–2245
Zhang Y, Wang H, Gao W, Wang F, Zhou N, Kammen DM, Ying X (2019) A survey of the status and challenges of green building development in various countries. Sustainability 11(19):5385

Part III
Urban Mental Well-Being

Chapter 11
Applications of Clinical Ecopsychology and Ecotherapy in Psychological Intervention and Well-Being: Scopes and Challenges

Sumona Datta, Tilottama Mukherjee, and Asmita Ganguly

Abstract Ecopsychology seeks to investigate the interaction and interdependence of human behavior with the natural environment as well as its implications on well-being and health. In the past couple of years, the conglomeration of Ecopsychology and Clinical Psychology has led to the emergence of subfields like Clinical Ecopsychology and Ecopsychiatry, both focusing on the applications of the principles of ecopsychology in health promotion and well-being. The concept of Ecotherapy is based on the understanding of ecopsychology, where the physical and mental health of an individual are studied in terms of the environment and natural ecosystem. The current chapter explores the implications of clinical ecopsychology and ecotherapy in providing psychological intervention and promoting well-being. The chapter provides an outline of the fundamental principles of clinical ecopsychology and ecotherapy as subfields of environmental psychology. The methodology follows a narrative review framework, exploring contemporary research on the implications of the basic tenets of ecopsychology. The discussion focuses on two chief aspects: (a) Therapeutic interventions based on ecopsychological principles, and (b) Implications in promoting health and well-being. Applications of ecotherapy are investigated in general health conditions and psychopathological conditions like stress-related disorders, attention-deficit/hyperactivity disorder, mood-related disorders, addictions, pain disorders, and other mental health conditions. Furthermore, the review explores the role of ecopsychological principles in nurturing psychological development and creative abilities among children as well as promoting well-being and satisfaction with life among adult individuals. The chapter concludes with

S. Datta (✉)
Department of Psychology, Government General Degree College, Singur, Hooghly, India
e-mail: sumonadatta.psych@gmail.com

T. Mukherjee
Department of Psychology, University of Calcutta, Kolkata, India

A. Ganguly
Clinical Psychology Centre, University of Calcutta, Kolkata, India

K. Mukherjee et al. (eds.), *Urban Climate*, The Urban Book Series,
https://doi.org/10.1007/978-3-032-17895-4_11

a note on the scopes and challenges of implementing the fundamental beliefs of ecopsychology in psychological intervention and endorsing well-being.

Keywords Ecopsychology · Ecotherapy · Ecopsychiatry · Psychopathology · Well-being

11.1 Introduction

We belong to a place, no matter where we go. At any point in time, we are always nested in some building, some locality, some city, some country, or some part of the globe, even when we are traveling, or in any natural environment. The interaction between "person" and "place" is, therefore, unending and interdependent. We shape our environment, just as our environment shapes us.

The concept of "Ecopsychology" connects Ecology with the understanding of psychology. Ecology (Haeckel 1866) is considered the scientific study of the association and interaction among individuals and their surrounding environment. It combines the understanding of ecology with the principles of psychotherapy, to nurture a healthy lifestyle among individuals, both ecologically and psychologically (Chowdhury And Jadhav 2012). Ecopsychology seeks to connect human nature with the natural environment and explore how the human-nature connection nurtures both the individual and his environment. Research on the interaction between nature and human behavior had begun early in the 1940s, with the works done by Hawley in his milestone work entitled "Ecology and the Human Ecology." In 2019, Park and Burgess defined human ecology as the "study of the spatial and temporal organization and relations of human beings with respect to the selective, distributive, and accommodative forces of the environment" (Park And Burgess 2019; Lawrence 2003, p.31). Another perspective comes from Lawrence (2001) who referred to human ecology as the study of the everchanging interrelationship between human beings and the biological, physical, and sociocultural features of their living environment. Here it is important to distinguish between personal mental space and the ethnocentric urban space. While the personal mental space refers to one's internal state of mind, comprising of the thoughts and feelings, the later refers to the surrounding environment of the individual, characterized by the physical, biological, and psychosocial factors. The study of human ecology, therefore, abridges the personal mental space with the ethnocentric urban space surrounding the individual.

Later works done by Wilkinson and O'Connor (1977, 1982) focused on how the concepts of clinical psychology can be combined with the understanding of ecopsychology in providing mental healthcare services to the community. Mental health issues were no longer considered to be the outcome of social processes, but also the product of physical and biological processes of the environment involved. A groundbreaking development in this context occurred in 1977 when the APA task force formed an "Eco-psychiatric Data Base" and introduced the term "Ecopsychiatry" to describe the human-nature interaction (Chowdhury And Jadhav 2012).

Steven Moffic (2008) opined that Ecopsychiatry must be considered a distinct discipline in clinical practice in the twentieth century (Chowdhury And Jadhav 2012), often defined as "Ecopsychiatry is a general approach in which the diverse sources of psychiatric knowledge are grouped to get a view of macro-reality emerging from a methodological integration of behavior sciences and a unified theory orientating to a realistic praxis in the normative" (Mariategui 1978). Although the concept of Ecopsychiatry was initiated with a focus on the physical environment of the individual, gradually it evolved and encompassed issues pertaining to ecology, evolution, economics, and the complex interaction of these with the mental health of the population. This led to the generalization of the concept of Ecopsychiatry to a much broader aspect and resulted in several interdisciplinary areas of research such as Ethnobiology (Berlin 1992), Ecological Anthropology, Applied Anthropology (Cohen 1997), Cultural Ecology (Steward 1972), and ecotherapy, denoting the interaction of ecopsychology and psychotherapy (Chowdhury And Jadhav 2012).

The current study aims to explore the implications of clinical ecopsychology and ecotherapy in providing psychological intervention and promoting well-being. This narrative review focuses on the therapeutic interventions based on ecopsychological principles, and their implications in promoting health and well-being.

11.2 Clinical Ecopsychology: Conceptual Framework

The impact of the environment on our mental health is enormous. There has been a growing body of research evidence suggesting the ill-impacts of toxic pollution as the leading cause of some health problems in India, specifically non-communicable diseases (NCDs) (Shetty 2024). Reports by Health Policy Watch suggested that 80% of the most polluted cities in the world are in India. Increased air pollution has been associated with worsened states of anemia (Chaudhary et al. 2022), high levels of cholesterol (Anand et al. 2024), hypertension (Prabhakaran et al. 2020), high diabetes (Mandal et al. 2023), and worsened mental health (Balakrishnan And Tsaneva 2023).

The discipline of Clinical Psychology deals with the diagnosis, assessment, and treatment of mental health problems and psychopathology. It encompasses studying the causal factors of a disorder as well as determining the diagnosis of a psychopathology. Another significant aspect includes developing intervention programs that can ensure a good prognosis for a mental health issue. Bodner (2024) emphasizes how environmental deterioration can exacerbate psychological problems like anxiety and depression and recognizes the connection between human mental health and environmental well-being. Clinical ecopsychology, which reflects a comprehensive understanding of the interaction between humans and nature, seeks to advance both individual mental health and environmental sustainability by integrating ecological awareness into therapy procedures. Thoma et al. (2021) identified the Vulnerability Stress Model as an explanatory model for explaining the impact of environmental issues on mental health. The model states that psychopathology in individuals is a product of the interaction between stress and an individual's susceptibility to it. Such vulnerability can arise from the complex interplay of biological, physical, and social factors. While biological issues can pertain to genetic, hormonal, disease, and immunological influences, social issues relate to social distancing and interpersonal discords. The personal factors encompass matters pertaining to behavioral, cognitive, and emotional dysfunctions. Thoma et al. (2021) pointed out that climatic and environmental factors impart potential ill effects on mental health through the triad of the biological, physical, and social factors affecting us at any point in time. There are numerous instances bearing evidence of this fact. Krefis et al. (2018) and Ma et al. (2019) found that a reduction in green space due to urbanization has ill effects on the mental health of individuals living in cities. Bojic (2018) noted that such increased stress produces more negative impacts on individuals since they are deprived of chances to apply nature-based strategies (Rezaei et al. 2021). Schell et al. (2012) described the situation in Seoul, South Korea, where due to rapid urbanization, and unplanned growth strategies in the late 1960s, there was a sudden fast transformation in the overall social and environmental milieu, leading to increased mental health problems. Rii and Ahn (2002) and Rahman and Ahmad (2012) reported the subsequent strategies undertaken by the South Korean government to mitigate the situation and enhance mental health problems. It can be noted that simple solutions such as providing greater access to green spaces (Rahman And Ahmad 2012) and green roof gardens (Currie And Bass 2008; Nowak et al. 1998) can enhance the psychological well-being of individuals dwelling in urban cities.

Research on the effects of nature on human health has expanded since Kaplan published some groundbreaking findings in the 1980s, demonstrating how spending time outdoors can help people regain their attention (Kaplan 1984; Kaplan 1995a). Even though people spend little time in nature these days, there is evidence that encountering it is good for one's health and well-being (Maller 2002). According to Douglas et al. (2017) and Nielsen (2011), post-modern urbanization (i.e., in urban studies, it typically refers to urban development patterns emerging from the late twentieth century onwards, characterized by decentralization, fragmentation, and increased complexity in urban landscapes) (Woodward And Jones 2008; Soja 1989) restricts access to outdoor recreation. Previous analyses examined the effects of nature interventions, particularly in the context of mental health and individuals with psychiatric diagnoses (Annerstedt et al. 2011; Bragg And Atkins 2016; Clatworthy et al. 2013). The prevalence of mental health disorders is rising, and estimates indicate that 15% of the world's healthcare costs in 2020 will be attributable to mental health disorders (Maller 2002).

11.2.1 Ecotherapeutic Interventions in Climate-Related Psychopathology

In the current decade, several novel terms have become existent in the literature of Ecopsychiatry and Ecopsychology. Budziszewska and Kalwak (2022) enlist some of these newer concepts, such as "Climate anxiety" and "Eco-anxiety," "Existential anxiety," "Climate fear," "Solastalgia," "Environmental Melancholia," "Climate Stress," and "Climate Depression." "Eco-anxiety" or "Ecological anxiety" refers to a state of obsession and irrational anxiety about diverse forms of environmental pollution. While individuals suffering from ecological anxiety nurture an irrational anxiety, those suffering from climate change anxiety often demonstrate a rational form of anxiety related to the consequences of climate change on society and lives (Budziszewska And Kalwak 2022). "Existential anxiety" refers to the anxiety associated with the existential crisis faced by human society as a whole in response to global climate change consequences. When we experience such emotion concerning the whole community or a particular group that we identify with, it becomes a collective emotion (Wohl et al. 2012). The term "Climate Fear" emerged out of the debate regarding the use of terms like fear, anxiety, and panic related to climate change in popular media (Budziszewska And Kalwak 2022). The term "Solastalgia" was introduced by the philosopher Glenn Albrecht (Albrecht 2005) to designate a painful and nostalgic feeling for losing a "well-known natural environment, such as the landscape, a meaningful place, or a sequence of seasons" (p.177, Budziszewska And Kalwak 2022). Another term "Environmental Melancholia" has been described (Lertzman 2015) as a deep sense of social indifference resulting from the gap between the real concern for environmental degradation and the absence of effort to demonstrate this concern in the form of actions. Such feeling of loss leads to apathy, less involvement, and pessimistic thinking, often resulting in melancholia. The term "Climate Stress" encompasses the psychological responses to situations of anticipated climate change (Clayton, Manning And Hodge 2014). Climate stress, when elevated to the level of excessive anxiety, is considered a "pre-PTSD" state. Finally, the concept of "Climate Depression" refers to the deep suffering and hopelessness, along with a feeling of helplessness in a state of climate crisis. Symptoms of climate depression often involve anxiety related to the climate crisis, hopelessness and helplessness, loss of meaning in life, fatalism, social withdrawal and loneliness, absence of power and passivity, along with aggression, rage, self-aggressive fantasies, and insomnia associated with persistent nightmares (Budziszewska And Kalwak 2022).

Hence, the concept of Ecopsychology integrates the psychological construal of the environmental impacts on our mental health and its clinical consequences on behavior. The newer forms of psychopathology present a complex scenario where the nature of psychopathology has transformed drastically. Hence, the implications of ecotherapeutic approaches in mitigating the risks of climate-related psychopathologies are significant.

11.3 Ecotherapy: Applications of Ecopsychological Principles

Ecotherapy has been traditionally considered as "applied ecopsychology" (Buzzell and Chalquist 2009), defined as "an umbrella term for nature-based methods of physical and psychological healing" (p. 18). The authors posit that the concept of ecotherapy is fundamentally based on the principles of ecopsychology, which advocates the disintegration of human beings from their environment and forms the basis of our pathologies. From the ecotherapeutic perspective, the physical and psychological well-being of individuals should be considered to pertain to environmental health and the ecosystem (Clinebell 1996; Swimme And Berry 1994). Ecotherapy seeks to reconnect individuals to their natural environment, and thereby facilitate coping with diseases and dysfunctions (Buzzell and Chalquist 2009). Applications of ecotherapeutic interventions can be found in approaches toward green views (Ulrich 1984; de Vries et al. 2003), horticultural therapy (Linden And Grut 2002), animal-assisted therapy (DeMayo 2009), art therapy (Degges-White And Davis 2010), green exercises (Pretty et al. 2005, 2007), wilderness therapy (Russell 2001), body therapy through movement (Clinebell 1996) and many others. More often, ecotherapeutic interventions rely on mindfulness practices (Ambrose-Oji 2013; Jordan And Hinds 2016) while implementing different forms of therapies.

Creating innovative methods to benefit from nature therapy is important not just for the psychiatric community but also could have significant clinical ramifications for general practice. Ulrich and colleagues were among the first to report that viewing nature through a window improved patient rehabilitation in a surgical unit. The patients experienced fewer complications following surgery, shorter lengths of stay in the hospital following surgery, and a decreased demand for moderate and powerful analgesics (Ulrich 1984). Green care encompasses a wide range of services, such as social rehabilitation and health promotion (Annerstedt et al. 2011; Bragg And Atkins 2016; Clatworthy et al. 2013). These interventions are also referred to as green prescription, ecopsychosocial intervention, outdoor adventure intervention, adventure therapy, and ecotherapy (Buckley And Brough 2017).

11.3.1 Ecotherapy: Implications on Stress Recovery

Numerous studies have demonstrated a protective relationship between an individual's perceived (de Vries et al. 2003) and actual (Maas et al. 2009; Mitchell And Popham 2008; Takano et al. 2002) health and the percentage of green space (GS) in their home. Stress can impact health by releasing stress hormones, which can lead to physiological adaptations (Cohen et al. 2007), or by increasing the tendency for behavioral risk factors for disease (Glantz And Schwartz 2008). Long-term activation of the stress response raises the possibility of permanent consequences, including the emergence of chronic illnesses and a weakened immune system (Cohen et al.

2007; Falagas et al. 2007; Ramirez et al. 1989). Elevated stress levels have been consistently associated with depression incidence, cardiovascular disease incidence and death, and HIV/AIDS progression (Cohen et al. 2007).

Stress recovery was more effective and rapid in the group that experienced natural settings, as determined by several physiological markers and a state-affect questionnaire (Ulrich et al. 1991a, b). Analogously, students seated in a room overlooking trees and subjected to a stressor experienced a quicker drop in diastolic blood pressure compared to their counterparts in a room without windows (Hartig et al. 2003). Furthermore, pupils exposed to forested surroundings showed a greater decrease in numerous physiological indices of stress compared to those exposed to urban environments devoid of vegetation (Lee et al. 2011). Greater percentages of green space in neighborhoods were associated with lower levels of chronic stress as determined by salivary cortisol levels and lower levels of self-reported stress based on measurements by the Perceived Stress Scale (Ward Thompson et al. 2012). Greener areas also showed fewer socioeconomic differences in mortality rates from all causes and from circulatory disorders, for which stress is a known factor (Mitchell And Popham 2008).

11.4 Dose–Response Analysis of Nature for Psychological Well-Being

According to emerging literature, exposure to nature at different doses has been linked to several health outcomes (Barton and Pretty 2010; Cox et al. 2017a, b; Shanahan et al. 2016). The term "dose of nature" refers to the amount of vegetation present and the frequency and length of exposure (Shanahan et al. 2015a, b; Sullivan et al. 2014). According to a recent experimental study, stress and the length of time spent in green spaces can have an impact on the dose of nature (Jiang et al. 2014). A dose–response analysis of nature revealed a correlation between decreased levels of depression and the weekly nature dose's minimal thresholds. High levels of nature are associated with better mental (White et al. 2013), physical (Mitchell And Popham 2008), and social health (Kingsley And Townsend 2006) as well as increased physical activity (Sugiyama et al. 2010) and a stronger sense of connection to nature (Lin et al. 2017; Shanahan et al. 2017). These findings are supported by population-level studies conducted in industrialized countries. People have the option to encounter nature immediately and conveniently through this "nearby nature" (Lachowycz And Jones 2013). A mix of public and private green spaces provides such nature. Additionally, gardens offer spaces where individuals may engage with other multisensory elements of nature, such as sunlight and fresh air, which can have positive health effects. The presence of vegetation in the surrounding area has been linked to improved mental health (Ulrich 1983; Kaplan 1995a; Berman et al. 2008), and brief exposure to natural settings can immediately lower blood pressure (Hartig et al. 2003) and increase sensations of mental renewal (Barton and Pretty 2010). Three aspects of natural exposure are distinguished by the nature dose–response paradigm (Shanahan et al. 2017, Jiang et al. 2016, Shanahan et al. 2015a, b, Sullivan et al. 2014): intensity (quality and quantity), frequency, and duration (Shanahan et al. 2017). A study conducted by Cox et al. (2016) revealed that there was a positive correlation between depression, perceptions of social cohesion, levels of physical activity, and nature orientation. Specifically, the study found that improvements in these domains were observed with increasing frequency (4–5 times per week) and duration of exposure (at least 30 min) to nearby natural areas. Additionally, it was discovered that higher degrees of nature relatedness and lower levels of mild or worse depression were linked to increased intensity of nature exposure. Those who spent comparatively less time outside during the survey week were more likely to exhibit depressive symptoms as well as inferior physical behavior, whereas those who spent comparatively more time outside demonstrated a stronger connection to nature. Living in greener neighborhoods was found to be substantially associated with higher levels of nature orientation and lower levels of depression (Cox et al. 2017a, b).

Hazer (2018) conducted a cross-sectional study and discovered that longer periods spent in green space, particularly when accessing it, are a reliable indicator of a statistically significant decrease in perceived stress in a non-hospitalized population in Baltimore going about their daily lives unhindered by experimental interference. The study also demonstrates that, after controlling for other variables that are known to predict stress, a statistically significant increase in the length of time spent physically and/or visually exposed to green space is predictive of incremental reductions in perceived stress. A similar study by Mennis et al. (2018) studied 9346 adolescents' level of psychological stress associated with exposure to green space. They found that with increased exposure to green space, psychological stress of the adolescents was lowered. They speculated that this association between green space with reduced stress is most likely explained by increased restoration of attention, and reduction of mental weariness with heightened exposure to vegetation and natural space (Mennis et al. 2018; Kaplan 1995a; Ulrich et al. 1991a, b). Another meta-analysis by Zhang et al. (2024) investigates the relationship between exposure to green spaces and the risk of depression and anxiety disorders. It was found in the study that a 10% increase in the proportion of green space was associated with a statistically significant reduction in the risk of depression as well as a reduced risk of anxiety. The meta-analysis supports the idea that higher exposure to green spaces is beneficial for mental health, especially in reducing the odds of depression. Although the evidence for anxiety is promising, it appears to be less definitive. The authors suggest that enhancing or preserving green spaces should be considered a promising public health intervention. These conclusions align with other research indicating that green spaces can decrease the risk of psychiatric disorders, including depression, anxiety, dementia, schizophrenia, and ADHD. In summary, the meta-analysis underscores the potential mental health benefits of increased green space exposure, advocating for urban planning strategies that incorporate more green areas to enhance public well-being.

These findings support the idea that green space exposure is part of a complex set of factors that predict how people respond to stress (Hartig et al. 2003; Lee et al. 2011; Mitchell And Popham 2008; Ulrich et al. 1991a, b; Ward Thompson et al. 2012).

The findings also offer additional evidence that the reported association between green space and health may be influenced by the amount of nature (de Vries et al. 2003; Maas et al. 2009, 2006; Takano et al. 2002). The study also emphasizes the possibility that an additional hour of weekly visual exposure could be the beneficial dose.

The population's health may be significantly impacted by behavioral interventions that promote exposure to surrounding nature and even the minimal amounts of neighborhood greening that are advised. Frequency of nature exposure was found to be a stronger predictor than duration of exposure across four self-reported health outcomes. Sport scientists have acknowledged that brief, repeated exposures are an efficient means to promote desired health effects (Gibala et al. 2006). Because nature orientation has been linked to increased life happiness (Capaldi et al. 2014, Zelenski And Nisbet 2014), decreased anxiety (Martyn And Brymer 2016), and environmental behavior (Restall And Conrad 2015), maintaining nature around the home may therefore be essential for both health and biological conservation.

11.5 Blue Space: Impact on Psychological Well-Being

Blue space refers to outdoor spaces, whether natural or man-made, that are primarily made up of water. They serve as valuable assets in fostering sustainable urban environments that support human health and well-being (Smith et al. 2022). In the past decade, there have been several researches on blue space being conducted in different urban hubs in India, such as Chandannagar in West Bengal (Ghosh And Singh 2022), English Bazar (Pal et al. 2023), Ahmedabad in Gujarat (Bruckner et al. 2022), and Kolkata in West Bengal (Gupta And De 2025). A comprehensive account of the global blue space hubs has been reported by Observer Research Foundation (2021) in major cities of Singapore, Canada, UK, Netherlands, United States, and China, which were all built with the intention of providing a blue space to their citizens and an enhanced psychological well-being consequently. Integrating blue spaces into urban planning can significantly reduce stress and depressive symptoms, contribute to better physical health, and promote social interactions (Wang And Qin 2011). Volker and Kistemann (2011) summarized the term "blue space" as all visible surface waters in space as an analogy to green space, not as a sub-category. Recent studies provide insight into the potential benefits of blue space, identifying that living near blue space is linked to significantly higher levels of physical activity (Georgiou et al. 2021), residential exposure to blue space was associated with well-being, and recreational visits to blue space (including inland and coastal waters) were associated with well-being, distress, anxiety, and depression (White et al. 2021). Living near the ocean was associated with greater mental health, according to one study (Pasanen et al. 2019). This was partially due to the residents' increased outdoor physical activity, such as walking. The relationship between living near a shore and mental health was significantly mediated by physical activity, while the presence of freshwater was directly linked to higher mental health and was not mediated by physical activity (Pasanen et al. 2019). This data suggests the possibility that, depending on the blue space setting (coastal versus inland), physical activity may have distinct mediation roles. Positive correlations between increased exposure to outdoor blue spaces and mental well-being (including happiness, optimism, positive thoughts, and affect) have been shown in numerous research that have looked at emotional components of mental well-being, such as happiness. In evaluating the potential benefits of living near a blue space for mental health, Murrin et al. (2023) found that living near a blue space was typically linked to increased well-being, such as positive thoughts and feelings and decreased anxiety symptoms, but not decreased depressive symptoms. Overall, the study's findings point to the potential benefits of blue space as a strategy for protecting against psychological distress and anxiety symptoms as well as for boosting positive mood, meaning, accomplishment, and pleasure.

Bratman et al. (2015) hypothesized that engaging with nature can foster awe experiences, which can elevate mood. Subsequent studies have demonstrated that exposure to coastal blue spaces increases emotions of serenity, relaxation, and renewal (Natural England. Monitor of Engagement with the Natural Environment: The National Survey on People and the Natural Environment—Happy and the Natural Environment 2013). It is possible that coastal blue spaces, in addition to physical activity, are linked to improved mental health through a feeling of awe.

11.6 Ecotherapeutic Approaches During the Pandemic: COVID-19

The impact on the mental health of frontline workers and the general public, as well as vulnerable populations like the elderly and those with chronic illnesses, is one of the most important "collateral damage" of COVID-19. These individuals must make significant lifestyle changes to adjust to the "new normal." (Tull et al. 2020; Wang et al. 2020; Holmes et al. 2020; Torales et al. 2023). The COVID-19 pandemic has been linked to psychological effects such as emotional disturbance, irritability, anxiety, stress, insomnia, depression, low mood, and emotional exhaustion, as well as socioeconomic distress brought on by economic loss (Reynolds et al. 2008; Wang et al. 2011; Pellecchia et al. 2015). Owing to its therapeutic qualities, home gardens can aid in people's relaxation and reduction of mental anguish. More and more research is showing that horticulture therapy—which includes gardening—offers significant health advantages to people, including lower rates of body mass index, anxiety, and depression as well as higher levels of life satisfaction, quality of life, and sense of community (Soga et al. 2017). Maintaining a home garden gives people the chance to unwind and enjoy the sunshine and fresh air, which has been shown to improve mood and sleeping habits, among other health advantages (Park et al. 2009). In addition, Maaset al. (2009) found that residents who had easier access to green spaces also had a stronger sense of social support and community, especially in metropolitan settings. Due to their therapeutic benefits, such as improved cognitive function and stress reduction, home gardens may be one of the most useful natural remedies for psychological stress, particularly for those who are confined to their homes and experiencing a difficult period as a result of COVID-19.

11.6.1 Open Green Spaces for Improvement in Mental Health

Having access to open areas helps reduce stress and anxiety as well as improve mood and inspire optimistic thinking. It increases psychological wellness and improves general well-being. Instant access to green space includes having a home garden, which has been shown to promote mood, enhance sleep quality, and improve cognitive performance. It is advantageous since it offers a chance to unwind (Basu et al. 2021). The association between emotional distress (apart from depression) and the number of days spent confined to one's home as a result of stay-at-home orders was considerably mitigated by the active interactions that occur when working in home gardens. Further benefits of being in green spaces include decreased rates of anxiety, mental exhaustion, aggression, and mortality as well as enhanced focus, self-discipline, social cohesiveness, and overall health (Chalquist 2009; Health Council of the Netherlands And RMNO 2004; Maller et al. 2002; MIND 2007; Newton 2007). According to the Disability Rights Commission (2006), persons who struggle with mental illness are also more likely to experience physical health issues and have a lower quality of life. It has been consistently observed that the prevalence of psychiatric illnesses, such as depression, bipolar disorder, and inpatients with serious mental disease, is much lower than the general population norm (Ware et al. 2007).

11.7 Ecotherapeutic Interventions in Clinical Setting

Ecotherapeutic interventions are highly effective in facilitating recovery among hospitalized patients. The presence of green foliage and flowering plants in hospital rooms has been associated with higher recovery among patients, lesser requirement of post-operative medications, more positive emotions, better physiological responses, and higher satisfaction with their rooms in comparison to control condition (Park And Mattson 2008, 2009a, 2009b). This finding has been supported by several other types of research, advocating the role of green space and plants inside hospital rooms in nurturing and fastening patient recovery (Bringslimark et al. 2009; Raanaas et al. 2010; Ulrich 1984; Ulrich, Dimberg And Driver 1991a, b).

11.7.1 Attention Deficit/Hyperactivity Disorder

Clinical case studies of children with Attention Deficit/Hyperactivity Disorder (ADHD) have been shown to benefit from ecotherapy, a treatment approach with roots in nature, according to the findings of the study by Altarawneh et al. (2023). People with ADHD can learn and improve their focus control in an environment that is both interesting and relaxing. According to Langberg et al. (2016), ecotherapy has demonstrated efficacy in mitigating symptoms of ADHD. This, in turn, creates a more cognitively engaging environment in the brain, resulting in improved grades and test scores. Reductions in impulsive tendencies and improvements in emotional regulation can lead to improvements in social interactions and behavior in both educational and extracurricular contexts. These improvements are often associated with a decrease in symptoms associated with ADHD (Pelham And Fabiano 2008). The extraordinary potential of ecotherapy to foster a collaborative effort culture and result in significant gains in the social maturity of young people is paramount. According to Jordan and Hinds (2016), ecotherapy is distinguished by its focus on cooperative learning opportunities in natural environments and group engagement (Altarawneh et al. 2023). In this context, the Attention Restoration Theory (Kaplan and Kaplan 1989) suggests that interaction with nature can lead to enhanced attention processes among children (Taylor et al. 2001) and adults (Kaplan 1995a; Summers And Vivian 2018). Exposure to green spaces and play in natural settings are extremely effective in reducing ADHD symptoms and enhancing concentration, motor abilities, and social abilities (Fjortoft And Sageie 2000; Fjortoft 2001 And 2004; Kuo And Taylor 2004; Taylor And Kuo 2009).

Exposure to natural settings and natural sounds has also been found to be effective in reducing pain symptoms (Malenbaum et al. 2008; Diette et al. 2003). Experiments involving patients being exposed to natural scenes and sounds reported greater tolerance to pain (Park et al. 2002) and higher thresholds in comparison to the control group (Tse et al. 2002). Walch et al. (2005) found that exposure to sunlight among spinal surgery patients led to decreased pain, reduced use of painkillers, lessened stress, and lower medical costs for pain medication (Walch et al. 2005; Summers And Vivian 2018).

11.7.2 Post-Traumatic Stress Disorder

Strength-based approaches in ecotherapy have been successfully implemented in Post-Traumatic Stress Disorder (PTSD), often characterized by severe anxiety and panic reactions after exposure to terrifying and traumatic stress situations like war, accidents, etc. (Summers And Vivian 2018). Such strength-based approaches such as recreation-based therapy are quite effective (Hawkins et al. 2016), typically emphasizing internal strengths (like interests, beliefs, skills, knowledge, etc.), external strengths (like social support, family support, community-based resources, etc.), and skill sets (for instance, character strength). Intervention involves focusing on the patient's aspirations, values, and hopes, instead of working on specific PTSD symptoms (Anderson And Hyen 2012; Hyen And Anderson 2012). Instances of effective recreation-based therapy include outdoor adventures and experiences, wilderness therapy, and green-space-based ecotherapy (Fredrickson And Anderson 1999; Ewert et al. 2001; Burls 2007; Dustin et al. 2011; Mowatt and Bennett 2011; Sibthorp And Jostad 2014).

11.7.3 Cognitive and Behavioral Problems

Ecotherapeutic approaches such as wilderness therapy and green space therapy are effective interventions for behavioral problems among adolescents (Russell et al. 1999; Hill 2007), specifically for psychological issues like adjustment problems, emotional issues, and addiction problems (Annerstedt And Wahrborg 2011; Summers And Vivian 2018). Moreover, other intervention techniques such as horticultural therapy, involving gardening activities, and interfacing with plants have been found to elevate mood by reducing stress (Wichrowski et al. 2005), enhance self-esteem and lessen depression (Lee et al. 2008; Son et al. 2004), elevate involvement and mood (Gigliotti et al. 2004; Gigliotti And Jarrott 2005), cognitive state and sleep (Lee And Kim 2008) in dementia. Furthermore, horticultural therapy and integrated adventure therapy are effective in dealing with substance abuse disorder (Richards And Kafami 1999; Bennett et al. 1998), as also in treating mood problems, schizophrenia, adolescent behavioral problems, anger management, and improving functionality at school, family, and social situations (Wilson And Lipsey 2000).

11.7.4 Ecotherapeutic Intervention: A Case Study by G.W. Burns

An interesting case study was reported by Burns (1998), showcasing the effectiveness of nature-guided therapeutic interventions like focusing on the sensual awareness of nature. Burns (1998) reported the case of Mary who presented chronic symptoms of depression along with suicidal ideation. Burns emphasized the works of Erickson, the founding father of the third wave of psychotherapy (Burns 1998), specifically an intervention based on nature-based therapy. The patient was asked to complete the Sensual Awareness Inventory (SAI), which required her to enlist 10–20 items indicating the activities from which she derives pleasure, enjoyment, and comfort for each of the sensual awarenesses like sight, sound, smell, taste, and touch. Burns reported that she was able to change her suicidal thoughts and sensations to ones of joy, laughing, and pleasure in a short period by just thinking about connecting with nature. Her expression of thankfulness for being allowed to work through the problems on her own reflects her sense of empowerment in accomplishing this. To help her meet two therapy objectives—first, to provide her with stimulation that would help her feel better and, second, to help her reach her other therapeutic aim of reducing weight—nature-based assignments or homework activities were proposed to her. As Mary planned to take a vacation with her spouse, Burns reported that holidays spent away from home are an effective way to start developing nature-guided experiential learning opportunities for clients while they're in new environments and are more likely to have various experiences. This helps the client establish expectations and create a goal. Mary on her visit to Uluru on vacation, described that, "I felt blessed to have had that time just there alone in such a spiritual place. Time stood still…the experience of standing alone in the vastness was truly a spiritual experience. The healing is complete" (p. 13; Burns 1998).

11.8 Challenges in Mainstreaming Ecotherapy in Clinical Practice

Albeit the advent of various ecotherapeutic interventions in clinical settings has gained enormous popularity and have been found to be extremely effective in reducing psychological distress, there have been several factors proving to be barriers in the large-scale implementation of ecotherapy in clinical practice. Isham et al. (2025) suggested that the relationship between nature and health is influenced by spiritual, cultural, gender, ethnic, and class-related issues, which in turn influences ecotherapeutic approaches. For instance, studies reported that activities involving exposure to outdoor environments are often considered to be a masculine practice while access to rural areas is often associated with the socioeconomic status of the individuals (Davies et al. 2020; Armstrong And Greene 2022). Studies also reported that individuals belonging to ethnic minorities often feel less included in social activities involving outdoor spaces (Armstrong And Greene 2022). Certain other physical barriers, such as accessibility to the space, particularly when transport is a concern, can pose a challenge to the implementation (Isham et al. 2025). Physical safety may also be an issue, where there are risks of physical injury involved (Shanahan et al. 2019). Another significant concern can be the emotional safety (King et al. 2023) involved in this case. Any form of psychotherapy conducted in a "closed" space ensures the privacy of the individual, while imparting a therapy in an open space might interfere with the confidentiality of the process involved and hence may lead to ethical concerns. Finally, another form of challenge comes from the fact that most of the healthcare professionals have only limited cognizance about ecotherapists and practitioners of nature-based therapies. This is further complicated by the uncertainty concerning the acceptance of ecotherapy as a potential therapeutic intervention by patients (Marx And More 2022).

11.9 Concluding Remarks: Implications of Ecotherapy in Promoting Well-Being

Buzzell and Chalquist (2009) proposed several recommendations for the inclusion of nature-based therapeutic interventions, such as actions that would involve protecting the natural environment, creating new associations with others in the context of the environment, and changing attitudes promoting concern not only for personal well-being but also for the well-being of the environment. The core idea behind the successful implementation of ecotherapeutic interventions lies in the fact that the nurturance involved is a two-way process. Only by protecting the natural ecosystem, human beings can preserve themselves. Hence, all the ecotherapeutic strategies involve the common act of nurturing, preserving, and synchronizing with nature in some way, and this, in turn, helps promote psychological well-being among us.

Considering these newly emerged forms of psychopathologies rooted in climate crisis, it is necessary to consider the applications of ecotherapeutic interventions. The above review of literature points to the fact that several forms of ecotherapeutic approaches exist and are widely being applied in different contexts. Such applications have been empirically validated, considering the vast literature already existing in the context of the climate crisis.

The future of research in ecopsychiatry and ecotherapeutic approaches concerns the advent of newer forms of nature-based ecotherapeutic approaches. There has been no dearth of studies suggesting the effectiveness of ecotherapeutic approaches in treating different psychopathologies (Coventry et al. 2021; Wilkie And Davinson 2021; Paakkolanvaara et al. 2023). For instance, Rueff and Reese (2023) compared the efficacy of ecotherapy and cognitive behavior therapy (CBT) in treating depression and anxiety by systematically reviewing 42 studies. Their study findings revealed that both therapeutic interventions were equally effective in depression, although CBT showed more efficacy in anxiety in comparison. In another instance, Wilkie and Davinson (2021) reviewed 52 research studies on nature-based therapeutic approaches and found that ecotherapeutic approaches facilitate a reduction in negative affect and mood, as well as improve cardiovascular functions and lessen stress levels.

However, the benefits of ecotherapy have not been limited to treating psychopathology, but also to enhance the well-being of human beings. Studies suggest that proximity to nature imparts potential benefits for the physical, psychological, and social well-being of individuals (Bratman et al. 2019). For instance, nature-based therapies have been found to enhance positive affect and reduce negative affect (Bratman et al. 2019; Paakkolanvaara et al. 2023), intensify sustained attention (Pasanen et al. 2018; Lee et al. 2015), improve social cohesion and social interactions (Jennings and Bamkole 2019), as well as sleep quality (Grigsby-Toussaint et al. 2015), and enhance cognitive flexibility and working memory capacity (Stevenson et al. 2018; Paakkolanvaara et al. 2023). In addition, it has been found that interaction with nature results in restoration, characterized by a significant reduction in stress (Paakkolanvaara et al. 2023) and major health benefits such as controlled blood pressure (Ideno et al. 2017) and cortisol levels (Roe et al. 2013).

With the advent of the third and fourth waves of psychotherapy, newer approaches toward ecotherapeutic interventions are being developed. For instance, Paakkolanvaara et al. (2023) reported the emergence of Flow with Nature (FWN) treatment as an effective mode of ecotherapeutic intervention for clinical depression. Such form of treatment has been influenced by several forms of newer interventions such as mindfulness-based cognitive theory (Hayes et al. 2012), Transtheoretical model of behavior change (Prochaska et al. 2020), evidence-based practice (Norcross And Wampold 2018), and creative art therapies (Zubala And Karkou 2018) and clinician's experience with nature-based therapeutic intervention (Paakkolanvaara et al. 2023).

Needless to mention, ecotherapeutic approaches are gaining popularity with each passing decade, owing to their wide applications in different settings, both for treating psychopathology as well as promoting psychological well-being. The success of the present chapter, therefore, lies in the fact that the current review presented the readers with a gainful insight into the integration of ecological principles with clinical psychology practices and their applications.

Authors Contribution Conceptual framework of the chapter was planned by SD and TM; Content of the chapter has been authored by all the authors.

Funding The research was not supported by any funding.

DataAvailability Data Availability Statement Not applicable for the current chapter.

Potential Conflicts of Interest On behalf of all authors, the corresponding author states that there is no conflict of interest.

Informed Consent Not applicable for the current chapter.

AI Discloser No AI tool has been used for writing the manuscript.

References

Albrecht G (2005) '*Solastalgia*'. A new concept in health and identity. PAN: Philosophy Activism Nature 3:41–55

Altarawneh ZSA, Al-Ajeely SA, Khasawneh MAS (2023) The impact of ecotherapy on academic performance and ADHD symptom severity in children. Migr Lett 20(5):1192–1204, ISSN: 1741–8984 (Print) ISSN: 1741–8992 (Online) www.migrationletters.com

Ambrose-Oji B (2013) Mindfulness practice in woods and forests: an evidence review. Research report for the Mersey forest, forest research. Alice Holt Lodge Farnham, Surrey

Anand K, Walia GK, Mandal S, Menon JS, Gupta R, Tandon N, Prabhakaran D (2024) Longitudinal associations between ambient PM2. 5 exposure and lipid levels in two Indian cities. Environ Epidemiol 8(2):e295

Anderson L, Heyne L (2012) Therapeutic recreation practice: a strengths approach. State College, PA: Venture Press

Annerstedt M, Wahrborg P (2011) Nature-assisted therapy: systematic review of controlled and observational studies. Scand J Public Health 39:371–388. https://doi.org/10.1177/1403494810396400

Armstrong A, Greene BT (2022) Sense of inclusion and race in a public, outdoor recreation setting: do place meanings matter? Soc Nat Resour 35:391–409. https://doi.org/10.1080/08941920.2022.2045413

Barton J, Pretty J (2010) What is the best dose of nature and green exercise for improving mental health—A multi-study analysis. Environ Sci Technol 44(10):3947–3955. https://doi.org/10.1021/es903183r

Basu M, DasGupta R, Kumar P, Dhyani S (2021) Home gardens moderate the relationship between Covid-19- induced stay-at-home orders and mental distress: a case study with urban residents of India. Environ Res Commun 3:105002. https://doi.org/10.1088/2515-7620/ac2ab2

Balakrishnan U, Tsaneva M (2023) Impact of air pollution on mental health in India. J Dev Stud 59(1):133–147

Bennett LW, Cardone S, Jarczyk J (1998) Effects of a therapeutic camping program on addiction recovery: the Algonquin Haymarket relapse prevention program. J Subst Abuse Treat 15:469–474. https://doi.org/10.1016/S0740-5472(97)00222-5

Berlin B (1992) Ethnobiological classification: principles of categorization of plants and animals in traditional societies. Princeton University Press, New Jersey

Berman MG, Jonides J, Kaplan S (2008) The cognitive benefits of interacting with nature. Psychol Sci 19:1207–1212

Bodnar S (2024) The fierce urgency of now: The case for including the environmental field in clinical work. Psychoanal Dialogues 34(6):795–811. https://doi.org/10.1080/10481885.2024.2416621

Bojić S (2018) Organic agriculture contribution to the rural tourism development in the north of montenegro. J Econ Bus 1:222–233

Bratman GN, Anderson CB, Berman MG, Cochran B, Vries D, Flanders J, Daily GC (2019) Nature and mental health: an ecosystem service perspective. Sci Adv 5(7):eaax0903. https://doi.org/10.1126/sciadv.aax0903

Bringslimark T, Hartig T, Patil GG (2009) The psychological; benefits of indoor plants: a critical review of the experimental literature. J Environ Psychol 29:422–433. https://doi.org/10.1016/j.jenvp.2009.05.001

Buckley RC, Brough P, (2017) Nature, eco, and adventure therapies for mental health and chronic disease. Front Public Health 5:220p

Bragg R, Atkins GJNECR (2016) A review of nature-based interventions for mental health care. Nat Engl Commissioned Rep 204:18p

Bratman GN, Daily GC, Levy BJ, Gross JJ (2015) The benefits of nature experience: Improved affect and cognition. Landscape Urban Plann 138:41–50

Brückner A, Falkenberg T, Kasturirangan U, Kistemann T (2022) Photovoice for enhanced healthy blue space research: an example of use from urban India. Cities Health 6(4):804–817

Budziszewska M, Kałwak W (2022) Climate depression. critical analysis of the concept. Psychiatr Pol 56(1):171–182

Burls A (2007) People and green spaces: promoting public health and wellbeing through Ecotherapy. J Public Mental Health 6(16):24–39. https://doi.org/10.1108/17465729200700018

Burns GW (1998) Nature-guided therapy: a case example of ecopsychology in clinical practice. J Outdoor Environ Educ 3(2):9–14

Chalquist C (2009) A look at the ecotherapy research evidence. Ecopsychology 1(2):1–10

Capaldi CA, Dopko RL, Zelenski JM (2014) The relationship between nature connectedness and happiness: A meta-analysis. Frontiers Psychol 5:92737

Chaudhary E, Dey S, Ghosh S, Sharma S, Singh N, Agarwal S, Jain S (2022) Reducing the burden of anaemia in Indian women of reproductive age with clean-air targets. Nat Sustain 5(11):939–946

Chowdhury AN, Jadhav S (2012) 52 eco-psychiatry: culture, mental health and ecology with special reference to India. In: Community mental health in India. Edited by Chavan BS, Gupta N, Úrun P, Sidana A, Jadhav S. Jaypee Brothers. India. pp 521–542. ISBN 978-93-5025-805-7

Clayton S, Manning CM, Hodge C (2014) Beyond storms and droughts: the psychological impacts of climate change. American Psychological Association and ecoAmerica, Washington, DC

Clinebell HJ (1996) Ecotherapy. Fortress Press, Minneapolis, MN

Cohen MJ (1997) Reconnecting with nature: finding wellness through restoring your bond with the earth. OR, Ecopress, Corvallis

Cohen S, Janicki-Deverts D, Miller GE (2007) Psychological stress and disease. JAMA: J Am Med Assoc 298(14):1685–1687

Coventry PA, Brown JVE, Pervin J, Brabyn S, Pateman R, Breedvelt J, Gilbody S, Stancliffe R, McEachan R, White PCL (2021) Nature-based outdoor activities for mental and physical health: systematic review and meta-analysis. Ssm–population Health 16:100934

Cox DTC, Shanahan DF, Hudson HL, Fuller RA, Anderson K, Hancock S, Gaston KJ (2017) Doses of nearby nature simultaneously associated with multiple health benefits. Int J Environ Res Public Health 14:172. https://doi.org/10.3390/ijerph14020172. www.mdpi.com/journal/ijerph

Cox D, Shanahan D, Hudson H, Fuller R, Anderson K, Hancock S, Gaston K (2017b) Doses of nearby nature simultaneously associated with multiple health benefits. Int J Environ Res Public Health 14(2):172

Cox EQ, Sowa NA, Meltzer-Brody SE, Gaynes BN (2016) The perinatal depression treatment cascade: Baby steps toward improving outcomes. J Clin Psychiatry 77(9):20901

Clatworthy J, Hinds J, MCamic P (2013) Gardening as a mental health intervention: A review. Ment Health Rev J 18(4):214–225

Currie BA, Bass B (2008) Estimates of air pollution mitigation with green plants and green roofs using the UFORE model. Urban Ecosyst 11(4):409–422

Davies J, McKenna M, Bayley J, Denner K, Young H (2020) Using engagement in sustainable construction to improve mental health and social connection in disadvantaged and hard to reach groups: a new greencare approach. J Ment Health 29:350–357. https://doi.org/10.1080/09638237.2020.1714001

de Vries S, Verheij RA, Groenewegen P, Spreeuwenberg P (2003) Natural environments: healthy environments? An exploratory analysis of the relationship between greenspace and health. Environ Plann A 35:1717–1731. https://doi.org/10.1068/a35111

Douglas O, Lennon M, Scott MJ (2017) Green space benefits for health and well-being: A life-course approach for urban planning, design. Douglas, Owen, Mick Lennon, and Mark J. Scott."Green Space Benefits for Health and Well-Being: A Life-Course Approach for Urban Planning, Design and Management." Elsevier, June 2017. https://doi.org/10.1016/j.cities.2017.03.011

Degges-White S, Davis N (eds) (2010) Integrating the expressive arts into counseling practice: theory-based interventions. Springer, New York, NY

DeMayo N (2009) Horses, humans, and healing, In: Ecotherapy: healing with nature in mind, eds Chalquist c, Buzzell L (San Francisco, CA: Sierra Club Books)

Diette GB, Lechtzin N, Haponik E, Devrotes A, Rubin HR (2003) Distraction therapy with nature sights and sounds reduces pain during flexible bronchoscopy: a complimentary approach to routine analgesia. Chest 123:941–948. https://doi.org/10.1378/chest.123.3.941

Disability Rights Commission (2006) Equal treatment: closing the gap. Disability Rights Commission, London

Dustin D, Bricker N, Arave J, Wall W, Wendt G (2011) The promise of river running as a therapeutic medium for veterans coping with post-traumatic stress disorder. Ther Recreat J 45:326–340

Ewert A, McCormick B, Voight A (2001) Outdoor experiential therapies: implications for TR practice. Ther Recreat J 35:107–122

Falagas M, Zarkadoulia E, Ioannidou E, Peppas G, Christodoulou C, Rafailidis P (2007) The effect of psychosocial factors on breast cancer outcome: a systematic review. Breast Cancer Res 9(4):R44. https://doi.org/10.1186/bcr1744

Fjortoft I (2001) The natural environment as a playground for children: the impact of outdoor play activities in pre-primary school children. Early Childhood Educ. J Environ Educ 29:111–117. https://doi.org/10.1023/A:1012576913074
Fjortoft I (2004) Landscape as playscape: the effects of natural environments on children's play and motor development. Child Youth Environ 14:21–44
Fjortoft I, Sageie J (2000) The natural environment as a playground for children: landscape description and analyses of a natural playscape. Landsc Urban Plann 20:83–97. https://doi.org/10.1016/S0169-2046(00)00045-1
Fredrickson LM, Anderson DH (1999) A qualitative exploration of the wilderness experience as a source of spiritual inspiration. J Environ Psychol 19:21–39. https://doi.org/10.1006/jevp.1998.0110
Georgiou M, Morison G, Smith N, Tieges Z, Chastin S (2021) Mechanisms of impact of blue spaces on human health: a systematic literature review and metaanalysis. Int J Environ Res Public Health 18(5):2486
Ghosh P, Singh KK (2022) Spatiotemporal dynamics of urban green and blue spaces using geospatial techniques in Chandannagar city, India. GeoJournal 87(6):4671–4688
Gibala MJ, Little JP, van Essen M, Wilkin GP, Burgomaster KA, Safdar A, Raha S, Tarnopolsky MA (2006) Short-term sprint interval versus traditional endurance training: similar initial adaptations in human skeletal muscle and exercise performance. J Physiol 575:901–911
Gigliotti CM, Jarrott SE (2005) Effects of horticulture therapy on engagement and affect. Can J Aging 24:367–377. https://doi.org/10.1353/cja.2006.0008
Gigliotti CM, Jarrott SE, Yorgason J (2004) Harvesting health: effects of three types of horticultural therapy activities for persons with dementia. Dementia 3:161–180. https://doi.org/10.1177/1471301204042335
Glantz K, Schwartz M (2008) Stress, coping, and health behavior. In: Glanz K, Rimer BK, Viswanath K (eds) Health behavior and health education: theory, research, and practice, 5. Jossey-Bass, San Francisco, CA, pp 211–236
Grigsby-Toussaint DS, Turi KN, Krupa M, Williams NJ, Pandi-Perumal SR, Jean- Louis G (2015) Sleep insufficiency and the natural environment: results from the US behavioral risk factor surveillance system survey. Prev Med 78:78–84. https://doi.org/10.1016/j.ypmed.2015.07.011
Gupta A, De B (2025) Blue space resilient urban planning to enhance severely distressed thermal environment. Phys Chem Earth, Parts a/b/c 137:103804
Haeckel EHPA (1866) Generelle Morphologie der Organismen. Allgemeine Grundzüge der organischen Formen-Wissenschaft, mechanische Begründet durch die von Charles Darwin reformirte Descendenz-Theorie. Volume I: Allgemeine Anatomie der Organismen. pp 32+574; vol 2: Alllgemeine Entwickelungsgeschichte der Organismen. pp 140+462. Georg Reimer, Berlin, Germany
Hartig T, Evans GW, Jamner LD, Davis DS, Garling T (2003) Tracking restoration in natural and urban field settings. J Environ Psychol 23(2):109–123
Hawkins BL, Townsend JA, Garst BA (2016) Nature-based recreational therapy for military service members: a strengths approach. Ther Recreat J 50:55–74. https://doi.org/10.18666/TRJ-2016-V50-I1-6793
Hayes SC, Strosahl KD, Wilson KG (2012) Acceptance and commitment therapy: the process and practice of mindful change (2nd Edition). The Guildford Press: New York, USA
Hazer M, Morley CP, Dieterlen S (2018) The relationship between self-reported exposure to greenspace and human stress in Baltimore, MD. Landsc Urban Plan 169(2018):47–56:0169-2046. https://doi.org/10.1016/j.landurbplan.2017.08.006
Health council of the Netherlands and RMNO (2004) Nature and health: the influence of nature on social, psychological and physical well-being. Health council of the Netherlands and Dutch advisory council for research and social planning: 2004/09E; RMNO publication No. A02ae
Heyne LA, Anderson LS (2012) Theories that support strength-based practice in therapeutic recreation. Ther Recreat J 46:106–128

Hill NR (2007) Wilderness therapy as a treatment modality for at-risk youth: a primer for mental health counselors. J Men Health Counsel 29:338–349. https://doi.org/10.17744/mehc.29.4.c6121j162j143178
Holmes EA et al (2020) Multidisciplinary research priorities for the COVID-19 pandemic: a call for action for mental health science. Lancet Psychiatry 7:547–60. https://doi.org/10.1016/j.ssmph.2021.100934
Ideno Y, Hayashi K, Abe Y, Ueda K, Iso H, Noda M, Suzuki S (2017) Blood pressure-lowering effect of Shinrin-yoku (Forest bathing): A systematic review and meta-analysis. (Report). BMC Complement Altern Med 17(1). https://doi.org/10.1186/s12906-017-1912-z
Isham A, Jefferies L, Blackburn J, Fisher Z, Kemp A (2025) Green healing: ecotherapy as a transformative model of health and social care. Curr Opin Psychol 102005
Jennings V, Bamkole O (2019) The relationship between social cohesion and urban green space: an avenue for health promotion. Int J Environ Res Public Health 16(3). https://doi.org/10.3390/ijerph16030452
Jiang B, Chang CY, Sullivan WC (2014) A dose of nature: tree cover, stress reduction, and gender differences. Landsc Urban Plan 132:26–36. https://doi.org/10.1016/j.landurbplan.2014.08.005
Jiang B, Li D, Larsen L, Sullivan WC (2016) A dose-response curve describing the relationship between urban tree cover density and self-reported stress recovery. Environ Behav 48(4):607–629
Jordan M, Hinds J (eds) (2016) Ecotherapy: theory, research and practice. London: Macmillan. https://doi.org/10.1057/978-1-137-48688-2
Kaplan R, Kaplan S (1989) The experience of nature: A psychological perspective. Cambridge university press
Kaplan S (1995) The restorative benefits of nature: toward an integrative framework. J Environ Psychol 15(3):169–182
Kaplan R (1984) Impact of urban nature: A theoretical analysis. Urban Ecology 8(3):189–197
Kaplan R (1985) Nature at the doorstep: Residential satisfaction and the nearby environment. J Architectural Plann Res, pp.115–127
King BC, Taylor CD, Garcia JA, Cantrell KA, Park CN (2023) Ethics and ecotherapy: the shared experiences of ethical issues in practice. J Adventure Educ Outdoor Learn 23:452–471. https://doi.org/10.1080/14729679.2022.2029512
Kingsley JY, Townsend M (2006) "Dig In" to social capital: community gardens as mechanisms for growing urban social connectedness. Urban Policy Res. 24:525–537
Krefis AC, Augustin M, Schlünzen KH, Oßenbrügge J, Augustin J (2018) How does the urban environment affect health and well-being? A systematic review. Urban Sci 2(1):21
Kuo FE, Taylor AF (2004) A potential natural treatment for attentiondeficit/ hyperactivity disorder: evidence from a national study. Am J Public Health 94:1580–1586. https://doi.org/10.2105/AJPH.94.9.1580
Lachowycz K, Jones AP (2013) Towards a better understanding of the relationship between greenspace and health: development of a theoretical framework. Landsc Urban Plan 118:62–69
Langberg JM, Dvorsky MR, Molitor SJ, Bourchtein E, Eddy LD, Smith Z, Evans SW (2016) Longitudinal evaluation of the importance of homework assignment completion for the academic performance of middle school students with ADHD. J Sch Psychol 55:27–38. https://doi.org/10.1016/j.jsp.2015.12.004
Lawrence RJ (2003) Human ecology and its applications. Landsc Urban Plan 65(1–2):31–40
Lawrence R (2001) Human ecology. In: Tolba MK (ed) Our Fragile world: challenges and opportunities for sustainable development, vol 1. Eolss Publishers, Oxford, pp 675–693
Lee J, Park BJ, Ohira T, Kagawa T, Miyazaki Y (2015) Acute effects of exposure to a traditional rural environment on urban dwellers: a crossover field study in terraced farmland. Int J Environ Res Public Health 12(2):1874–1893
Lee J, Park BJ, Tsunetsugu Y, Ohira T, Kagawa T, Miyazaki Y (2011) Effect of forest bathing on physiological and psychological responses in young Japanese male subjects. Public Health 125(2):93–100

Lee S, Kim MS, Suh JK (eds) (2008) Effects of horticultural therapy on selfesteem and depression of battered women at a shelter in Korea. Acta Horticult 790:139–142. https://doi.org/10.17660/ActaHortic.2008.790.19
Lee Y, Kim S (2008) Effects of indoor gardening on sleep agitation and cognition in dementia patients—A pilot study. Int J Geriatr Psychiatry 23:485–489. https://doi.org/10.1002/gps.1920
Lertzman R (2015) Environmental melancholia: psychoanalytic dimensions of engagement. Routledge, New York
Lin BB, Gaston KJ, Fuller RA, Wu D, Bush R, Shanahan DF (2017) How green is your garden? Urban form and socio-demographic factors influence yard vegetation, visitation, and ecosystem service benefits. Landsc Urban Plan 157:239–246
Linden S, Grut JJ (2002) The healing fields: working with psychotherapy and nature to rebuild shattered lives. London: Frances Lincoln Publishing
Ma B, Zhou T, Lei S, Wen Y, Htun TT (2019) Effects of urban green spaces on residents' well-being. Environ Dev Sustain 21:2793–2809
Maas J, Verheij RA, de Vries S, Spreeuwenberg P, Schellevis FG, Groenewegen PP (2009) Morbidity is related to a green living environment. J Epidemiol Community Health 63(12):967–973
Maas J, Verheij RA, Groenewegen PP, de Vries S, Spreeuwenberg P (2006) Green space, urbanity, and health: How strong is the relation? J Epidemiol Community Health 60:587–592. Retrieved from http://www.ncbi.nlm.nih.gov/pubmed/16790830
Malenbaum S, Keefe FJ, Williams A, Ulrich R, Somers TJ (2008) Pain in its environmental context: implications for designing environments to enhance pain control. Pain 134:241–244. https://doi.org/10.1016/j.pain.2007.12.002
Maller C, Townsend M, Brown P, St. Leger L (2002) Healthy parks healthy people: the health benefits of contact with nature in a park context—A review of current literature. Melbourne, Australia: Deakin University/Parks Victoria
Mandal S, Jaganathan S, Kondal D, Schwartz JD, Tandon N, Mohan V, Narayan KV (2023) PM2. 5 exposure, glycemic markers and incidence of type 2 diabetes in two large Indian cities. BMJ Open Diabetes Res Care 11(5)
Mariategui J (1978) Ecology and psychiatry. Acta Psiquiqtricq y Psicologica De America Latina 24:101–108
Martyn P, Brymer E (2016) The relationship between nature relatedness and anxiety. J Health Psychol 21(7):1436–1445
Marx V, More KR (2022) Developing Scotland's first green health prescription pathway: a one-stop shop for nature-based intervention referrals. Front Psychol 13:817803. https://doi.org/10.3389/fpsyg.2022.817803
McMahan EA, Estes D (2015) The effect of contact with natural environments on positive and negative affect: a meta-analysis. J Posit Psychol 10(6):507–519. https://doi.org/10.1080/17439760.2014.994224
Mennis J, Mason M, Ambrus A (2018) Urban greenspace is associated with reduced psychological stress among adolescents: a geographic ecological momentary assessment (GEMA) analysis of activity space. Landsc Urban Plan 174:1–9
MIND (2007) Ecotherapy the green agenda for mental health. London
Mitchell R, Popham F (2008) Effect of exposure to natural environment on health inequalities: an observational population study. Lancet 372(9650):1655–1660
Moffic S (2008) Is ecopsychiatry a speciality for the 21st century? Psychiatr News 43:37
Mowatt RA, Bennett J (2011) Veteran stories, PTSD effects and therapeutic fly-fishing. Ther Recreat J 45:286–308
Murrin E, Taylor N, Peralta L, Dudley D, Cotton W, White RL (2023) Does physical activity mediate the associations between blue space and mental health? A cross-sectional study in Australia. BMC Public Health 23:203. https://doi.org/10.1186/s12889-023-15101-3
Newton J (2007) Well-being and the natural environment: a brief overview of the evidence. Retrieved 11 May 2010 www.sustainabledevelopment.gov.uk/what/documents/WellbeingAndTheNaturalEnvironmentReport.doc

Nielsen M (2011) Inverse governmentality: The paradoxical production of peri-urban planning in Maputo, Mozambique. Critique Anthropol 31(4):329–358

Norcross JC, Wampold BE (2018) A new therapy for each patient: evidence-based relationships and responsiveness. J Clin Psychol 74:1889–1906. https://doi.org/10.1002/jclp.22678

Nowak DJ, McHale PJ, Ibarra M, Crane D, Stevens JC, Luley CJ (1998) Modeling the effects of urban vegetation on air pollution. Air pollution modeling and its application XII. Springer, Boston, MA, pp 399–407

Pal S, Singha A, Mondal S, Debanshi S (2023) Ecological consequences of urban blue space transformation. Environ Sci Pollut Res 30(12):34115–34134

Park SA, Shoemaker CA, Haub MD (2009) Physical and psychological health conditions of older adults classified as gardeners or nongardeners. Hort Sci 44:206–210

Park RE, Burgess EW (2019) The city. University of Chicago press

Park S-H, Mattson RH (2009) Ornamental indoor plants in hospital rooms enhanced health outcomes of patients recovering from surgery. J Alternat Compliment Med 15:975–980. https://doi.org/10.1089/acm.2009.0075

Park S-H, Mattson RH (2009b) Therapeutic influences of plants in hospital rooms on surgical recovery. HortTechnology 44:102–105

Park S-H, Mattson RH, Kim E (2002) Pain tolerance effects of ornamental plants in a simulated hospital room. Acta Horticult 639:241–247

Park S-H, Mattson RH (2008) Effects of flowering and foliage plants in hospital rooms on patients recovering from abdominal surgery. HortTechnology 18:563–568

Pasanen TP, White MP, Wheeler BW, Garrett JK, Elliott LR (2019) Neighbourhood blue space, health and wellbeing: the mediating role of different types of physical activity. Environ Int 131:105016

Paakkolanvaara JV, Hyvönen K, Salonen K, Muotka J, Korpela K (2023) Profiles of depression and restoration in nature-based group therapy. Ecopsychology 15(3):231–243

Pasanen T, Johnson K, Lee K, Korpela K (2018) Can nature walks with psychological tasks improve mood, self-reported restoration, and sustained attention? Results from two experimental field studies. Front Psychol 9:2057. https://doi.org/10.3389/fpsyg.2018.02057

Pelham WE Jr, Fabiano GA (2008) Evidence-based psychosocial treatments for attention-deficit/ hyperactivity disorder. J Clin Child Adolesc Psychol 37(1):184–214

Pellecchia U, Crestani R, Decroo T, Van den Bergh R, Al-Kourdi Y (2015) Social consequences of Ebola containment measures in Liberia. PLoS ONE 10:e0143036

Prabhakaran D, Mandal S, Krishna B, Magsumbol M, Singh K, Tandon N (2020) GeoHealth hub study investigators, COE-CARRS study investigators. Exposure to particulate matter is associated with elevated blood pressure and incident hypertension in urban India. Hypertension 76(4):1289–1298

Pretty J, Peacock J, Hine R, Sellens M, South N, Griffin M (2007) Green exercise in the UK countryside: effects on health and psychological wellbeing, and implications for policy and planning. J Environ Plann Manage 50:211–231. https://doi.org/10.1080/09640560601156466

Pretty J, Peacock J, Sellens M, Griffin M (2005) The mental and physical health outcomes of green exercise. Int J Environ Health Res 15:319–337. https://doi.org/10.1080/09603120500155963

Prochaska JO, Norcross JC, Saul SF (2020) Generating psychotherapy breakthroughs: trans theoretical strategies from population health psychology. Am Psychol 75(7):996–1010. https://doi.org/10.1037/amp0000568

Raanaas RK, Patil GG, Hartig T (2010) Effects of an indoor foliage plant intervention on patient well-being during a residential rehabilitation program. HortScience 45:387–392

Rahman SR, Ahmad H (2012) Green roofs as urban antidote: a review on aesthetic, environmental, economic and social benefits. Paper presented at the Sixth South East Asian Technical Consortium in King Mongkut University of Technology Thonbur, Bangkok, Thailand

Ramirez JA, Craig TK, Watson JP, Fentiman IS, North WR, Rubens RD (1989) Stress and relapse of breast cancer. BMJ (Clin Res Ed) 298(6669):291–293

Reynolds DL, Garay JR, Deamond SL, Moran MK, Gold W, Styra R (2008) Understanding, compliance and psychological impact of the SARS quarantine experience. Epidemiol Infect 136:997–1007

Rezaei M, Kim D, Alizadeh A, Rokni L (2021) Evaluating the mental-health positive impacts of agritourism; a case study from South Korea. Sustainability 13(16):8712

Richards HJ, Kafami DM (1999) Impact of horticultural therapy on vulnerability and resistance to substance abuse among incarcerated offenders. J Offender Rehabil 29:183–193. https://doi.org/10.1300/J076v29n03_11

Rii HU, Ahn JS (2002) Urbanization and its impact on Seoul, Korea. 2001 Forum on Urbanizing World and UN Human Habitat II. Columbia University, New York, pp 83–100

Rueff M, Reese G (2023) Depression and anxiety: a systematic review on comparing ecotherapy with cognitive behavioral therapy. J Environ Psychol 102097

Roe JJ, Thompson CW, Aspinall PA, Brewer MJ, Duff EI, Miller D, Mitchell R, Clow A (2013) Green space and stress: Evidence from cortisol measures in deprived urban communities. Int J Environ Res Public Health 10(9):4086–4103

Restall B, Conrad E (2015) A literature review of connectedness to nature and its potential for environmental management. J Environ Manage 159:264–278

Russell K (2001) What is wilderness therapy? J Exp Educ 23:170–176. https://doi.org/10.1177/105382590102400203

Russell KC, Hendee JC, Phillips-Miller D (1999) How wilderness therapy works: an examination of the wilderness therapy process to treat adolescents with behavioral problems and addictions. In: Proceedings: wilderness science in a time of change, eds Cole DN, McCool SF (2000. Proceeding RMRS-P-000). Ogden, UT: U.S. Department of Agriculture, Forest Service, Rocky Mountain Research Station

Schell L, Cotton S, Luxmoore M (2012) Outdoor adventure for young people with a mental illness. Early Interv Psychiatry 6:407–414

Shanahan DF, Astell-Burt T, Barber EA, Brymer E, Cox DT, Dean J et al (2019) Nature–based interventions for improving health and wellbeing: the purpose, the people and the outcomes. Sports 7:141. https://doi.org/10.3390/sports7060141

Shanahan DF, Bush R, Gaston KJ, Lin BB, Dean J, Barber E, Fuller RA (2016) Health benefits from nature experiences depend on dose. Sci Rep 6:28551. https://doi.org/10.1038/srep28551

Shanahan DF, Fuller RA, Bush R, Lin BB, Gaston KJ (2015a) The health benefits of urban nature: how much do we need? Bioscience 65(5):476–485. https://doi.org/10.1093/biosci/biv032

Shanahan DF, Cox DTC, Fuller RA, Hancock S, Lin BB, Anderson K, Bush R, Gaston KJ (2017) Variation in experiences of nature across a gradient of tree cover in compact and sprawling cities. Landsc Urban Plan 157:231–238

Shanahan DF, Lin BB, Bush R, Gaston KJ, Dean JH, Barber E, Fuller RA (2015b) Toward improved public health outcomes from urban nature. Am J Public Health 105:470–477

Shetty (2024) From Anaemia to mental health—Growing body of Indian research links polluted air to range of chronic conditions. Health Policy Watch. https://healthpolicy-watch.news/growing-body-of-data-links-indias-polluted-air-to-worsening-health/ Accessed 10 March 2025

Sibthorp J, Jostad J (2014) The social system in outdoor adventure education programs: present and future. J Exp Educ State Know 37:60–74. https://doi.org/10.1177/1053825913518897

Smith N, Foley R, Georgiou M, Tieges Z, Chastin S (2022) Urban blue spaces as therapeutic landscapes: a slice of nature in the city. Int J Environ Res Public Health 19(22):15018. https://doi.org/10.3390/ijerph192215018

Soga M, Gaston KJ, Yamaura Y (2017) Gardening is beneficial for health: a meta-analysis. Prev Med Rep 5:92–9

Soja EW (1989) Postmodern geographies: the reassertion of space in critical social theory. Verso. Retrieved from https://mars1980.github.io/Space/resources/postmoderngeographies.pdf

Son KC, Um SJ, Kim SY, Song JE, Kwack HR (2004) Effect of horticultural therapy on the changes of self-esteem and sociality of individuals with chronic schizophrenia. Acta Horticult 639:185–191. https://doi.org/10.17660/ActaHortic.2004.639.23

Stevenson M, Schilhab T, Bentsen P (2018) Attention restoration theory II: a systematic review to clarify attention processes affected by exposure to natural environments. J Toxicol Environ Health Part B 21(4):227–268. https://doi.org/10.1080/10937404.2018.1505571

Steward JH (1972) Theory of culture change: the methodology of multilinear evolution. University of Illinois Press

Sugiyama T, Francis J, Middleton NJ, Owen N, Giles-Corti B (2010) Associations between recreational walking and attractiveness, size, and proximity of neighborhood open spaces. Am J Public Health 100:1752–1757

Sullivan WC, Frumkin H, Jackson RJ, Chang CY (2014) Gaia meets ascelpius: creating healthy places. Landsc Urban Plan 127:182–184

Summers JK, Vivian DN (2018) Ecotherapy—A forgotten ecosystem service: a review. Front Psychol 9:1389. https://doi.org/10.3389/fpsyg.2018.01389

Swimme B, Berry T (1994) The universe story: from the primordial flaring forth to the ecozoic era—A celebration of the unfolding of the cosmos. Harper, San Francisco, CA

Takano T, Nakamura K, Watanabe M (2002) Urban residential environments and senior citizens' longevity in megacity areas: the importance of walkable green spaces. J Epidemiol Community Health 56(12):913–918

Taylor MJ, Edmonds GE, McCarthy G, Allison T (2001) Eyes first! Eye processing develops before face processing in children. NeuroReport 12(8)1671–1676

Taylor AF, Kuo FE (2009) Children with attention deficits concentrate better after walk in the park. J Atten Disord 12:402–409. https://doi.org/10.1177/1087054708323000

Thoma MV, Rohleder N, Rohner SL (2021) Clinical ecopsychology: the mental health impacts and underlying pathways of the climate and environmental crisis. Front Psychiatry 12:675936. https://doi.org/10.3389/fpsyt.2021.675936

Torales J, O'Higgins M, Castaldelli-Maia JM, Ventriglio A (2020) The outbreak of COVID-19 coronavirus and its impact on global mental health. Int J Soc Psychiatry 66:317–320002076402091 5000

Torales J, Vilallba-Arias J, Bogado JA, O'Higgins M, Almirón-Santacruz J, Ruiz Díaz N, García O, Amarilla-Salvioni D, Castaldelli-Maia JM, Ventriglio A, Barrios I (2023) Satisfaction with Telepsychiatry during the COVID-19 pandemic: Patients' and psychiatrists' report from a University Hospital. Int J Soc Psychiatry 69(1):156–160

Tse MM, Ng JK, Chung JW, Wong TK (2002) The effect of visual stimuli on pain threshold and tolerance. J Clin Nurs 11:264–269. https://doi.org/10.1046/j.1365-2702.2002.00608.x

Tull MT, Edmonds KA, Scamaldo K, Richmond JR, Rose JP, Gratz KL (2020) Psychological outcomes associated with stay-at-home orders and the perceived impact of COVID-19 on daily life. Psychiatry Res 113098

Ulrich RS (1984) View through a window may influence recovery from surgery. Science 224:420–421. https://doi.org/10.1126/science.6143402

Ulrich RS, Dimberg U, Driver BL (1991) Psychophysiological indicators of leisure benefits. In: Benefits of leisure, eds Driver BL, Brown PJ, Peterson GL (State College, PA: Venture)

Ulrich RS, Simons RF, Losito BD, Fiorito E, Miles MA, Zelson M (1991b) Stress recovery during exposure to natural and urban environments. J Environ Psychol 11(3):201–230

Ulrich RS (1983) Aesthetic and affective response to natural environment. In: Behavior and the natural environment. Altman I, Wohlwill JF, Eds. Plenum Press: New York, NY, USA, pp 85–125

Völker S, Kistemann T (2011) The impact of blue space on human health and well-being–Salutogenetic health effects of inland surface waters: A review. Int J Hyg Environ Health 214(6):449–460

Walch JM, Rabin BS, Day R, Williams JN, Cho K, Kang JD (2005) The effect of sunlight on postoperative analgesic medicine use: a prospective study of patients undergoing spinal surgery. Psychos Med 67:156–163. https://doi.org/10.1097/01.psy.0000149258.42508.70

Wang Y, Xu B, Zhao G, Cao R, He X, Fu S (2011) Is quarantine related to immediate negative psychological consequences during the 2009 H1N1 epidemic? Gen Hosp Psychiatry 33:75–77

Wang X, Qin Y (2011) A water quality monitoring system based on wireless sensor network and cloud computing. Procedia Environ Sci 10:2479–2486. https://doi.org/10.1016/j.proenv.2011.09.385

Ward Thompson C, Roe J, Aspinall P, Mitchell R, Clow A, Miller D (2012) More green space is linked to less stress in deprived communities: evidence from salivary cortisol patterns. Landsc Urban Plan 105(3):221–229

Ware JE, Kosinski M, Turner-Bower DM, Gandeck B (2007) User's manual for the SF-12v2TM health survey. Lincoln, Quality Metric Incorporated

Wang C, Pan R, Wan X, Tan Y, Xu L, McIntyre RS, Choo FN, Tran B, Ho R, Sharma VK, Ho C (2020) A longitudinal study on the mental health of general population during the COVID-19 epidemic in China. Brain Behav Immun 87:40–48

White MP, Elliott LR, Grellier J, Economou T, Bell S, Bratman GN et al (2021) Associations between green/blue spaces and mental health across 18 countries. Sci Rep 11(1):1–12

White MP, Alcock I, Wheeler BW, Depledge MH (2013) Would you be happier living in a greener urban area? A fixed-effects analysis of panel data. Psychol Sci 24:920–928

Wichrowski M, Whiteson J, Haas F, Mola A, Rey MJ (2005) Effects of horticultural therapy on mood and heart rate in patients participating in an inpatient cardiopulmonary rehabilitation program. J Cardiopulm Rehabil 25:270–274. https://doi.org/10.1097/00008483-200509000-00008

Wilkie S, Davinson N (2021) Prevalence and effectiveness of nature-based interventions to impact adult health-related behaviours and outcomes: a scoping review. Landsc Urban Plan 214:104166

Wilkinson CB, O'Connor WA (1982) Human ecology and mental illness. Am J Psychiatry 139:985–990

Wilkinson CB, O'Connor WA (1977) An ecologic approach to mental health: introduction and overview. Psychiatr Ann 7:10–15

Wilson SJ, Lipsey MW (2000) Wilderness challenge programs for delinquent youth: a meta-analysis of outcome evaluations. Evaluat. Program Plann. 23:1. https://doi.org/10.1016/S0149-7189(99)00040-3

Wohl MJ, Squires EC, Caouette J (2012) We were, we are, will we be? The social psychology of collective angst. Soc Personal Psychol Compass 6(5):379–391

Woodward K, Jones JP (2008) The condition of postmodernity (1989): David Harvey. https://doi.org/10.4135/9781446213742.n14

Zhang Y, Wu T, Yu H, Fu J, Xu J, Liu L, Tang C, Li Z (2024) Green spaces exposure and the risk of common psychiatric disorders: a meta-analysis. Ssm–population Health 25:101630. https://doi.org/10.1016/j.ssmph.2024.101630

Zelenski JM, Nisbet EK (2014) Happiness and feeling connected: The distinct role of nature relatedness. Environ Behav 46(1):3–23

Zubala A, Karkou V (2018) Arts therapies in the treatment of depression. Routledge, Abingdon, UK

Chapter 12
Collage Urbanism: Creating a Framework for Citizen-Centric Urban Eco-Psychology

Deepashree Choudhury

Abstract Urban design plays a critical role in shaping the ecological, cultural, and psychological fabric of cities, influencing the wellbeing of their inhabitants. The principles of Collage Urbanism and Eco-Psychology offer valuable frameworks for integrating these diverse dimensions into urban planning. While Collage Urbanism advocates for multi functionality, cultural layering, and spatial diversity, Eco-Psychology emphasizes the importance of human-nature connections for mental health and resilience. Despite the shared goals of these two frameworks, their integration into practical urban design remains underexplored, particularly in terms of creating spaces that are not only ecologically sustainable but also psychologically restorative and culturally enriching. This study proposes an original evaluative matrix for integrating Collage Urbanism and Eco-Psychology, enabling measurable assessment of psychological wellbeing, ecological balance, and cultural inclusivity across urban environments, a framework previously underdeveloped in literature. The intersection of Collage Urbanism and Eco-Psychology is explored through the analysis of three different urban stretches of Park Street, Kolkata, a major metropolis in Eastern India; each demonstrating varying degrees of integration of these principles. The findings suggest that the integration of Collage Urbanism and Eco-Psychology principles can serve as a blueprint for future urban development that prioritizes environmental sustainability, cultural richness, and psychological wellbeing. The chapter contributes to the growing discourse on urban design by offering a comprehensive understanding of how these principles can provide actionable insights for urban planners, architects, and policymakers seeking to create urban environments that balance functionality with the wellbeing of their inhabitants.

Keywords Urban design · Collage urbanism · Eco-psychology · Urban wellbeing · Mental health · Sustainability practices

D. Choudhury (✉)
Professor and Head, Interior Design Department, Budha Group of Institutions, Karnal, India
e-mail: ar.deepachoudhury@gmail.com

K. Mukherjee et al. (eds.), *Urban Climate*, The Urban Book Series,
https://doi.org/10.1007/978-3-032-17895-4_12

12.1 Introduction

Urban design has evolved into a powerful discipline with the potential to influence the ecological, cultural, and psychological wellbeing of urban dwellers. Historically, cities have served as centers of economic activity, cultural exchange, and social interaction. However, the rapid urbanization of the twentieth and twenty-first centuries has brought complex challenges, including overcrowding, environmental degradation, and socio-spatial inequalities. Conventional urban planning approaches, often centered on functional zoning and infrastructural efficiency, have struggled to address the diverse and evolving needs of urban populations (Kumar et al. 2024).

Collage Urbanism, an urban design approach that draws from the artistic concept of collage, and as the art of collage unites the many disparate elements, sometimes molded through multiple mediums to represent a unified finished product capable of a single interpretative and functional theme, collage urbanism emphasizes the integration of diverse spatial, cultural, and functional elements into a cohesive urban fabric (Choudhury 2020). This approach advocates for multifunctionality, inclusivity, cultural layering, and citizen engagement, offering a flexible and adaptive alternative to traditional urban planning paradigms (Rowe and Koetter 1987). Similarly, Eco-Psychology highlights the intrinsic human need to connect with nature, emphasizing the critical role of green spaces and ecological diversity in fostering mental and emotional wellbeing (Hu and Chen 2018).

Although Collage Urbanism and Eco-Psychology appear to address different aspects of urban environments, they share a common underlying objective: promoting sustainability and enhancing the quality of urban life. If the core principles of both these domains are synthesized, they can collectively contribute to building a citizen centric, sustainable urban future that prioritizes the quality of life for all residents.

12.2 Data and Method

12.2.1 Research Aim and Objectives

The primary aim of this research is to develop and apply an evaluative framework that integrates the principles of Collage Urbanism and Eco-Psychology, thereby assessing how urban spaces can simultaneously support spatial diversity, ecological vitality, and psychological wellbeing. The study seeks to demonstrate how the synthesis of these two approaches can contribute to the creation of citizen centric and mentally restorative urban spaces, ultimately promoting more sustainable urban futures. Research Objectives of the study are listed below:

1. To explore how the synthesis of urban design strategies advocated collage urbanism and eco-psychology approaches can contribute to the creation of citizen-centric and mentally restorative urban spaces, ultimately promoting more sustainable urban futures.
2. To formulate an integrated evaluative framework that combines spatial, ecological, and psychological indicators derived from the principles of Collage Urbanism and Eco-Psychology for assessing urban environments at a micro-urban scale.
3. To apply the proposed framework to the selected urban stretches (S1, S4, and S5 of Park Street, Kolkata) to correlate the presence of integrated principles with the degree to which these stretches embody citizen-centric and mentally restorative urban spaces.
4. To qualitatively explore spatial patterns and indicative interrelationships between different principles of integrative framework in order to propose recommendations for the design of more citizen centric, sustainable, and psychologically supportive urban spaces.

12.2.2 Significance of the Study

Even with major improvements in urban planning, fragmentation, socio-spatial isolation, and decreased urban liveability are still major problems in many cities today. Standardized, efficiency-driven design methodologies that give functional zoning precedence over the sociocultural and psychological aspects of urban living are frequently the cause of these difficulties. Consequently, urban areas often become inflexible and alienating, technically useful but devoid of the depth, variety, and emotional resonance required to promote a fulfilling human experience. The increased concern about mental health in urban settings exacerbates this problem. Urban dwellers' rates of stress, anxiety, and depression are rising as a result of urban stressors such as noise pollution, crowding, and the loss of accessible green places.

However, traditional urban planning methods continue to ignore the psychological and emotional effects of the built environment, concentrating only on physical infrastructure without addressing the deeper human need for connection with nature, culture, and community. Collage Urbanism, which draws inspiration from the art form of collage, offers a potential solution to these issues by going beyond traditional planning models to support more resilient and vibrant urban experiences, while eco-psychology focuses on how urban environments can either promote or undermine psychological health depending on the level of human-nature reciprocity ingrained in urban design.

This study fills this critical gap by examining how a synthesized application of Collage Urbanism and Eco-Psychology can contribute to the creation of citizen-centric, ecologically vibrant, and psychologically supportive urban environments. It attempts to provide an evaluative framework that measures not just spatial functionality but also the emotional, cultural, and ecological vitality of urban spaces.

Integrating the principles of Collage Urbanism and Eco-Psychology can be a very powerful way to promote sustainability and enhanced quality of life, but conscious effort toward this is still largely lacking among theorists and practitioners alike, and the effects of such amalgamation in real-world urban contexts remain underexplored.

12.2.3 Data Set and Methodology

This chapter presents a literature study which focuses on theoretical underpinnings of collage urbanism and Eco-Psychology and their interrelationship. Academic publications such as journal articles, conference papers, and review papers were given priority. Over the course of three months, from November 2024 to January 2025, this review was carried out. Relevant literature was identified through systematic keyword searches across major academic and institutional databases, including ScienceDirect, Web of Science, JSTOR, Scopus, and SAGE Journals. Keywords such as Collage Urbanism, Eco-Psychology and Urban Design, Environmental Psychology and Mental Health, Citizen-centric Urban Design, Urban Green Spaces and Wellbeing, Multifunctionality in Urban Public Spaces, Cultural Layering in Urban Planning, Urban Biodiversity and Psychological Health, Public Space Adaptability and Mental Wellbeing, Human-Nature Reciprocity in Cities, and Sustainable Urban Morphology were used to guide the searches. In addition to database-driven searches, backward and forward snowballing (nonprobability sampling technique) was employed for identifying additional relevant studies (Wohlin et al. 2022; Mengist et al. 2020).

Specific repositories such as the Urban Design Research Institute (UDRI, Mumbai), the World Resources Institute (WRI), the World Green Building Council (World-GBC), and specialized platforms like DELNET and ShodhGanga were consulted for region-specific research and global perspectives. Resources from the WHO, particularly the Urban HEART Tool (World Health Organization 2010), along with SAGE Journals and the American Psychological Association (APA) databases, provided crucial insights into the psychological dimensions of urban design. Sustainability frameworks like those developed by UN-Habitat and the Rockefeller Foundation's 100 Resilient Cities program were also studied. Initially, in the first phase, a pool of 120 research papers was identified. After rigorous screening to eliminate duplications, irrelevance, and ensure alignment with the study's aims and objectives, 40 high-quality publications were selected for detailed analysis. Select working papers and books were also reviewed. From the literature study, the evaluative framework consisting of 9 parameters were deduced as per the objective stated.

Case Study Selection: Park Street, Kolkata, was divided into six stretches, each approximately 350–500 m long, to facilitate systematic study and data collection. Among these, three stretches—S1, S4, and S5—were selected for detailed primary data collection and the application of the integrated evaluation framework developed through the literature review. The selection of S1, S4, and S5 was guided by a stratified purposive sampling approach. This involved deliberately identifying subgroups or "strata" that are meaningfully different based on critical urban variables such as land use, spatial configuration, cultural expression, ecological integration, and patterns of public engagement. The objective was to capture inherent internal spatial consistency within each stretch, consistency in land use types, edge conditions, and plot configurations, rather than uniformity across all stretches. This consistency is defined as "homogeneity" in this study. These characteristics were mapped and verified through physical surveys and available cartographic documents. Based on this mapping, S1, S4, and S5 were found to exhibit the maximum degree of internal homogeneity while still representing distinct typological diversity necessary for comparative analysis. The three selected stretches represent distinct spatial typologies: S1: High commercial intensity, strong cultural layering, and relatively higher green space integration; S4: Mixed-use character with institutional and commercial dominance, moderate ecological richness, and moderate cultural identity; S5: Dense residential commercial mix with low green coverage and relatively weaker cultural expression. Simultaneously, macro-level homogeneity was maintained: all selected stretches belong to the same historical corridor of Park Street, developed during a similar period, subjected to the same climatic, cultural, and administrative influences, and anchored in a common urban morphological framework characterized by major arterial connectivity and colonial-era spatial patterns. This consistent broader context ensures that any observed differences across the stretches arise from intrinsic urban factors rather than from external environmental disparities, enhancing the internal validity of the comparative analysis.

Data Collection Strategy: Data was collected through semi-structured interviews and structured questionnaires. Face-to-face interviews with 30 key informants from each stretch, including urban planners, local government officials, business owners, community leaders, residents, and visitors to the area provided in-depth insights. Field observations, site survey, activity mapping, and building audit were other methods applied to collect data about selected indicators. Smart Maps, the first base map of the city prepared by Colonel Smart in 1905, drawn at a scale of 1:600 (Dasgupta 2018), and GIS maps were sourced from Kolkata Municipal Corporation and used for triangulating site surveys, field visits, and building audits. The evaluation of both mental health indicators and perceived connection to nature was conducted using a Likert scale survey, where participants rated their agreement with specific statements on a scale of 1 to 5. 30 participants from each stretch rated predefined statements, and average scores were calculated. The statements addressed stress reduction, emotional wellbeing, and the restorative impact of public spaces, while for perceived connection to nature, they focused on feelings of closeness to nature, enjoyment of green spaces, and satisfaction with natural elements. The formula applied for analysis in both cases was:

$$\text{Average Score} = \text{Total of 5 Ratings Provided by Respondent/Number of Ratings}$$

For instance, if a participant provided ratings of 4, 4, 3, 5, and 2 for five statements, the total score would be 18, with an average score of $18/5 = 3.618/5 = 3.618/5 = 3.6$. This process was repeated for all 30 participants per stretch, and the aggregated average scores were compared across the three stretches to assess differences in mental health benefits and perceived connection to nature.

The study adopted an exploratory sequential mixed-methods approach, starting with qualitative field observations and semi-structured interviews to identify patterns, followed by structured surveys using Likert scales to assess perceptions across predefined parameters. The use of both qualitative and quantitative tools allowed cross-validation of findings (triangulation), ensuring greater analytical depth.

12.2.4 Scope and Limitations

Although convenience sampling was used for primary data collection because of practical limitations such as time, access allowances, and respondent availability in dynamic urban environments, attempts were made to diversify the respondent base by age, occupation, and socio-economic background to reduce inherent sampling biases. Bias was also reduced by respondent stratification across demographic dimensions and through triangulation of survey evidence with field observation and secondary urban data. Since this is an exploratory, framework-building study, convenience sampling is acceptable for preliminary stage conceptualization, not for population generalization. Since this study's main purpose was exploratory, to build and use an integrated evaluative framework and not to statistically generalize to the larger population, convenience sampling was deemed suitable for identifying initial trends and relations.

Although the research aims to correlate the occurrence of integrated Collage Urbanism and Eco-Psychology principles with the level of citizen centricity and psychological restorative capacity of urban spaces, no systematic statistical correlation techniques were used. This is due to the fact that the main purpose of the research was exploratory and diagnostic in nature, aimed at qualitative and semi-quantitative assessment and not statistical validation. The integrated framework conceived is based on interpretive evaluation using structured field observation, perception surveys, and thematic analysis of spatial characteristics. Next studies, utilizing larger datasets and standardized variable control, can utilize formal statistical methods to confirm the exploratory findings outlined here.

12.3 Collage Urbanism

Collage Urbanism emerged as a critical response to the homogenizing tendencies of modernist urban planning. As a concept, it draws inspiration from the artistic notion of a collage—an assembly of diverse elements that coexist harmoniously within a single framework. The idea originated in the mid-twentieth century, particularly through the work of architects and urban theorists like Colin Rowe and Fred Koetter, who introduced it in their book Collage City (Rowe and Koetter 1987). They argued against rigid, modernist approaches to urban planning, proposing instead an adaptable and incremental urbanism that allows for diversity, layering, juxtaposition of diverse urban elements, embracing complexity, multiplicity, and temporal layering in urban design (Daglioglu 2016). Unlike the monolithic planning paradigms of modernism, collage urbanism embraces the complexities of urban environments, acknowledging the multiplicity of needs, identities, and histories present in cities. Its emphasis on pluralism and contextual adaptability makes it particularly relevant in contemporary urban scenarios characterized by rapid change and diverse populations (Davis 2017).

12.3.1 Principles of Collage Urbanism

1. The Concept of Layering and Integration: Welcomes a mosaic of contrasting urban components, mixing old and new, built and natural, functional and aesthetic, and attempts to produce an interesting interplay between past context and current requirements.
2. The Concept of Adaptability and Flexibility: Encourages urban structures that are adaptable and resilient to alteration over time, supporting changing societal, environmental, and technological requirements.
3. Pluralism and Inclusivity: Enjoys cultural, social, and spatial diversity in the city's fabric, promoting inclusiveness for everyone in the community and inviting designs to respond to the identities and hopes of diverse user groups.
4. Contextuality: Values the local context, geography, climate, culture, and community needs in guiding design and planning decisions.
5. Fragmentation as Strength: Sees fragmentation not as a constraint but as a chance to build a richly textured urban context to advance creativity and engagement.
6. Interconnection and Continuity: Creates connections among contrasting urban components through green corridors, routes, and common public areas, encouraging connectivity and cohesion.
7. Human-Centric Design: Centers around human experience, making spaces accessible, livable, and supportive of social interaction, mental wellbeing, and productivity.
8. Sustainability and Ecology: Comprehends green infrastructure and sustainable practices to increase environmental quality and urban resilience and promotes coexistence of built and natural environments for ecological equilibrium.
9. Participatory Processes and Citizen-Focused Urbanism: Engages stakeholders, such as inhabitants and neighborhood communities, in the design and planning process to make spaces appealing and functional for them and align with their aspirations. When people are involved proactively, they become more attached and own the spaces more, leading to long-term success and sustainability of urban initiatives.
10. Artistic Expression and Innovation: incorporates creative and artistic elements in urban design, fostering a sense of identity, belonging, and visual appeal.

Collage Urbanism is a versatile and inclusive approach that integrates diverse elements into a cohesive urban framework, prioritizing adaptability, sustainability, and human wellbeing (Choudhury 2024).

12.3.2 Contemporary Interpretations and Applications: Explorations Through an Integrated the Lens of Collage Urbanism Principles

Contemporary urban practitioners and theorists have built upon Collage Urbanism's original principles, incorporating them into modern-day design practices. Design technique of collage has been used multiple times in urban contexts, highlighting its ability to refurbish spaces through the superposition of cultural, historical, and functional aspects (Daglioglu 2016). The application of collage techniques within urban design helps to integrate disparate urban elements toward the creation of a harmonious urbanscape that is not only visually diverse but socially inclusive as well (Choudhury 2024).

The real-world implementation of Collage Urbanism theories can be observed in various worldwide urban projects incorporating heterogeneous elements in order to achieve dynamic, inclusive, and sustainable spaces. The projects demonstrate the ways in which the principles of Collage Urbanism, including layering, flexibility, inclusivity, contextuality, fragmentation, interconnectedness, human-oriented design, sustainability, participatory mechanisms, and artistic expression, may be translated in actual environments.

Barcelona's Superblocks strategy remodels urban streets into pedestrian spaces, diminishing car traffic and stimulating community engagement. The strategy is reflective of layering by incorporating dwellings, businesses, and parks in existing urban form. It reflects adaptability by employing a versatile design that supports a range of activities for the community, and inclusivity by repurposing streets as public realms. The strategy is courteous to the city's grid history, exhibiting contextuality, and builds connections with pedestrian networks, adhering to interconnection principles (Nieuwenhuijsen et al. 2024). By emphasizing quality-of-life improvement and the integration of green infrastructure, it also responds to human-oriented design and sustainability. The development process incorporated community participation, emphasizing participatory processes, and promotes creative public space use, reflecting artistic expression.

The New York City High Line converts a derelict railway into a raised linear park, combining industrial heritage with natural landscapes and modern design, demonstrating layering and integration. Its adaptive reuse of outmoded infrastructure demonstrates adaptability, and retaining historical aspects demonstrates contextuality. The linear design produces distinctive urban experiences, making fragmentation an asset. Through joining different neighborhoods together, it boosts interconnection and provides a serene area for relaxation, focusing on human-centric design (Kim and Chae 2018). The park includes rotating artworks, mirroring artistic expression. Nevertheless, critiques emphasize limitations in multifunctionality and community adaptability, suggesting areas of improvement in inclusivity and cultural expression.

HafenCity in Hamburg redevelops former port lands into a mixed-use urban quarter combining old warehouses with contemporary design, showing integration and layering. Its architecture supports changing urban requirements, showing adaptability, and provides varying housing and cultural spaces, ensuring inclusivity. It shows Hamburg's maritime culture heritage, supporting contextuality, and integrates well into existing infrastructure, adding interconnection (Eleftheriou and Knieling 2017). Prioritizing public spaces and pedestrian access emphasizes human-centered design, while the inclusion of environmentally friendly practices addresses sustainability. Stakeholder engagement in planning processes emphasizes participatory processes, and architectural diversity promotes artistic expression.

Bogota's Ciclovia converts urban streets into vehicle-free spaces each Sunday, promoting recreational activities and socialization. This short-term restructuring emphasizes adaptability and flexibility, and accommodates participants of all socio-economic status, supporting inclusivity. By fulfilling local recreation needs, it demonstrates contextuality, and uses multiple city areas to develop a network of activities, making fragmentation its strength. As the network of activities extends beyond the city center, it gains contextual depth, becoming more inclusive (Montero 2017). Linking neighborhoods along active transportation pathways increases interconnectivity, promoting health and happiness highlights human-scaled design. Minimizing vehicle emissions within events supports sustainability, and community engagement highlights participatory processes. Arts performances and exhibits during the events highlight artistic expression.

Delhi's Raahgiri Day, motivated by Bogota's Ciclovia, encourages non-motorized travel and community participation by temporarily taking back streets for public space. It encourages participation from all segments of society, emphasizing inclusivity, and adapts activities to local cultural environments, emphasizing contextually. Mobilizing multiple city areas simultaneously converts fragmentation into a strength and promotes sustainable mobility, increasing interconnection. By enhancing public health and social cohesion, it focuses on human-centered design, promoting environmentally friendly modes of transportation, and addressing sustainability. Its focus on enabling citizens to influence urban experiences highlights participatory processes, while cultural performances and creative workshops symbolize artistic expression. Its transient nature restricts long-term ecological effects and infrastructure development (Global Designing Cities Initiative, n.d.).

Community gardens in Detroit represent grassroots initiatives transforming vacant lots into productive green spaces, enhancing food security and community well-being. Integrating agriculture into urban settings demonstrates layering and integration, while utilizing underused spaces for evolving community needs illustrates adaptability. Engaging diverse populations in urban farming promotes inclusivity, and responding to local economic challenges reflects contextuality. Re-zoning isolated vacant lots into community assets makes fragmentation a virtue, and establishing networks of green space throughout neighborhoods maximizes connection.

Enhancing mental wellbeing and social interaction prioritizes human-scale design, and encouraging environmental stewardship works toward sustainability (Hu and Chen 2018). Led by community initiative, these projects highlight processes that are participatory in nature, and include creative landscaping and community art initiatives, highlighting artistic expression.

Although it is theoretically rich and has practical uses, Collage Urbanism has no universally applied method of measurement for its use in urban spaces. Current urban evaluation tools, like those emphasizing sustainability indicators, tend not to reflect the subtlety principles of Collage Urbanism, such as cultural layering or adaptive reuse. To fill this void, there is a critical need for a structured evaluative method that formalizes Collage Urbanism's qualitative principles into quantifiable parameters. Such a framework would allow urban designers, planners, and researchers to systematically evaluate the existence and efficacy of collage elements in urban environments.

12.3.3 Development of the Assessment Framework

To address this need, the following framework of assessment has been formulated based on the tenets of Collage Urbanism and informed by recent urban design literature. The framework demarcates significant parameters representative of collage principles, including spatial variety, cultural diversity, and ecological incorporation. Every parameter is linked to particular indicators and measures, allowing for a complete assessment of urban areas based on Collage Urbanism. Table 12.1 shows this framework, providing a tool to measure the applications and effect of Collage Urbanism in urban areas. Table establishes 15 parameters and their respective indicators with rational behind their selection.

12.4 Literature Review

12.4.1 Eco-Psychology in Urban Contexts

Eco-psychology is a cross-disciplinary framework which brings together ecology, science of organism relationships and biophysical environment with Psychology; science of mind, behavior, and emotion processes (Thompson 2009; Fridgen 2005). The discipline deals with the dynamic relationship between man and environment, recognizing that change in one will necessarily impact the other (Gifford 2001; O'Neill and Teka 2023). This dual lens is important at a time when cities, with increased investment in physical infrastructure such as parks, streets, and public spaces, are still facing ecological deterioration and deteriorating mental health of residents (Pelletier et al. 2008).

Table 12.1 Assessing integration of collage urbanism in an environment. *Source* Author

	Collage urbanism principle	Parameter	Indicator	Indicator description	Measurement methodology	Collage urbanism integration
1	Layering and integration	Spatial layering	Percentage of premises showing effective integration of historical and contemporary elements	Evaluates old and new integration	Physical surveys, community surveys and interviews	Evaluates the synthesis of past and present, demonstrating how historical continuity and modern interventions create a layered urban narrative, reflects blending of timescales
2	Adaptability and flexibility	Temporal flexibility	No. of multi-use spaces	Spaces adapting over time (day, week, seasons)	Site observations over different times, interviews, usage analysis	Reveals the adaptability of spaces across different times and uses, supporting a flexible, evolving urban landscape inherent to collage principles, shows multifunctionality of space
3	Pluralism and inclusivity	Cultural inclusivity	Percentage of premises representing and accommodation of diverse cultural identities in any form including architecture, events, activities	Heritage, art, hubs incorporated	Physical surveys, activity mapping, community surveys and interviews	Assesses the visibility and accommodation of multiple cultural identities, reinforcing urban diversity and inclusivity essential to collage urbanism, celebrates diversity and identity
4	Contextuality	Contextual Responsiveness	Percentage of premises showing significant alignment with geography, climate, culture	Response to environmental/ social context	Physical surveys, cultural profiling of spaces, community surveys and interviews	Measures the urban fabric's responsiveness to local conditions, ensuring rootedness and relevance, key for contextual collage integration, ensures design sensitivity

(continued)

Table 12.1 (continued)

	Collage urbanism principle	Parameter	Indicator	Indicator description	Measurement methodology	Collage urbanism integration
5	Fragmentation as strength	Fragmentation utilization	Percentage of premises showing Innovative use of fragmented spaces	Creative reuse of underutilized spaces	Physical surveys, community surveys and interviews, consultations with experts	Highlights how disjointed or leftover spaces are creatively reimagined to enrich the urban texture, a hallmark of collage urbanism, showcases creativity
		Fragmentation balance	Degree of linkage in fragmented spaces (both visual and spatial), Physical linking features (bridges, corridors)	Distribution of cohesive vs fragmented areas	Physical surveys, connectivity analysis, community surveys and interviews	Examines the controlled coexistence of fragmentation and cohesion, embodying collage urbanism's embrace of spatial complexity, merging urban diversity
6	Interconnection and continuity	Walkability	Walkability scores	Establishing links through pedestrian pathways	Walkability index scoring, pedestrian counts, physical surveys, community surveys and interviews	Indicates the permeability and continuity of the urban fabric, ensuring fragmented elements are meaningfully linked, assesses accessibility
		Connectivity	Quality and extent of Physical linkages	Measures accessibility in terms of roads and access and their quality	Connectivity analysis after physical survey, Observation, community surveys and interviews	Assesses the strength of networks connecting diverse spaces and users, supporting the holistic integration of fragmented urban elements
		Permeability	Ease of movement through space	Open and interconnected layouts		Measures how freely users can traverse spaces, critical for creating fluid, collage-like urban experiences, freedom of circulation

(continued)

Table 12.1 (continued)

	Collage urbanism principle	Parameter	Indicator	Indicator description	Measurement methodology	Collage urbanism integration
7	Human-centric design	Human experience	User comfort and engagement	Psychological and emotional wellbeing	community surveys and interviews, behavioral mapping, emotional wellbeing scales	Evaluates how urban environments prioritize human needs and interactions, enhancing inclusivity and liveability within collage urbanism
8	Sustainability and ecology	Ecological integration	Percentage of premising accommodating green infrastructure and sustainability practices	Extent of green features		Reflects how natural systems are integrated into the urban matrix, ensuring environmental resilience within a layered urban fabric
9	Participatory urbanism	Community participation	Level of stakeholder engagement	Resident involvement in planning	Public meeting records of local municipality, community surveys and interviews	Captures how collective input shapes urban spaces, fostering community ownership and authentic collage creation
		Social interaction spaces	No. of community spaces in a specified area	Public plazas, parks, gathering spaces	Physical surveys, studying available maps, field observations	Measures the facilitation of social interaction and cohesion, vital for a vibrant and inclusive collage urbanism environment
		Technology integration	Percentage of public spaces with wi-fi, other smart city features	Digital infrastructure presence	Physical survey with specifically designed survey instrument, community surveys and interviews	Assesses the digital inclusivity of spaces, ensuring technology enhances the layered, interconnected fabric of the urban landscape

(continued)

Table 12.1 (continued)

	Collage urbanism principle	Parameter	Indicator	Indicator description	Measurement methodology	Collage urbanism integration
		Economic inclusivity	Percentage of commercial areas devoted to Informal and small-scale economic opportunities	Space for vendors	Physical survey with specifically designed survey instrument, Interviews	Demonstrates support for economic diversity, integrating various socio-economic layers within the urban collage, economic resilience and inclusivity
10	Artistic Expression and innovation	Artistic and innovative expression	Presence of creative elements	Artistic identity and innovation in urban fabric	Site surveys, documentation of art installations, community surveys and interviews	Highlights the role of creativity in enriching the urban tapestry, reinforcing identity and innovation within collage urbanism

The idea of Eco-Psychology was first put forward by Theodore Roszak (Roszak 2021), and it was he who brought forward the term ecological unconscious: a profound, yet repressed, human connection with nature. Roszak thought that reclaiming access to this unconscious realm is critical to discovering solutions to collective psychological and ecological crises (Igor 2020). Eco-psychology engages a number of significant dimensions: emotional response to environments in nature, psychological effect of environmental harm (for instance, harm triggered by natural catastrophes or global change), and the transpersonal experience of environment-related identity (Kellert and Speth 2009). New metropolitan work supports such arguments, demonstrating how experience in nature sites significantly benefits mental wellness indicators (Jiang et al. 2024). For instance, in Kolkata, a survey found that 41% of respondents felt relaxed in green spaces (Biswas and Sen 2020). These observations coincide with global policy efforts such as the United Nations Sustainable Development Goals (SDG 11.7) that require the universal access to green public spaces that are inclusive by the year 2030 (Sorooshian 2024).

Aside from tangible physical health benefits, research indicates that urban biodiversity and visibility of nature, even from one's home, are closely correlated with decreased stress, aggression, and mental fatigue (Alcock et al. 2013). Conservation psychology, a branch of environmental psychology, also underscores the importance of establishing emotional connections with nature in biodiversity conservation as well as environmental sustainability (Elsadek et al. 2019; Gaekwad et al. 2022).

Grounded on these theoretical foundations, the ensuing core principles of eco-psychology can be extracted:

- Human-Nature Connection: Reinforcing everyday engagements with nature to develop psychological resilience
- Therapeutic Approaches: Applying nature exposure as a mechanism for emotional recovery and wellbeing (Hu and Chen 2018).
- Holistic Integration: Merging ecological and psychological issues in urban planning to ensure sustainability (Moutel 2016).
- Awakening Environmental Reciprocity: Reviving the intrinsic human tendency to care for and live alongside the natural world (Nieuwenhuijsen et al. 2024).
- Enhancing Ecological Wellbeing and Biodiversity: Ensuring that urban ecosystems are resilient and service providing (Kawamura and Brady 2022).

Therefore, an urban design paradigm that incorporates eco-psychological principles should target measurable aspects such as accessibility to nature, environmental stewardship, biophilic urban design, social connectivity, and psychological wellbeing.

Therefore, drawing on the theoretical findings outlined, Table 12.2 consolidates main Eco-Psychology principles into quantifiable parameters and indicators. This systematic framework allows for a methodical assessment of how urban spaces can facilitate psychological wellbeing, ecological wellness, and human-nature exchange at a micro-urban level. Each parameter is directly related to main eco-psychological principles, thereby ensuring conceptual clarity and operational utility for urban evaluation. These are encapsulated in Table 12.2:

Table 12.2 Eco-psychology principles and corresponding urban evaluation metrics. *Source* Author

Eco-psychology principle	Parameter	Indicator	Indicator description	Measurement methodology	Importance of integrating this eco-psychology indicator in urbanscape
Human-nature connection	Human-nature interaction	Time spent in natural settings	Measures the frequency and duration of individual interactions with natural environments (parks, forests)	Surveys, behavioral diaries tracking hours/week or visits/month	Promotes mental restoration, reduces urban stress, and nurtures emotional resilience
Accessibility to nature	Green spaces availability	% of green cover in urban areas, proximity to public green spaces	Assesses the spatial distribution and accessibility of green areas in urban settlements	spatial analysis, accessibility mapping within 300 m buffer zones	Enhances physical and psychological access to restorative natural environments
Environmental stewardship	Environmental engagement	Participation in environmental programs, volunteer hours	Captures citizens' active participation in ecological activities and advocacy	Surveys, volunteer program records, participation counts	Cultivates environmental ethics and collective urban stewardship
Psychological resilience	Psychological wellbeing	Stress reduction, Mood improvement after nature exposure	Examines emotional and psychological improvements post nature interaction	Psychological scales (Likert survey), physiological tests	Highlights the critical mental health benefits of nature integration in dense urban environments
Biophilic urbanism	Integration in urban design	Presence of biophilic design elements, Tree density	Evaluates how nature is intentionally integrated into urban form and architecture	Physical audits of biophilic features	Strengthens sensory connections with nature, fostering emotional and ecological ties
Education and awareness	Educational initiatives	Number of eco-psychology workshops, Enrollment rates	Assesses the availability and uptake of educational activities related to Eco-Psychology and sustainability	Institutional reports	Builds long-term ecological literacy and supports eco-conscious citizenship

(continued)

Table 12.2 (continued)

Eco-psychology principle	Parameter	Indicator	Indicator description	Measurement methodology	Importance of integrating this eco-psychology indicator in urbanscape
Social connectivity	Social cohesion and interaction	Community activities in natural settings, Sense of belonging	Measures the role of green public spaces in fostering community ties and social wellbeing	Event logs, community survey scores on belonging (scale 1–10)	Reinforces urban social networks and contributes to inclusive, resilient communities
Ecological health	Biodiversity and ecosystem services	Native species richness, air purification, pollination	Assesses the richness and functionality of urban ecosystems and their service provision	Biodiversity field surveys, ecosystem service evaluations	Promotes resilient ecosystems, ensuring ecological services critical to human survival

12.4.2 Integration of Collage Urbanism and Eco-Psychology: Toward a Unified Framework

Though Collage Urbanism and Eco-Psychology have their origins in distinct traditions of discipline; urban design and environmental psychology respectively, their shared hope for citizen-oriented, inclusive, resilient, and sustainable urban living is profoundly conjoint. Both call for urban environments that promote diversity, flexibility, emotional comfort, environmental health, and participatory government (Dawes and Ostwald 2017). Aware of these convergences, this research suggests that their unification can be systematized by a series of synthesized principles, bringing theoretical commonalities to practical urban design principles. This shift from abstract alignment to practical frameworks serves an important research gap by providing a systematic approach to assessing and improving urban areas with spatial, ecological, and psychological perspectives all at once (Agrawal 2022).

12.4.3 Proposed Evaluative Framework for Integration

In order to operationalize this transdisciplinary synthesis, the research presents an innovative evaluative system that integrates the theoretical issues of Collage Urbanism and Eco-Psychology into empirical dimensions. Contrary to earlier research that has addressed urban form, ecology, and psychology separately or qualitatively, this system facilitates comparative, reproducible, and context-specific analysis at a micro-urban scale.

The structure (seen in Table 12.3 is organized across ten fundamental parameters), each one of them has definite indicators to it, measurement procedures, and ranking methodology. To what gives this structure the edge is the fact that it is holistic in approach and transdisciplinary in approach; Parameters such as Mental Health Score and Sense of Connection with Nature are produced from ecological psychological studies but are measured with spatial design evaluation and user feedback surveys; Quantifiable such as Multifunctionality and Cultural Layering embody the Collage Urbanism's spatial logics while establishing connection to sociopsychological wellbeing outcomes.

12.5 Park Street, Kolkata

Kolkata's Park Street is one of India's most iconic roads, embodying both the city's resurgence and the complexities of its development. Park Street serves as a physical and symbolic link between colonial Calcutta and modern-day Kolkata (Refer to Fig. 12.1). It reflects the city's journey from a glorious past to a more modest present, marked by the replacement of colonial mansions and branded stores with humbler,

Table 12.3 Proposed evaluative framework for integration of collage urbanism and eco-psychology principles. *Source* Author

Integrated principle	Rooted in collage urbanism	Rooted in eco-psychology	Indicator	Indicator description	Measurement methodology	Remarks—how indicator measurement helps achieve objective
Layered spatial integration	Emphasis on mosaicking old and new functions	Recognition of environmental layers supporting wellbeing	Diversity of land use within walkable radius	% mix of residential, commercial, cultural, and ecological elements	Physical survey, secondary land use records	Shows integration of built and natural elements and fosters diverse citizen engagement
Multifunctional nature-culture spaces	Layering, pluralism, fragmentation	Human-nature reciprocity, holistic integration	Diversity of space use	Number and type of simultaneous ecological, cultural, and social activities supported	Physical survey, activity mapping	Aids by evaluating multifunctionality; Helps by assessing the experiential diversity in real cases
Human-nature reciprocity	Urban elements designed to connect built and green	Psychological health through interaction with nature	Perceived mental health score	Average Likert scale rating (1–5) of stress reduction/calmness	Structured user survey	Measures ecological exposure and psychological health directly
Adaptive public realm	Flexible, changing, multi-use urban spaces	Psychological comfort through adaptable environments	Flexibility of public spaces	Number of different activities spaces can support across seasons	Field observations, physical survey, public event calendars	Demonstrates space adaptability and resilience

(continued)

Table 12.3 (continued)

Integrated principle	Rooted in collage urbanism	Rooted in eco-psychology	Indicator	Indicator description	Measurement methodology	Remarks—how indicator measurement helps achieve objective
Connectivity and continuity	Linking fragmented urban spaces through corridors	Continuity strengthens sense of belonging and ecology	Green and physical linkages	Percentage of roads, street, trails connecting different premises and giving access to them in the selected area	physical survey, Available GIS MAPS	Reveals ecological and experiential continuity
				Percentage of shaded connections mapped		
Participatory design	Citizen involvement as a design force	Empowerment fosters psychological ownership	Citizen participation index	% of citizens involved in design/maintenance events	Interviews	Measures degree of citizen-centrism
Eco-functional sustainability	Green infrastructure as part of urban fabric	Ecological balance nurtures long-term wellbeing	Sustainable practices implementation	No. of green infrastructure interventions (solar, rainwater harvesting, etc.)	Site audits, secondary documents, Interviews	Links sustainability goals to eco-psychological outcomes
Emotional place attachment	Spaces designed to nurture emotional bonds	Rooting emotional wellbeing in physical place identity	Sense of belonging score	Average Likert score (1–5) on place identity and pride	Resident surveys and interviews	Measures depth of human-place relationships
Temporal adaptability and resilience	Acceptance of temporality and urban change	Dynamic urban experience supports mental agility	Seasonal/temporal space use variations	Documentation of different uses across time	Longitudinal field observations	Shows time-resilience of urban collage

Fig. 12.1 Park Street, Kolkata. *Source* Author

deteriorating establishments. This transformation mirrors Kolkata's history, transitioning from its colonial prominence to the post-independence era of refugee influx and urban decline. Stretching approximately 3.5 km from Chowringhee Road in the west to Park Circus Station in the east, Park Street showcases a gradient of social and spatial usage, from the grandeur of colonial heritage to the neglected urban fabric near Park Circus. Emerging from the old colonial town, it is surrounded by diverse districts: Salt Lake City, developed in the 1960s, and the rapidly expanding Rajarhat to the east, with Kolkata International Airport to the northeast and South Kolkata's main residential areas to the south. Renamed Mother Teresa Sarani in 2001, Park Street continues to symbolize Kolkata's evolving urban landscape.

For this study, Park Street is divided into six distinct stretches (S1 to S6) (Refer to Fig. 12.2). This study includes three stretches namely S1, S4, and S5 as case studies. Rather than focusing on analyzing urbanscape purely through historical or architectural lenses, this analysis applies an integrated eco-psychological and collage urbanism matrix to measure spatial quality and citizen experience.

12.5.1 *Integration of Collage Urbanism and Eco-Psychology: Toward a Unified Framework*

The first stretch, designated as S1, is primarily characterized as a retail and entertainment hub, comprising seven prominent premises. This area is marked by landmark

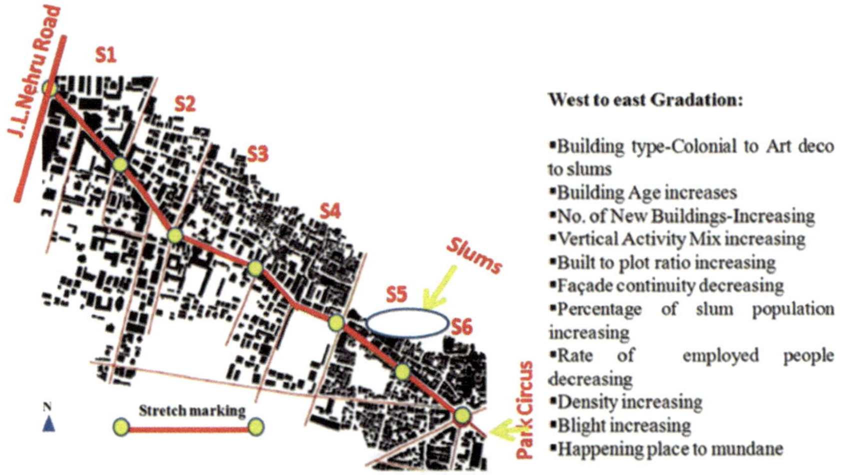

Fig. 12.2 Stretch division in Park Street, Kolkata, and description of changing urban characteristics. *Source* Author

buildings with significant historical associations, showcasing the influence of European architectural styles from the colonial era. The structures, ranging from G + 1 to G + 4 stories, occupy large premises spanning between 1 to 4 acres. These buildings include a mix of institutional structures, mixed-use developments with offices and residences, and a 5-star hotel. The stretch reflects a sense of exclusivity and status inherited from colonial times, emphasized by its grand buildings, courtyards, plazas, wide roads, and spacious footpaths. It features excellent connectivity at both city and local scales, pedestrian-friendly infrastructure with shaded footpaths, inviting meeting spaces, active edges, and a strong visual identity. Four premises within this stretch consist of perimeter blocks arranged around courtyards, with transparent retail facades at the ground level. These house clubs, bars, showrooms of international brands, arcades with street-front shops, and vibrant activities spill over onto the pavements. Additionally, vendors selling books and imported items enhance the liveliness of the area (Refer to Figs. 12.3, 12.4, and 12.5).

Stretch S4 spans approximately 400 m and features plot sizes ranging from 0.3 acres to 10 acres. This segment accommodates more mid and high-rise buildings compared to the previous stretches and primarily serves as an institutional and retail zone. The area is characterized by a diverse mix of building types, including multi-story structures with retail spaces on the ground floor and offices or residences on upper levels, apartment blocks, a continuous line of temporary roadside eateries, public institutions, parks, and garages. Major landmarks in this stretch include a hospital, an old church, a missionary school, and a 300-year-old cemetery renowned worldwide for its remarkable collection of tombs, which serves as a unique historical reference point in this otherwise modernized area. However, other historical connections are minimal. The stretch has a low percentage of street-front retail and transparent facades, with many buildings set back from the road behind boundary

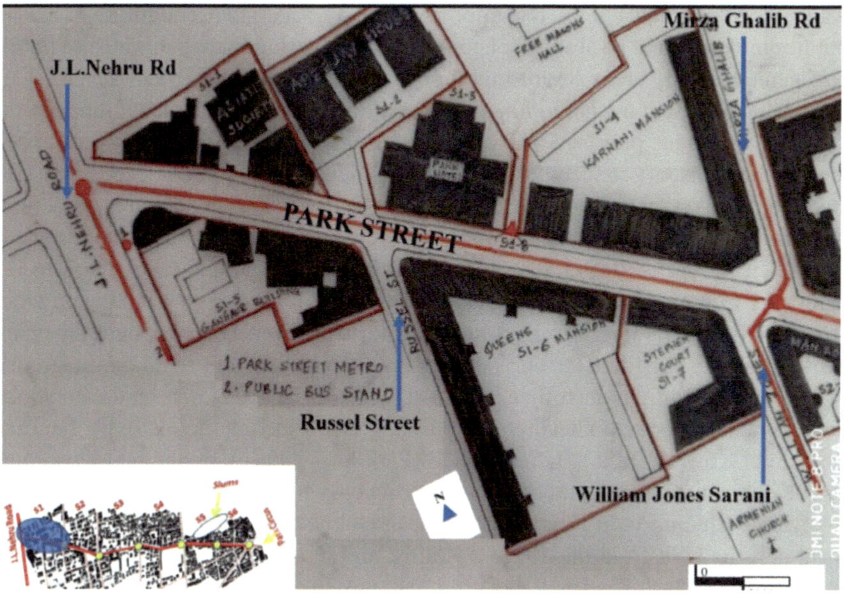

Fig. 12.3 Figure-ground map of Stretch 1, Park Street, Kolkata. *Source* Author

Fig. 12.4 Part Elevation of Street, Stretch 1, Park Street, Kolkata. *Source* Author

Fig. 12.5 A few buildings of Stretch 1, Park Street, Kolkata. *Source* Author

walls. This, coupled with a reduction in publicly accessible buildings, a lack of public amenities, and an absence of comfortable gathering spaces, diminishes public engagement in the area. Architectural styles here range from Indo-Deco to sleek modern designs, but the predominance of institutional use limits the vibrancy of the stretch. Despite the presence of Parkomat, a multi-story parking facility, and excellent public transport connectivity making it an ideal entry point to Park Street, the stretch fails to retain visitors for extended periods due to the lack of lively attractions (Refer to Figs. 12.6, 12.7 and 12.8).

Stretch S5 spans 300 m. The plot sizes in this stretch range from 0.03 to 0.2 acres and accommodate more mid and high-rise buildings compared to the previous stretches. It is primarily a mixed-use zone, featuring a diverse range of building types. Most structures are three to four stories high, with street-front retail on the ground floor, small institutions and offices like coaching centers and clinics on the first floor, and residences above. Other notable building types include hotels (e.g., Jameson Inn, Hotel Royal Garden), a cinema hall (Park Showhouse), public buildings (e.g., Park Circus Post Office), and mosques. Architecturally, the stretch showcases a mix of styles, including Indo-Deco and European architectural elements combined with local designs built from the 1920s onwards, and modern glass structures from the twenty-first century. The small plot sizes and closely packed buildings with minimal front and side setbacks, along with continuous street-front retail, contribute to a dense

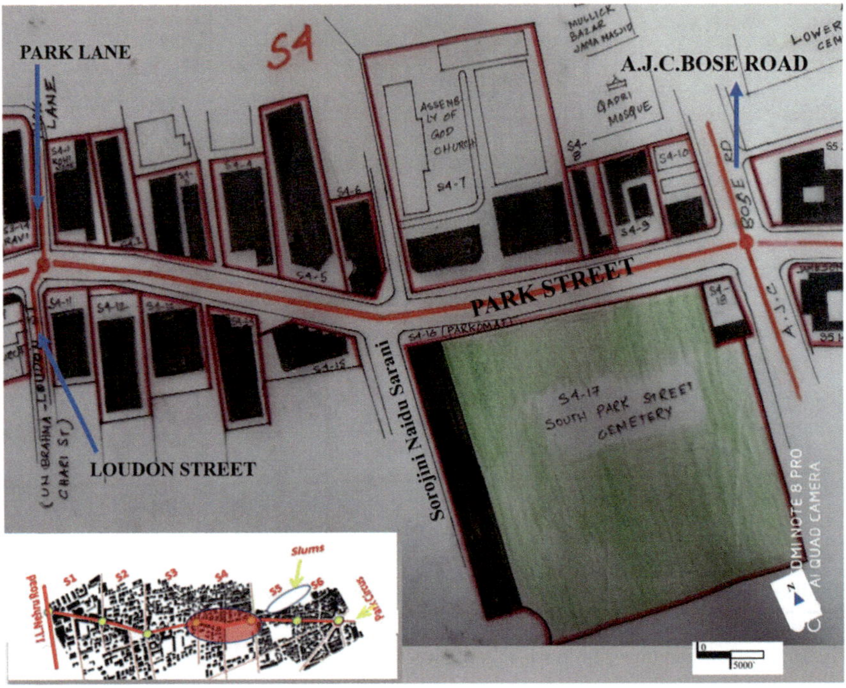

Fig. 12.6 Figure-ground map of Stretch 4, Park Street, Kolkata. *Source* Author

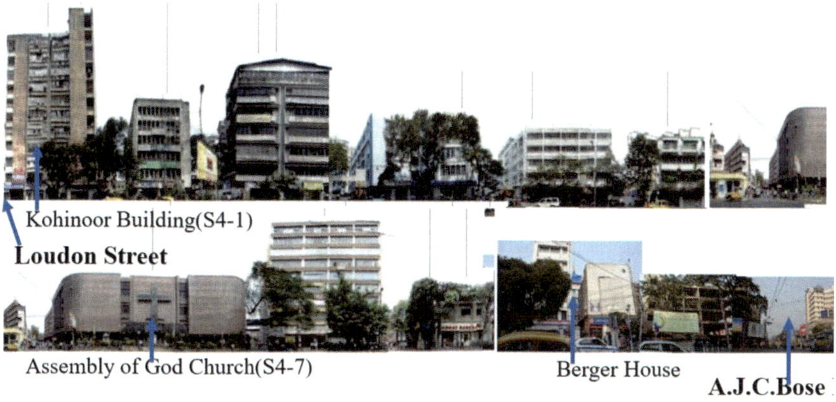

Fig. 12.7 Part Elevation of Street, Stretch 4, Park Street, Kolkata. *Source* Author

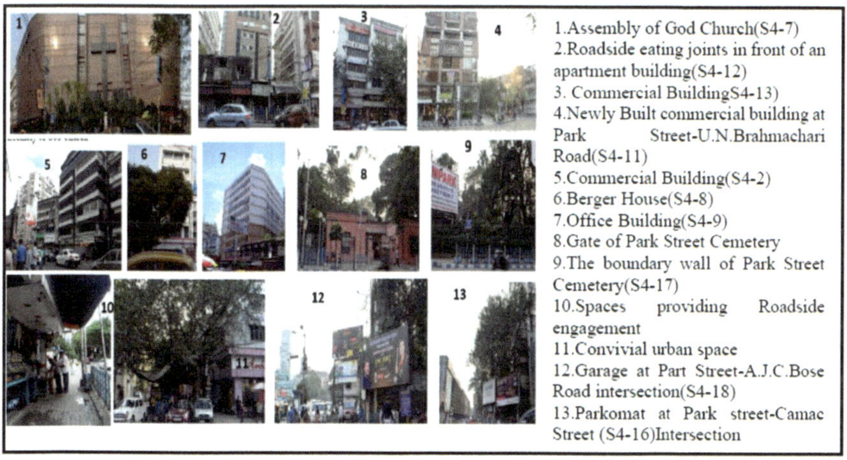

Fig. 12.8 A few buildings of Stretch 4, Park Street, Kolkata. *Source* Author

urban environment. This, combined with informal activities, everyday landmarks, and numerous convivial urban spaces, fosters high levels of public engagement. However, many buildings in this stretch are structurally unsound and in urgent need of renovation. The presence of squatter settlements and encroachments, particularly near Park Circus, obstructs both visual and spatial access to urban spaces. This hinders the ability to congregate and socialize, as the street's narrow and congested environment, coupled with the encroachments, diminishes the area's overall livability and deteriorates its environmental conditions (Refer to Figs. 12.9, 12.10 and 12.11).

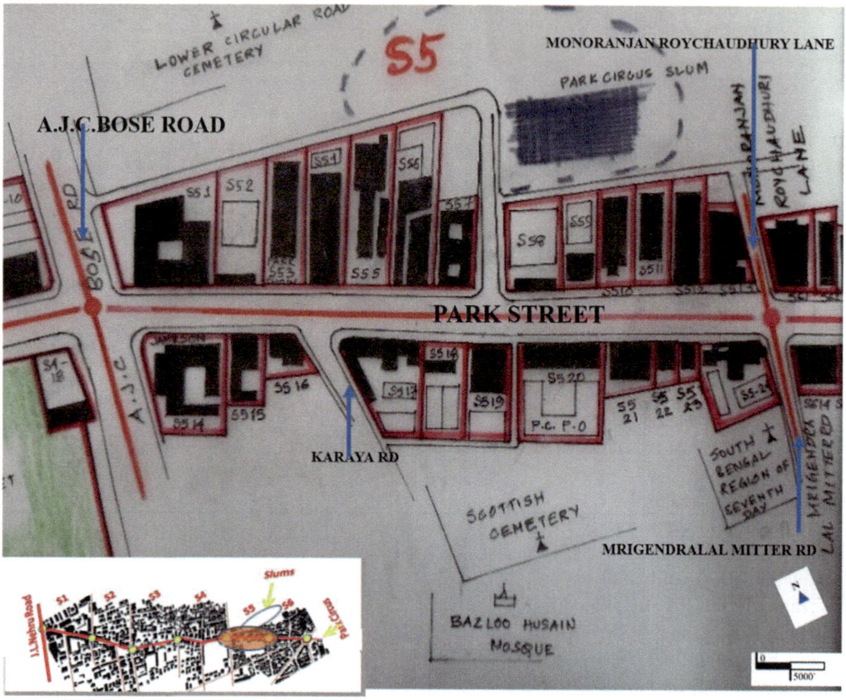

Fig. 12.9 Figure-ground map of Stretch 5, Park Street, Kolkata. *Source* Author

Fig. 12.10 Part Elevation of Street, Stretch 5, Park Street, Kolkata. *Source* Author

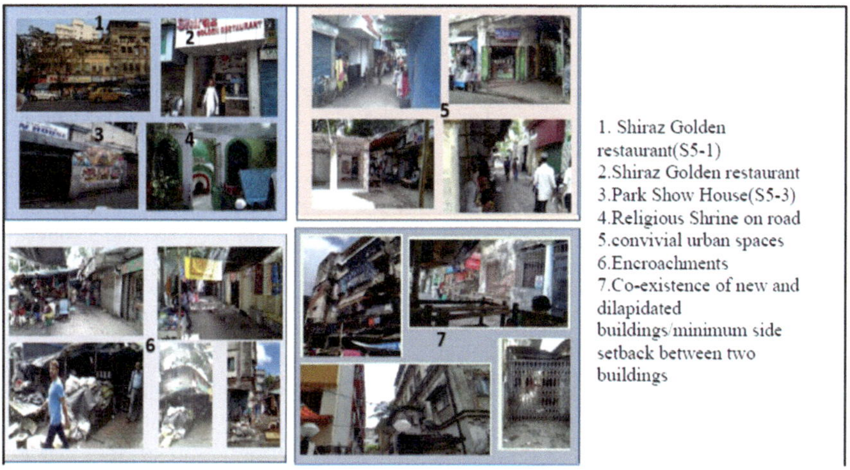

Fig. 12.11 A few buildings of Stretch 5, Park Street, Kolkata. *Source* Author

12.6 Findings

Table 12.4 depicts the findings arrived at by applying the integrated framework listed in Table 12.3 on case study stretches.

12.6.1 Integration of Collage Urbanism and Eco-Psychology: Toward a Unified Framework

The developed framework (combining Collage Urbanism and Eco-Psychology principles) was successfully applied across three selected urban stretches: S1, S4, and S5. A structured evaluation of 9 parameters revealed measurable differences between the stretches. Key Findings per Stretch are listed below:

S1:

- Strong multifunctionality (diverse commercial, social, recreational uses).
- 95% public accessibility, high visual identity, moderate sustainability practices.
- High emotional connection (average mental health score of 8/10).

S4:

- Mixed institutional commercial fabric.
- Only 45% public accessibility, fragmented pedestrian experience, moderate emotional connection (6/10).

S5:

- High multifunctionality (street-level vibrancy, social spaces).

Table 12.4 Findings. *Source* Author

Sl. no	Integrated principle	S1: retail and entertainment hub	S4: institutional and retail zone	S5: mixed-use dense zone	Ranking criteria (high/ moderate/ low)	Inference regarding citizen-centric and mentally restorative urban spaces
1	Layered spatial integration	High	Moderate	High	High: multiple overlapping urban functions	S1 and S5 offer rich urban experiences through layered land uses, supporting vibrant, diverse citizen activities
2	Multifunctional nature-culture spaces	Moderate	Low	High	High: different functions integrated with nature	S5 excels with its multifunctionality supporting social and psychological engagement
3	Human-nature reciprocity (mental health score)	High	Moderate	Low	High: mental health score > 7/10	S1 contributes to psychological restoration; S5 struggles due to lack of greenery and environmental stressors
4	Adaptive public realm	Moderate	Low	High	High: spaces support multiple changing uses	S5's flexibility supports dynamic, community-driven spaces enhancing inclusivity
5	Connectivity and continuity	High	Moderate	Moderate	High: continuous walkability and street life	S1 enhances mental wellbeing through seamless pedestrian movement and sociability
6	Participatory design	Low	Low	Moderate	High: visible citizen engagement	Only S5 shows moderate citizen-driven spaces; S1 and S4 lack active public participation

(continued)

Table 12.4 (continued)

Sl. no	Integrated principle	S1: retail and entertainment hub	S4: institutional and retail zone	S5: mixed-use dense zone	Ranking criteria (high/moderate/low)	Inference regarding citizen-centric and mentally restorative urban spaces
7	Eco-functional sustainability	Moderate	Low	Low	High: presence of green technology and shading	Sustainability features are weak across all, though S1 has marginally better shading infrastructure
8	Emotional place attachment	High	Moderate	High	High: strong identity and pride indicators	Both S1 and S5 generate emotional connectedness crucial for mental wellbeing and belonging
9	Temporal adaptability and resilience	Moderate	Low	High	High: space adapts seasonally/temporally	S5's resilience ensures lively, evolving spaces aligned to citizens' changing needs

- 80% public accessibility, strong adaptability, but poor environmental quality, low emotional connection (4/10).

S1 emerged as the most balanced and citizen-centric, scoring highest across most indicators. S5 showed high flexibility and social engagement but environmental degradation affected mental wellbeing. S4 lagged in ecological richness and public engagement.

12.7 Discussion

Framework Efficacy: The framework was found to be effective in differentiating levels of citizen-centricity and psychological supportiveness, confirming the use of spatial, ecological, and emotional measures as a strong assessment method.

12.7.1 Interpretation of Stretch Performance

S1's layering of history, green courtyards, shaded pedestrian infrastructure, and cultural integration are the best examples of effective Collage Urbanism integrated with Eco-Psychology principles, which result in high mental health scores. it has the highest level of integration of Collage Urbanism and Eco-Psychology principles. Its high cultural layering, strong visual identity, comprehensive pedestrian infrastructure, and moderate ecological features provide a holistically citizen-focused and psychologically nurturing environment.

S5's strong multifunctionality fosters social interaction (a Collage Urbanism advantage), yet environmental degradation and weak maintenance erode psychological resilience (Eco-Psychology issue). Extend S5 possesses strong adaptability and multifunctionality, active ground-level action, and relaxed interaction, conducive to effective social interaction. But environmental degradation (weak maintenance, encroachments) erodes its effectiveness as a fully psychologically restorative area.

S4's institutional supremacy and physical severance (boundary walls, minimal open public space) lead to decreased accessibility, connectivity, and emotional connection, detracting from collage and eco-psychological hopes. Stretch S4 exemplifies ecologically and socially fragmented integration, with institutional uses taking over land and limited public-realm vitality. The stretch falls short of layered richness and participatory nature to foster mental welfare, suggesting radical retrofitting toward citizen-oriented and eco-psychological urbanism.

Key Insights:

- Citizen participation, visual connectedness to nature, and adaptive multifunctional spaces are essential to obtaining psychologically restorative and socially inclusive urban spaces.
- Environmental quality (biodiversity, shaded areas, green cover) is highly associated with perceived emotional attachment and mental health gains.
- Fragmentation, if imaginatively layered (as in S1), enhances urban vibrancy, but if institutionally inflexible (as in S4), it heightens disconnection.

Broader Implications: Urban design systems have to transcend functional and aesthetic considerations to actively incorporate ecological and psychological health into planning choices.

The interwoven principles of Collage Urbanism and Eco-Psychology present a compelling, cross-disciplinary framework through which to formulate future citizen-based cities.

12.8 Conclusion

This research highlights the importance of combining Collage Urbanism and Eco-Psychology principles to design urban spaces that are multifunctional, ecologically diverse, and psychologically nurturing. Of the three urban stretches that were examined, stretch 1 (S1) is found to be best in balancing these principles and having high rates of urban green cover, accessibility, biodiversity, and cultural blend. The results prove that S1's multilayered strategy of urban planning is conducive to mental health and environmental sustainability, in accordance with the fundamental principles of both theories.

Stretch 4 (S4), although moderately successful, identifies accessibility and adaptability challenges, implying the necessity for strategic interventions to improve inclusivity and functionality. Stretch 5 (S5), however, exhibits limited integration, indicating the detrimental effect of poor green coverage and cultural representation on ecological and psychological health.

The study is also of practical relevance to urban planners, architects, and policymakers. It highlights the need for a balanced strategy in urban design that employs ecological and cultural diversity as means of enhancing mental health and wellbeing. Substantial future research can apply this framework to various urban settings using more precise instruments to assess ecological and psychological indicators. This would further optimize strategies for creating resilient and inclusive cities.

12.9 Recommendation

On the basis of the assessment of Stretches S1, S4, and S5, a set of strategic propositions is made with a view to developing citizen-centered, resilient, and sustainable urban spaces. To begin with, it is important to extend cultural layering in new buildings through bringing in local elements of heritage, public art exhibitions, and culture events. This method, significantly responsible for the emotional attachment and sense of place evident in S1, should become a ubiquitous practice in urban interventions. Furthermore, ecological diversity should be prioritized beyond basic green coverage. Developing biodiverse environments that include trees, shrubs, and indigenous species would actively enhance cognitive and emotional renewal, filling shortcomings highlighted particularly in S5.

Public space planning must also focus on flexibility, with movable furniture, multipurpose areas, and dynamic programming that can adapt to shifting social requirements, emulating the multifunctionality advantage seen in S5. At the same time, measures to mitigate environmental degradation are a must. In sections such as S5, urban redevelopment measures and anti-encroachment drives need to be taken to revive accessibility, safety, and visual appeal, thus preserving the psychological advantages of public spaces.

Moreover, reconnecting isolated institutional areas like those in S4 to the surrounding public space is essential. This can be done by providing shared open spaces, accessible green corridors, and cultural activities that overcome physical and social barriers. Local communities need to be addressed as well through participatory greening, engaging communities in actively creating, managing, and scheduling green and cultural spaces, enhancing ownership, mental health, and ecological stewardship. Lastly, a comprehensive planning checklist based on the integrated framework established in this research should be institutionalized in the initial stages of urban projects to guarantee that both ecological psychological and spatial-cultural parameters are incorporated systematically into urban policymaking and design. These suggestions form the foundation for developing more resilient and inclusive urban spaces; yet, given the constrictions experienced in this research, there is much room for further study to further illuminate and improve understanding of integrated urban ecological, psychological dynamics.

12.10 Further Research Scope

Although this research presents useful insights into the application of Collage Urbanism and Eco-Psychology principles at a micro-urban level, it also indicates some areas of potential future research development. For instance, this research was mostly based on qualitative and perceptual analysis; in the future, statistical methods like correlation, regression analysis, or structural equation modeling can be used to empirically confirm the relationships between cultural layering, ecological diversity, adaptability, and mental health outcomes. Second, convenience sampling was adequate to study exploratory factors, but subsequent research should use stratified random sampling from varied demographic groups to enhance representativeness and the solidity of conclusions. Moreover, extending the application of the integrated framework to various urban typologies, e.g., peri-urban, planned townships, or slums, would challenge its applicability and flexibility outside of Park Street, Kolkata. More advanced spatial analytics, including GIS-based analysis, remote sensing, and 3D modeling, can also be added in future studies to provide more accurate measures of urban vegetation, biodiversity, connectivity, and socio-spatial interactions. Additionally, longitudinal designs monitoring mental health outcomes before and after urban design intervention will enable researchers to make more robust causal inferences about the psychological effects of urban spaces. Incorporating climate resilience metrics, including heat island mitigation and flood absorption, would further enhance the ecological aspect of the framework. Finally, creating a standardized and validated assessment tool based on the suggested evaluative framework would provide planners, designers, and policymakers with a replicable approach to systematically audit and improve citizen-focused and psychologically supportive urban environments.

Funding The research was not supported by any funding.

DataAvailability Data Availability Statement The datasets used during and/or analyzed during the current study are available from the author on reasonable request.

Potential Conflicts of Interest The author states that there is no conflict of interest.

Informed Consent Informed consent was obtained from all individual participants included in the study.

AI Discloser No AI tool has been used for writing the manuscript.

References

Agrawal N (2022) Environmental sustainability through ecopsychology. Int J Innov Res Anal 2(4):248–256

Alcock I, White MP, Wheeler BW (2013) Longitudinal effects on mental health of moving to greener and less green urban areas. Environ Sci Technol 48(2):1247–1255

Biswas R, Sen S (2020) Urban eco-psychological attitude during COVID-19 'Lockdown': a survey. Int J Creat Res Thoughts 8(7)

Choudhury D (2020) Collage city urbanism: the case of Kolkata. Thesis. Delhi: school of planning and architecture. http://hdl.handle.net/10603/382544

Choudhury D (2024) Chapter 28–Collage urbanism: creating sustainable cities through equitable urban resources for good health, Editor(s): Sivaramakrishnan L, Dahiya B, Sharma M, Mookherjee S, Karmakar R, Developments in environmental science, Elsevier, (15), vol 15:595–630. https://doi.org/10.1016/B978-0-443-21948-1.00028-5

Daglioglu EK (2016) Karl popper's architectural legacy: an intertextual reading of collage city (1). METU J Fac Archit 33(1):107–119. https://doi.org/10.4305/METU.JFA.2016.1.6

Davis CL (2017) The postwar postscripts of collage city: exploring the rhetorical integrations of race and geometry in 'Contextualism,' 1963–1978. India documents. Retrieved 4 June 4, from https://fdocuments.in/document/the-postwar-postscripts-of-collage-city.html

Dasgupta K (2018) Mapping Calcutta. Kolkata: Centre for studies in social sciences

Dawes MJ, Ostwald MJ (2017) Christopher alexander's a pattern language: analysing, mapping and classifying the critical response. City, Territ Arch 4(1):1–10. https://doi.org/10.1186/s40410-017-0073-1

Eleftheriou V, Knieling J (2017) The urban project of HafenCity. Today's urban and traffic profile of the area. Executive summary of methodology and traffic research conducted in the region. Transp Res Procedia 24(1):73–80. https://doi.org/10.1016/j.trpro.2017.05.070

Elsadek M, Liu B, Lian Z (2019) The influence of urban roadside trees and their physical environment on stress relief measures: a field experiment in Sangh. Urban For Urban Green 42(1):51–60. https://doi.org/10.1016/j.ufug.2019.05.007

Fridgen C (2005) The voice of the earth: an exploration of ecopsychology, In: 2nd edition. Environ Pract 7(1):62–63. https://doi.org/10.1017/S1466046605280055

Gaekwad JS, Moslehian AS, Roös PB, Walker A (2022) A meta-analysis of emotional evidence for the biophilia hypothesis and implications for biophilic design. Front Psychol, pp 1–17. https://doi.org/10.3389/fpsyg.2022.750245

Global designing cities initiative (GDCI) (n.d.) Case study: Raahgiri Day; Gurgaon, India. Retrieved from Global street design guide: https://globaldesigningcities.org/publication/global-street-design-guide/streets/special-conditions/temporary-street-closures/case-study-raahgiri-day-gurgaon-india/

Gifford R (2001) Environmental psychology: principles and practice. Optimal Books, London

Hu M, Chen R (2018) A framework for understanding sense of place in an urban design context. Urban Sci 2(2):34–40. https://doi.org/10.3390/urbansci2020034
Igor B (2020) Implementation of ideas of ecopsychology in urban environment. Ecopoiesis: Eco-Hum Theory Pract 1(1):41–45
Jiang X, Wang X, He L (2024) Effects of virtual exposure to urban greenways on mental health. Front Psych 2(1):1–17. https://doi.org/10.3389/fpsyt.2024.1256897
Kawamura S, Brady C (2022) Climate change resilience in the built environment. World Green Building Council, New York
Kellert SR, Speth JG (2009) The coming transformation: values to sustain. Forestry & Environmental Studies Publications Series, Yale, p 5
Kim HJ, Chae BK (2018) Exploring public space through social media: an exploratory case study on the High Line New York City. Urban Design Int 23(2)
Kumar A, Suresh H, Mall S, Maliwar S (2024) India forum for nature-based solutions: annual summit. New Delhi: WRI India
Mengist W, Soromessa T, Legese G (2020) Method for conducting systematic literature review and meta-analysis for environmental science research. MethodsX 7:100777. https://doi.org/10.1016/j.mex.2019.100777
Montero S (2017) Worlding Bogotas Ciclovia: from urban experiment to international "best practice." Lat Am Perspect 44(2):111–131. https://doi.org/10.1177/0094582X16668310
Moutel N (2016) From ecocriticism to ecopsychology in theodore roszak's frankenstein. Caliban-Fr J Engl Stud, pp 87–100.https://doi.org/10.4000/caliban.3297
Nieuwenhuijsen M, Nazelle AD, Pradas MC (2024) The superblock model: a review of an innovative urban model for sustainability, live ability, health and wellbeing. Environ Res 251(1):118550. https://doi.org/10.1016/j.envres.2024.118550
O'Neill P, Alavi S, Yadav P, Chen R, Teka N (2023) Saving the future by looking to the past: personal relationships to the environment. J Ecopsychology 3(2):1–10
Pelletier LG, Lavergne KJ, Sharp EC (2008) Environmental psychology and sustainability: comments on topics important for our future. Can Psychol 49(1):304–315
Roszak T (2021) Voice of the earth: an exploration of ecopsychology. Phanes Press, New York
Rowe C, Koetter F (1987) Collage city. MIT Press, Cambridge
Sorooshian S (2024) The sustainable development goals of the United Nations: a comparative midterm research review. J Clean Prod 453(1):142272. https://doi.org/10.1016/j.jclepro.2024.142272
Thompson M (2009) Reviewing ecopsychology research: exploring five databases and considering the future. Ecopsychology 1(1):32–37
Wohlin C, Kalinowski M, Felizardo KR, Mendes E (2022) Successful combination of database search and snowballing for identification of primary studies in systematic literature studies. Inf Softw Technol 147:106908. https://doi.org/10.1016/j.infsof.2022.106908
World Health Organization, The WHO Centre for health. (2010) Urban HEART: urban health equity assessment and response tool. Kobe: World Health Organization, The WHO Centre for Health

Part IV
Urban Environmental Management and Resilience

Chapter 13
Catalyzing Awareness on Urban Climate: A Critical Study on Narrative Elements and Reception of Sustainability Advertising

Sayak Pal, Nitesh Tripathi, Sharmila Kayal, and Swati Agarwal

Abstract Urban areas are prone to face the consequences of climate change like "warmer temperatures," "altered wind," "urban dryness," "lower levels of humidity," "slower wind speed," "complications occurring due to excessive rainfall," etc., which sometimes cause massive distress for the residents, forcing the government to take drastic and immediate actions. There are numerous campaigns addressing climate change and projects that aim at sustainability. However, the responses of the Indian citizens are still nascent, leading to negligence in action toward the prevention of climate change. Advertisements play a significant part in generating awareness among people and leading their actions toward sustainability. Thomas Gieryn, in 1999, called sustainability a "boundary term" where the primary concern is to build epistemic communities with shared understanding while sharing common commitments across the nations. The study selects and analyzes a list of audio-visual advertisements to promote sustainability and increase the consciousness of sustainable practices among the audiences. Those advertisements are also exposed to the selected audiences to understand their perception and reactions. The outcome suggests that many brands adopt sustainable advertising to become more relatable. Also, there has been a significant change in the awareness level of the viewers subjected to sustainable advertising.

S. Pal (✉)
Department of Liberal Arts, Humanities and Social Sciences (DLHS), Manipal Academy of Higher Education (MAHE), Manipal, India
e-mail: palsayak01@gmail.com

N. Tripathi
Department of Mass Communication, School of Liberal Studies, Pandit Deendayal Energy University, Gandhinagar, India

S. Kayal
Department of Mass Communication, St. Xavier's University, Kolkata, India

S. Agarwal
Department of Journalism and Mass Communication, Adamas University, Kolkata, West Bengal, India

K. Mukherjee et al. (eds.), *Urban Climate*, The Urban Book Series,
https://doi.org/10.1007/978-3-032-17895-4_13

Keywords Urban climate · Sustainability · Sustainable advertisement · Audience perception · Sustainable practices · Climate change

13.1 Introduction

13.1.1 Urban Climate Change

Urban areas often face the consequences of climate change, and sometimes, these changes massively hit the residents, forcing the government to restrict the further spread of contamination. A study by Valéry Masson, Aude Lemonsu, Julia Hidalgo, and James Voogt on "urban climate and climate change" explains the factors like "warmer temperature", "altered wind", "urban-modified cloud and precipitation", "over" and "downwind of the city" that determines the changes of the climate of the cities like London and Paris. The study also explains the climate differences between the daytime and night, which often vary across the geographical location of the urban settlement, as it differs across the inland city, coastal city, a city in the valley, a polar city, etc. (Masson et al. 2020). Preventive measures are necessary to undertake as the percentage of urban settlements will increase over time. By 2050, the urban population is projected to increase by up to 2.5 billion (Grimm et al. 2008; Seto et al. 2017). Improving habitability, therefore, has to be one of the foremost essential initiatives that need to be addressed, especially for urban settlements. This is because of the higher rate of vulnerability in urban areas due to climate change. Sustainability has to be paired with economic development to prolong the lifespan of urban habitability (Bader et al. 2018). Urban populations are also prone to exposure to additional heat stress because of the increased surface temperature compared to the surrounding areas. The "land-usage and land-cover changes" (LULCC) also cause the heat to increase, and optimizing the LULCC would not only reduce the temperature of the urban climate but also gradually help in slowing the global temperature rise (Cugnon et al. 2019; IPCC 2019; Troeyer et al. 2020). The Sustainable Development Goals (SDGs) enacted by the United Nations also stress the improvement of human settlements as a part of their 2030 agenda (Kuang 2020).

There are other complications that are common across urban areas, like urban dryness, lower levels of humidity, slower wind speed, and complications occurring due to excessive rainfall, etc. (Jia et al. 2020; Hamdi et al. 2011). According to the global climate modeling groups, urban land covers are considered negligible compared to the global weather as the cities cover only about 0.2% of land area; therefore, the sensible and latent heat fluxes are also lesser for the urban areas (0.12 W $m-^{2}$ and 0.08 W m^{-2}) (Lamptey 2009; Zhang et al. 2013; Chen et al. 2016). However, the minimum mean temperature rise is observed to be much higher in the urban areas. A study found that in the USA, the minimum temperature rise is measured between 14 and 21% (Hausfather et al. 2013). The rise in temperature in urban areas is closely linked with industrialization in countries like China (Song et al. 2016). Similar records are also found in Europe, Japan, Puerto Rico, and Saudi Arabia (Hamdi et al. 2020). In India, urban areas are also prone to multiple climate crises like cyclones, coastal surges, water stresses, temperature increases, heat waves, and spreading diseases (Sahay 2017). By 2050, there are projected to be 400 million people living in urban areas (United Nations Division of Economic and Social Affairs (UN DESA) 2014). The increasing rate of urban population is primarily because of the continuous increase in GDP contribution from the urban population. The urban settlements are also increasing their capacity to support this economic and population growth (Kumar et al. 2010). The causes of the increased surface temperature in the urban areas in India are interlinked with multiple local stressors like population growth, changes in land usage, and industrialization (Lele et al. 2018).

There are campaigns addressing climate change, and projects that aim at sustainability are also being implemented. However, the responses of the Indian citizens are still nascent, leading to negligence in action toward the prevention of climate change. Therefore, it is crucial to link the climate goals with local development (Khosla and Bhardwaj 2019). Similar to other industries, the advertising industry also bears the responsibility to address this crisis of climate change, while a conscious advertisement can play a crucial role in preventing the planet's climate change and environmental degradation (Sheehan and Morrison 2018).

13.1.2 Sustainability and Its Goals

Sustainability comes from the Latin word "sustinere," which means "to support." In the 1610s, the word "bearable," now almost obsolete for this purpose, formed the word sustainable by joining "sustain +" and "–able." The word is attested from 1845: "defensible, capable of being upheld." In 1965, the word also formed a meaning "capable of being continued or maintained at a certain level" (Dictionary 2024). The most common definition of sustainability comes from the "United Nations Commission on Environment and Development" which defines the word as "sustainable development is development that meets the needs of the present without compromising the ability of future generations to meet their own needs." Another definition by UCLA illustrates the word: "The integration of environmental health, social equity, and economic vitality in order to create thriving, healthy, diverse and resilient communities for this generation and generations to come. The practice of sustainability recognizes how these issues are interconnected and requires a systems approach and an acknowledgment of complexity" (UCLA 2024).

Thomas Gieryn, in 1999, called sustainability a "boundary term" as we have sustainability spread across multiple sectors like the economy, cities, resource management, livelihood, business, and development (Gieryn 1999). The major concern is to build epistemic communities with shared understanding while, at the same time, sharing common commitments across them, which should be practiced across the world. Sustainability draws out a large amount of money for forging networks, diverse actors, institutes, organizations, projects, and many more to support the ventures. The term "sustainability" was coined by the German forester Hans Carl von Carlowitz (in his 1712 text Sylvicultura Oeconomica). By 1980, the term received a larger audience. The contemporary movement on the environment was born in the 1960s and 1970s (Scoones 2007).

The 1992 Rio Conference on sustainability and sustainable development, attended by 178 governments and convened by the United Nations, NGOs, campaign groups, and civil societies, found its place on the political agenda (Holmberg et al. 1991). The conference outcome resulted in multiple high-level convention processes focused on biodiversity, climate change, and desertification, all aimed at addressing environmental issues. The agendas focused on the community-led processes, including the contributions from community groups, local government, and citizens (Selman 1998). The seventeen goals that were adopted by the UN are "no poverty," ensuring the eradication of poverty everywhere, and bringing the poor under the "social protection system"; "zero hunger," to achieve food surety by ending hunger, improving nutrition and promoting sustainable agriculture; "good health and well-being," for promoting healthy lives and well-being across ages; "quality education," for everyone, involving equitable, inclusive quality education and lifelong education. "gender equality," with a special focus on empowering girls and women across the member nations; "clean water and sanitation," aims at making the clean water available to everyone and ensuring sanitization across the member nations; "affordable and clean energy," access to everyone, a step forward to achieve sustainability by switching to modern energy; "decent work and economic growth," to promote sustainable economic growth; "industry, innovation, and infrastructure," to promote innovation and sustainable industrialization; "reduce inequalities," among the member nations and within their societies; "sustainable cities and communities," to make human settlements safe, sustainable and resilient; "responsible consumption and production," for the member nations; "climate action," to combat the impacts of climate changes; "life below water," ensuring sustainability across, seas, oceans and marine life; "life on land," to protect territorial ecosystem; "peace, justice, and strong institutions," to amend justices across the member nations; and "partnerships for the goals" to encourage global partnership (Nations 2024; Pal et al. 2023).

13.1.3 Urban Climate and Sustainable Infrastructure

The majority of greenhouse gasses (GHG) are emitted from urban settlements, where twenty-five percent of the GHG are generated from urban transportation, thirty-two percent comes from the built environment in the urban area, five percent from the municipal waste, which is a significant source of GHG from the urban areas across the world (Martos et al. 2016). Urban infrastructure (UI) is a combination of engineering perspectives that involves the irrigation system, urban sanitization, energy consumption, education and healthcare centers, etc., and municipal utilities that include electricity consumption, telecommunication facilities, urban transportation, etc. Therefore, UI comprises not only the intervention of local government but also issues related to climate change, economic growth, and municipal waste for an urban settlement (Hodson and Marvin 2010; El-Diraby and Osman 2011). While concerned about the sustainability of the urban settlements, the principle of sustainability applies to understanding whether an urban area is moving towards sustainability or not. According to the Brundtland report, consumption and production should be compared using the equation: $(dP/dt)-(dC/dt) \geq 0$. The rate of change in production should be equal to or greater than the rate of change in consumption to maintain sustainability (Derrible 2018).

13.1.4 Advertisement on Sustainability

Sustainability is portrayed through advertisements for many brands, and one of the primary reasons is the demand from the consumer, who is constantly looking for sustainability to be reflected through their product offerings, initiatives, and practices (Rathee and Milfeld 2024). Studies suggest that the packaged products that are sustainably marketed (almost one out of five products) are growing faster than their traditional counterparts (Crawford 2020). Studies also suggest that around eighty-five percent of the world's population has shifted toward sustainable product consumption (Wire 2021; Xu et al. 2021). This changing nature of consumers is encouraging marketers to redesign their advertisements with a sustainable approach (Kong et al. 2021; Simpson and Radford 2012). The following audio-visual advertisements are found to be aimed at promoting sustainability through both direct and indirect approaches to increase consciousness of sustainable practices among the audiences.

13.1.5 Scope of the Study

The study aims to understand the existing literature on sustainability with respect to urban climate change. Audio-visual advertisements play a prominent role in educating people to bring about positive behavioral changes. The study evaluates several audio-visual social cause advertisements aiming at urban climate change. The research aims to help advertisers understand the audience's perception of sustainable social advertisements on urban climate change.

13.1.6 Theoretical Framework

The study utilizes social cognitive theory in principle to explain how behavior change is facilitated through awareness messages from media platforms. In this study, advertisements, which are a potent way to disseminate sustainability and environmental awareness, have been taken into consideration and further dissected to evaluate their effectiveness in aiding behavioral change on a mass level. The theory focuses on how human behavior is shaped through observational learning and social experience. The theory also posits that people learn new behaviors by observing behaviors of others and the consequences of their behavior (In this case, the characters in the advertisements become role models who promote appropriate behavior, which is later learned by the viewers as well). In addition to this, people also tend to copy the behavior that is being rewarded and refrain from those that are being punished (in the advertisements). This theory helps in confirming that viewers don't just observe but also learn environment-appropriate behavior. The same concept was utilized in the research and became the basis for screening advertisements to the participants of focused group discussion, who were later subjected to questions related to the effectiveness of the advertisement, change in awareness, and adoption of new behaviors (if any).

13.2 Data and Method

13.2.1 Objectives of the Study

The study was conducted with the intention of analyzing the sustainability-related messages embedded in audio-visual advertisements. In addition, the study also aimed to understand and assess the perception of audio-visual advertisements that promote sustainability among viewers. The study adopted qualitative research methods to understand the selected audio-visual advertisements to increase the consciousness of sustainable practices among the audiences.

13.2.2 Methods Used

In the first phase, the advertisements that deal with sustainability were identified and later analyzed using narrative analysis. The audio-visual advertisements selected for the study were from 2017 to 2023, as the Sustainable Development Goals for 2030 were adopted by the UN in 2015. A narrative analysis was conducted on these advertisements to understand concepts like urban climate change, sustainability, and its goals, the connection between urban climate and sustainable infrastructure, and the requirements of advertisements on sustainable infrastructure. Narrative analysis is a qualitative method that often helps to know the softer means, latent, and hidden metaphorical, in semi-structured audio-visual narratives (Berger 2019).

In the second phase, a focused group discussion was conducted to understand pre and post awareness among the respondents about sustainability and sustainable practices. Focus group discussion is a qualitative approach often used to gain an in-depth understanding of a social issue, which makes the process extensive in collecting qualitative data from individuals (Nyumba et al. 2018). For this study, snowball sampling was applied, and the participants were chosen based on the *k* method. Leo A. Goodman has defined snowball sampling in his study "Snowball Sampling," where the process of sampling is defined as the selection of samples randomly from the finite population, and each one of the chosen samples is asked a name *k* (specified integer) to be considered as a sample (Goodman 1961). The participants for the study selected as a sample were faculty members and students from five reputed universities and colleges in Kolkata, India. Overall, 10 focused group discussions were conducted (two from each institute, one with faculty members and the other with students). Each focused group discussion had at least eight participants, which is the minimum requirement for conducting the FGD. For the students' FGD, a mix of postgraduate and undergraduate students was taken to gain a diversity of opinions.

As the students in the universities and colleges of Kolkata hail from various parts of West Bengal and the entire country, the sample can be considered appropriate for eliciting opinions on the topic. However, it must be stated here that the authors haven't attempted to make any form of generalization as the reception of audio-visual content such as advertisements varies according to the background, geography, culture, religion, and literacy level of the person. The reception of such content may be as diverse as the population itself. For each FGD, the first general awareness of sustainability was elicited. Later, advertisements on sustainability were screened on the projector, and then questions specifically dealing with perception toward the advertisements, appeals and elements of the advertisements, and change in awareness level among the respondents were gauged through several questions.

13.3 Results and Discussion

13.3.1 Narrative Analysis of Audio-Visual Advertisements

#BringBackOurStars (Oct 13, 2017)—Airtel India created this ad, and later on, the same was uploaded to YouTube from Airtel India, which was indirectly aimed at reducing pollution in urban areas. The ad starts with a caption showing children looking for stars but unable to find them in the sky. The same is shown throughout the ad, where children are shown decorating the ceiling with glowing stars, smiling at a pair of gloves with stars embroidered on them, smiling at the golden stars on the grading paper at school, decorating a Christmas tree with an electric star, wearing star-shaped fun glasses, drawing star and moon on the desk with crayon, looking at star lantern, smiling on a tiny star sticker on a mirror, making a hollow star shape out of paper and holding them towards the night sky in search of stars but eventually got disappointed for not being able to find any real stars on the sky.

A man came to rescue and arranged the light-fitted drones to fly them away to the night sky to imitate the stars while forming constellations and shooting stars. The ad ended with a request to give the children back the sky so they can see the stars by reducing the pollution in urban areas, with a pledge to fight against pollution on the auspicious occasion of Diwali.

Have You Given Your Contribution—Green Good Deed—MoEFCC (June 07, 2018)—On June 7, 2018, the Ministry of Environment, Forest and Climate Change released an advertisement featuring Amitabh Bachchan advocating for human intervention to prevent environmental degradation. The ad starts with a woman buying vegetables from a street vendor, who delivers them in a plastic bag. Mr. Amitabh Bachchan comes to the scene with a bunch of kids wearing t-shirts with the "Green Good Deeds" logo and accusing her of not loving the earth. He then shows the woman a video where a fisherman explains the harm done to the ocean, its creatures, and the fishermen whose livelihoods are based on the ocean. The woman then returned the plastic bag to the street vendor, and the children started rejoicing. In the end, Mr. Bachchan urges everyone to engage in "Green Good Deeds," where they would do something for nature. The ad focuses on the adverse effect of plastic waste, which largely comes from urban areas, and directly calls for behavioral changes among the people.

Breathe Free | Environment Protection Video | Volvo Auto India (October 08, 2018)—An ad from Volvo Auto India was uploaded on YouTube on October 08, 2018, directly differentiating the atmospheric changes between a polluted urban area and a landscape full of green plants, lakes, and mountains. The 2.34-min ad starts with a WHO report that states that the polluted environment causes 1.7 million children's deaths every year. The following scene shows a child waking up and sitting at the breakfast table with a gloomy face. The child then goes to school with other children wearing masks to protect themselves from the pollution and dust that makes the urban weather look dull. Once the child comes back from school, he stares at the empty part where he used to play with other kids. He also misses the stars in the sky as they are no longer visible due to the pollution. This is when his father notices his despair and takes him on a ride where they spend a day amidst the lush and green nature. The family returns to their home carrying many plants to be handed to their society for plantation. The ad urges the viewers to amend their responsibility to help sustain clean air for the atmosphere.

HUL Start a Little Good—Water (December 20, 2018)—Hindustan Unilever shared this wonderful ad on YouTube on December 20, 2018, to showcase the massive water shortage among certain areas and disparities in using these valuable earthly resources across people from different geographical backgrounds. The ad starts with a female voice in a narrative of a phenomenon where a "bath shower panel" was placed in a village in a desert. A farmer locates the "bath shower panel" on his way to the field when the farmer opens the shower and starts drinking the water. Gradually, other villagers followed the farmer and started drinking the water from the showers and collecting it in their utensils for later usage. The ad ends with a message from the same narrator saying that while half of the villagers could quench their thirst with the water, an urban person was not done taking a bath. The message also floats at the end of the ad, explaining the initiatives taken by Hindustan Unilever to save billions of liters of water and urging people to save water as well.

#CleanAirEveryday (Dec 02, 2020)—The ad from Star भारत was particularly designed for the Indian urban population, where air pollution in the urban area caused by carbon emission from cars is a pressing issue. Air pollution can cause significant damage to the brains and lungs of people, especially to children. The ad starts with a busy urban road filled with cars and the smoke emanating from them. The following scene is of a household where a mother is helping her child to revise general knowledge by asking her questions, and the father is getting ready for the office. The child could not answer the question, and the parents were shocked at her performance, which was significantly better than the previous year. The child answered in a harsh and haggard voice that she tried hard to remember but could not remember the information. The parent turned their attention and found that the child looked like an old woman (metaphor). A voiceover then asks to restrict the driving of private vehicles for a day every week, which would reduce the air pollution that can cause severe damage to the lungs and brain of a child.

Bin Boy (May 04, 2022)—This advertisement from Hindustan Unilever targets urban pollution caused by household waste. The ad starts with a knock on the front door of an urban house when a woman opens the door and finds out the cleaner is pointing out the dustbin. The woman was shocked to find out that her son was sitting inside the garbage bin. She tried to pull him out but failed and called her husband for help. His father also tried to reason with the boy but failed. Slowly, people started gathering around to find out the reason behind this unusual behavior. Finally, the boy's grandmother came, sat beside him, and asked him the reason. The boy responded that despite multiple warnings to his family to separate water waste from dry waste, the family remained indifferent, which would lead to damage to his future. Next, the boy was seen coming out of the garbage bin, and his mother and father were busy separating wet waste from dry waste. The ad ends with the message of the dire consequences of not managing waste properly and sends appeals to people to keep dry waste from wet waste.

Take a #GoodnessResolution with Mamaearth|Plastic Positive (Dec 27, 2022)—On December 27, 2022, Mamaearth shared a campaign on social responsibility towards urban waste management. The ad starts with two women coming to a park for morning exercise and finding out that the park is filled with mostly plastic waste from last night's party. They also spotted a cleaner on her job clearing the place of waste; the women started criticizing the people who littered the place with plastic leftovers and mentioned that they were not solely responsible for this mess, even being part of the last night's celebration. Suddenly, a girl came to the park and found the cleaner working her way to clean the heavily littered place. The girl looked around her surroundings and started picking up trash to put in the waste bins. The two women suggested to each other that the cleaning could be easier and faster if people joined the process when they spotted the girl helping out the cleaner by picking out the trash from the place. The ad ends with a message where people are asked to take part in keeping the environment clean.

Tata Tea Jaago Re—To fight Climate Change—Nursery Rhymes (Jun 03, 2023)—Tata Tea shared this ad on YouTube with the help of popular nursery rhymes but with twisted and different meanings, aiming to gain awareness about sustainability. The ad shows a group of children performing on stage during a school function. They started with the poem "Jack and Jill" with the lyrics that tell the story of Jack and Jill being unable to fetch the water because all the water had dried up, which could be the situation for the kids one day. The next popular rhyme, "Machli jal ki rani hai," was twisted into "Machli jal ki thi rani," they suffered and slowly died because of the pollutants thrown in the water. The third rhyme was "Twinkle twinkle little star," which was twisted into "Twinkle twinkle little star, how I wonder what you are, up above the smoggy sky, I have never seen you, I wonder why?" that talked about the air pollution that causes breathing issues. The kids then brought in a projector to show their parents the popular nursery rhymes that might change if the pollution is uncontrolled, requiring human intervention. The ad ends with a request to visit the brand's website to find out how we can fight against climate change.

The Journey to Zero Pollution with Shoonya|#ShoonyaKiShakti (Jul 19, 2023)—Shoonya India, by NITI Aayog, promotes "zero pollution mobility" across the nation, shared this ad on July 19, 2023, on YouTube. The ad shows a change in surface transportation that usually causes massive air pollution because of the burning of fossil fuels for transportation. The change from conventional to sustainable forms of surface transportation can lead to reduced CO_2 emission in the atmosphere and a significant reduction in air pollution, a pertinent issue, especially in urban areas. The electric vehicle from Shoonya India promotes the sustainable usage of surface transportation.

The nine advertisements were carefully selected for the study to address the areas that require significant development and improvement. The campaign from Airtel India: "#BringBackOurStars," aimed at addressing air pollution, along with the advertisement from Star भारत: "#CleanAirEveryday," Volvo Auto India: "Breathe Free," and another from NITI Aayog: "#ShoonyaKiShakti." Two advertisements aimed at reducing the usage of harmful plastics were from the Ministry of Environment, Forest and Climate Change: "Green Good Deed" and from Mamaearth: "#GoodnessResolution." One advertisement on limiting water wastage from Hindustan Unilever: "Water," another one from the same organization on waste management, "Bin Boy," and the last one advertisement from Tata Tea aiming at multiple issues promoting sustainable practices and preventing environmental degradation, by raising awareness on the issues regarding water shortages as well as contamination, temperature rise, contamination of water bodies, excessive use of harmful plastics, improper disposal of waste, air pollution, health degradation due to pollutants, and the climate change. The advertisements aim at the components that are the root cause of urban environment degradation to raise awareness and promote sustainable practices (Table 13.1).

The sustainable goals comprise a total number of seventeen goals that were adopted by the United Nations, which are "no poverty," "zero hunger," "good health and well-being," "quality education," "gender equality," "clean water and sanitation," "affordable and clean energy," "decent work and economic growth," "industry, innovation, and infrastructure," "reduce inequalities," "sustainable cities and communities," "responsible consumption and production," "climate action," "life below water," "life on land," "peace, justice, and strong institutions," and "partnerships for the goals." However, when the advertisements focus only on the

Table 13.1 Consolidated table that indicates the audio-visual advertisements that are connected with the sustainable development goals (SDGs)

Audio-visual advertisements	Sustainable development goals (SDGs)
#Bring back our stars	Industry, innovation, and infrastructure Sustainable cities and communities Climate action
Have you given your contribution—green good deed—MoEFCC	Good health and well-being Clean water and sanitation Sustainable cities and communities Climate action Life below water
Breathe free I environment protection video I volvo auto India	Good health and well-being Sustainable cities and communities Climate action
HUL start a little good—water	Clean water and sanitation Responsible consumption and production
#Clean air every day	Good health and well-being Sustainable cities and communities Climate action
Bin boy	Sustainable cities and communities Life below water Life on land
Take a #Goodness resolution with mamaearthIplastic positive	Responsible consumption and production Life on land
Tata tea Jaago Re—to fight climate change—nursery rhymes	Good health and well-being Responsible consumption and production Climate action Life on land
The journey to zero pollution with ShoonyaI#ShoonyaKiShakti	Affordable and clean energy Industry, innovation, and infrastructure Sustainable cities and communities Responsible consumption and production Climate action

specific areas or goals, all nine advertisements touched nine goals with different frequencies: "Sustainable cities and communities" covered six different advertisements, as well as "climate action" with the same number of frequencies, followed by "good health and well-being" and "responsible consumption and production" in four advertisements, "life on land" in three advertisements, all three "industry, innovation, and infrastructure", "clean water and sanitation" and "life below water" in two advertisements each, and "affordable and clean energy in one advertisement (Table 13.1).

13.3.2 Reception of Audio-Visual Advertisements Among the Respondents

The focus group discussion was conducted to determine the selected participants' general understanding of sustainable practices, which will help evaluate sustainable practices among the urban population. The participants were asked to illustrate their understanding of sustainable practices before they were exposed to the advertisements. The responses came in the form of related keywords and key phrases like "green products," "eco-friendly products," "not using leather," "using products that are beneficial to the conservation of the environment," and "not using plastic products." Afterward, the participants were exposed to the nine selected advertisements focusing on sustainable practices in the urban landscape to increase the consciousness of sustainable practices among the audiences. The same question was again asked of the participants, enquiring about their understanding of sustainable practices. This question was repeated to find out the difference in understanding of sustainable practices among the same audience group after exposure to the advertisements. The answers to this question, after the exposure of the advertisements, came with a detailed list of keywords and phrases like "saving water," "controlling pollution," "placing different dustbins in every locality so that people do not litter," "spreading awareness," "activities such as segregating waste," "ensuring proper waste disposal," "go with renewable fuels," "reducing carbon emission," and "use resources in such a way that we have its content now and we can save it up for later for our future generations."

The third question was asked whether they had witnessed any practices in their localities that may go against sustainable practices. Participants brought in a list of issues that they encounter in their locality, starting from drainage systems that carry contaminated water to the waterbodies where people use the water for daily usage, immersion of idols in the waterbodies that end up polluting the water, lack of mindset among the people to practice sustainability, excessive usage of cars that are causing a greater amount of air pollution, lacking in waste management system by failing to separate dry waste from wet waste, and spitting in the lift. They also mentioned the importance of working toward sustainability and making such efforts toward a sustainable future.

The participants were then asked to share their experiences after exposure to the advertisements. While they referred to the advertisements as nice and effective in raising awareness, they could also recall particular advertisements, like the "Hindustan Unilever" ad, which were more impactful than the rest. There are adverse reactions found among the participants as well where they have identified the advertisement like "Airtel" ad and quoted "Airtel tried to raise awareness about the lack of stars, but sending 100 drones to the sky is not actually sustainable". One of the participants also identified the advertisement from "Volvo" and quoted, "For a car company like Volvo, showing the people help in growing the greenery that shows that only the rich can buy cars and go into the green, but the poor will suffer the prolusion." Another of the participants identified the ad on electric vehicles from "ShoonyaKiShakt" and quoted "The ShoonyaKiShakt ad emphasizes electric vehicles. Sunya means zero, which means if you buy an electric vehicle, it produces zero pollution, but most electric vehicles get their charges from coal power (electricity generated from burning fossil fuels). The electric batteries can cause massive pollution in the environment". They also questioned the government's awareness of such issues, which could have concentrated more on sustainability.

The motive of sustainability advertisements is not only to make people aware of the concept and offer sustainable products but also to bring permanent changes in their behavior. The fifth question asked the participants was about the behavioral changes they would undertake. They came up with answers: "I wish to close the tap when I am brushing," "segregating waste is something that we can try and we should be doing," "not using plastics," "stop littering waste," "I tend to forget and sometimes do not find dustbins; therefore, I throw garbage anywhere" which the participant repents on, "I will not throw plastics outside," "I will use eco-friendly products from now," "I will carry jute bags with me so that I can put products in that." The influence over the people on their behavioral changes would be at its prime when they would not only take the initiative to change their behavior but also encourage others to participate. The last question asked the participants was the actions they would like to take to encourage others toward sustainable practice. The participants not only came up with options like "preventing people from using plastic bags," "encouraging the usage of eco-friendly bags," "reducing the wastage of water," and "preventing people from contaminating water bodies" but also questioned the government's role in creating awareness and taking actions toward sustainability.

Focus group discussion helped understand the selected advertisements' impact on the audience. The first two questions remained the same; however, one of them was asked before, and the other one was after the exposure to the advertisements. The participants were found to be aware of the concept of sustainable practices but with very little information and understanding. The participants were found to be much more enlightened on sustainable practices after being exposed to the advertisements and came up with a list of sustainable practices that are shown in the advertisements. This shows the recall of information about sustainable practices that they could retrieve from the advertisements. There are some contradictions that are found during the conversations as well. For example, they mentioned the need for companies and corporations to act in favor of promoting sustainability, which might encourage changes in the behavior of the citizens. A participant also mentioned about the class differences that are liked with the sustainable practices, where the awareness and implementation would differ across different groups of people belonging to different classes. Another one of the participants put forward the need to carefully control our usage to save resources that are scarce for future generations.

The participants were able to recall about the unsustainable practices that they have witnessed in their localities. Surprisingly, many of those practices are found to be linked to the advertisements that they are exposed to as a part of this experiment. The possible reason behind this behavior is that the participants recalled the unsustainable practices they had seen in the advertisements because they could link such activities and identify the follies in the people's behavior. Although the participants agreed on the positive impact of the advertisements, some advertisements were identified as not fruitful as those ads highlighted more product promotion than focusing on the cause (sustainability). Sometimes, these advertisements also miss details like carbon footprints and related power consumption that cause more pollution, even though the product claims to be sustainable.

When the participants were asked to mention their behavioral changes, they came up with a lot of practices that they wished to undertake, including using jute bags, using eco-friendly products, preventing littering waste, restricting the usage of plastics, etc. Such consent indicates the initiation of change towards sustainable practices, which is one of the main objectives of creating and sharing sustainability advertisements. The participants were willing to encourage others to adopt sustainable practices like better waste management and reduced water wastage. They were also found to be interested in informing the role of local representatives about unsustainable practices as well as demanding government intervention in preserving and promoting sustainable practices.

13.4 Conclusions

13.4.1 Conclusions of the Study

The "Sustainable cities and communities" and "Climate action" are the two most prominent SDGs that are relevant to promoting sustainable practices in the urban landscape, followed by "Good health and well-being" and "Responsible consumption and production," therefore, the audio-visual advertisements have managed to keep their focus on the most relevant issues concerning sustainable practices while these advertisements are primarily targeted for the urban audiences.

The focus group discussion was conducted to assess the audience's perception of the audio-visual social cause advertisements promoting sustainable practice. The participants were also tested based on their perception of sustainable practices by pre and post-exposure to the advertisements. The changes were observed in their perception after exposure to the advertisements, which mentioned more components that are essential in sustainable practices. The participants were also found to mention areas like saving resources for future generations, the intervention of companies and organizations, and class differences that are interlinked and essential for sustainable development. The motive of the ads is not only to bring changes in the viewers' attitudes but also to encourage them to take the initiative in helping others practice sustainability. The exposure to the advertisements not only probed them to plunge into practice more toward sustainability but also made them try to change others' behavior towards sustainable practices. While the participants liked the ads aiming at different sustainable practices, they also became concerned about the excess promotion of brands that sometimes overshadow the root cause of such advertisements, like the ads from Airtel, Volvo, and NITI Aayog.

13.4.2 Recommendations

Based on the study, it can be said for sure though the instances of sustainability advertising are very less but those few instances are very impactful and innovative. No wonder, they have been liked by most of the viewers (and also the participants) that can be confirmed from the user engagement. This is something that is encouraging to advertisers, companies, and governments across the globe. They should take a hint from this and join the bandwagon of disseminating messages related to sustainability and environmental issues. By doing so, they will become more relevant. They would also enjoy praise from the public, who would appreciate their gesture and deem these companies/organizations environment-friendly. The study also hints at the use of solution-based advertising vouching for innovative solutions to sustainability and climate change-related issues rather than a mere statement of the grim situation of the environment and climate.

AI Discloser No AI tool was utilized for writing the manuscript.

References

Bader DA, Blake R, Grimm A, Hamdi R, Kim Y, Horton R, Rosenzweig C, Alverson K, Gaffin S, Crane S (2018) Climate change and cities. Second Assess Rep Urban Clim Chang Res Netw 27–60. https://doi.org/10.1017/9781316563878.009

Berger AA (2019) Brands and cultural analysis. Springer International Publishing

Chen H, Zhang Y, Yu M, Hua W, Sun S, Li X, Gao C (2016) Large-scale urbanization effects on eastern Asian summer monsoon circulation and climate. Clim Dyn 47:117–136. https://doi.org/10.1007/s00382-015-2827-3

Crawford E (2020) Sustainably-marketed products drove more than half of all growth across CPGs in past five years. https://www.foodnavigator-usa.com/Article/2020/07/23/Sustainably-marketed-products-drove-more-than-half-of-all-growth-across-CPGs-in-past-five-years

Cugnon G, Caluwaerts S, Duchêne F, Hamdi R, Termonia P, Top S (2019) Climate sensitivity to land use changes over the city of Brussels. Geogr Pannonica 23(4):269–276

Derrible S (2018) An approach to designing sustainable urban infrastructure. MRS Energy Sustain 5:E15

Dictionary OE (2024) Sustainable. Etymonline. https://www.etymonline.com/word/sustainable

El-Diraby TA, Osman H (2011) A domain ontology for construction concepts in urban infrastructure products. Autom Constr 20(8):1120–1132. https://doi.org/10.1016/j.autcon.2011.04.014

Gieryn T (1999) Cultural boundaries of science: credibility on the line. Chicago University Press, Chicago, IL

Goodman LA (1961) Snowball sampling. Ann Math Stat 148

Grimm NB, Faeth SH, Golubiewski NE, Redman CL, Wu J, Bai X, Briggs JM (2008) Global change and the ecology of cities. Science 319(5864):756–760. https://doi.org/10.1126/science.1150195

Hamdi R, Kusaka H, Doan QV, Cai P, He H, Lup G, Kuang W, Caluwaerts S, Duchêne F, Schaeybroek BV, Termonia P (2020) The state-of-the-art of urban climate change modeling and observations. Earth Syst Environ 4:631–646. https://doi.org/10.1007/s41748-020-00193-3

Hamdi R, Termonia P, Baguis P (2011) Effects of urbanization and climate change on surface runoff of the Brussels capital region: a case study using an urban soil-vegetation-atmosphere-transfer model. Int J Climatol. https://doi.org/10.1002/joc.2207

Hausfather Z, Menne MJ, Williams CN, Masters T, Broberg R, Jones D (2013) Quantifying the effect of urbanization on U.S. historical climatology network temperature records. J Geophys Res Atmos 118:481–494. https://doi.org/10.1029/2012JD018509

Hodson M, Marvin S (2010) Can cities shape socio-technical transitions and how would we know if they were? Res Policy 39(4):477–485. https://doi.org/10.1016/j.respol.2010.01.020

Holmberg J, Bass S, Timberlake L (1991) Defending the future: a guide to sustainable development. Earthscan, London

IPCC (2019) Climate change and land: an IPCC special report on climate change, desertification, land degradation, sustainable land management, food security, and greenhouse gas fluxes in terrestrial ecosystems. In: Shukla PR, Skea J, Buendia EC, Masson-Delmotte V, Pörtner H-O, Roberts DC (eds) (In Press)

Jia TB, Ren G, Zhao X, Zhang P, Li X (2020) Half-century urban drying in Shijiazhuang city. Environ Res Commun 2:075006

Khosla R, Bhardwaj A (2019) Urban India and climate change. India Warm World Integr Clim Chang Dev 459–476. https://doi.org/10.1093/oso/9780199498734.003.0025

Kong HM, Witmaier A, Ko E (2021) Sustainability and social media communication: how consumers respond to marketing efforts of luxury and non-luxury fashion brands. J Bus Res 131:640–651. https://doi.org/10.1016/j.jbusres.2020.08.021

Kuang W (2020) National urban land-use/cover change since the beginning of the 21st century and its policy implications in China. Land Use Policy 97:104747. https://doi.org/10.1016/j.landusepol.2020.104747

Kumar S, Kapoor R, Rewal R, Seth S, Walia A (2010) Developing an energy conservation building code implementation strategy in India. ACEEE Summer Study Energy Effic Build 8:209–224

Lamptey B (2009) An analytical framework for estimating the urban effect on climate. Int J Climatol 30(1):72–88. https://doi.org/10.1002/joc.1873

Lele S, Srinivasan V, Thomas BK, Jamwal P (2018) Adapting to climate change in rapidly urbanizing river basins: insights from a multiple-concerns, multiple-stressors, and multi-level approach. Water Int 43(2):1–24. https://doi.org/10.1080/02508060.2017.1416442

Martos A, Pacheco-Torres R, Ordóñez J, Jadraque-Gago E (2016) Towards successful environmental performance of sustainable cities: intervening sectors. A review. Renew Sustain Energy Rev 57:479–495. https://doi.org/10.1016/j.rser.2015.12.095

Masson V, Lemonsu A, Hidalgo J, Voogt J (2020) Urban climates and climate change. Annu Rev Environ Resour 45:411–444

Nations U (2024) The 17 goals. https://sdgs.un.org/goals

Nyumba TO, Wilson K, Derrick CJ, Mukherjee N (2018) The use of focus group discussion methodology: insights from two decades of application in conservation. Methods Ecol Evol 9(1):20

Pal S, Kayal S, Tripathi N, Agarwal S (2023) Deciphering sustainability messages in audio-visual advertisements toward achieving the SDGs. Liberal Stud 8 (1):71–95. https://doi.org/10.5281/zenodo.8170807

Rathee S, Milfeld T (2024) Sustainability advertising: literature review and framework for future research. Int J Advert 43(1):7–35

Sahay S (2017) Urban adaptation to climate sensitive health effect: evaluation of coping strategies for dengue in Delhi India. Sustain Cities Soc 37:178–188. https://doi.org/10.1016/j.scs.2017.11.017

Scoones I (2007) Sustainability. Dev Pract 17(4/5):589–596. https://doi.org/10.1080/09614520701469609

Selman P (1998) Local agenda 21: substance or spin? J Environ Plan Manag 41(5):533–553. https://doi.org/10.1080/09640569811443

Seto KC, Golden JS, Alberti M, Turner BL II (2017) Sustainability in an urbanizing planet. Proc Natl Acad Sci 114:8935–8938. https://doi.org/10.1073/pnas.1606037114

Sheehan K, Morrison D (2018) Advertising leadership and climate change: the efficacy of industry professionals to address climate issues. Advert Soc Q 19:1. https://doi.org/10.1353/asr.2018.0009

Simpson BJ, Radford SK (2012) Consumer perceptions of sustainability: a free elicitation study. J Nonprofit Public Sect Mark 24(4):272–291. https://doi.org/10.1080/10495142.2012.733654

Song Y, Liu H, Wang X, Zhang N, Sun J (2016) Numerical simulation of the impact of urban non-uniformity on precipitation. Adv Atmos Sci 33:783–793

Troeyer KD, Bauwelinck M, Aerts R, Profer D, Berckmans J, Delcloo A, Hamdi R, Schaeybroeck BV, Hooyberghs H, Lauwaet D, Demoury C, Nieuwenhuyse AV (2020) Heat related mortality in the two largest Belgian urban areas: a time series analysis. Environ Res 188:109848. https://doi.org/10.1016/j.envres.2020.109848

UCLA (2024) What is sustainability? sustain.ucla.edu. https://www.sustain.ucla.edu/what-is-sustainability/

United Nations Division of Economic and Social Affairs (UN DESA) (2014) World urbanization prospects: the 2014 revision, highlights. Population Division, UN DESA, New York

Wire B (2021) Recent study reveals that more than a third of consumers are willing to pay more for sustainability as demand grows for environmentally-friendly alternatives. https://www.businesswire.com/news/home/20211014005090/en/Recent-Study-Reveals-More-Thana-Third-of-Global-Consumers-Are-Willing-to-Pay-More-for-Sustainability-as-Demand-Grows-for-Environmentally-Friendly-Alternatives

Xu W, Jin X, Fu R (2021) The influence of scarcity and popularity appeals on sustainable products. Sustain Prod Consum 27:1340–1348. https://doi.org/10.1016/j.spc.2021.03.014

Zhang GJ, Cai M, Hu A (2013) Energy consumption and the unexplained winter warming over northern Asia and North America. Nat Clim Chang 3:466–470. https://doi.org/10.1038/nclimate1803

Chapter 14
Artificial Intelligence and Urban Disasters Resilience: Building a Conceptual Model Using Literature Review

Syed Tabrez Hassan

Abstract Enhancing the technological and procedural components of calamities and disaster research stands out as an important and pivotal process within disaster management. The application of artificial intelligence (AI) presents a spectrum of applications, including but not limited to tracking as well as mapping, geospatial evaluation, remote sensing methods, drone technology, robotics, network and telecom services, machine learning, analysis of accident hotspots, smart city planning, transport and route planning, and environmental impact evaluation. These technology-based facets signify communal change and bear profound implications for studying societal responses to disasters. Researchers have adeptly employed diverse methodologies and techniques, adopting disciplinary, multidisciplinary, and interdisciplinary approaches to investigate hazards and disasters. Researchers have integrated both qualitative data and quantitative data collection and analysis strategies. This research offers an outline of AI's applications in disaster prevention and management across its phases, emphasizing AI's integral role in expediting and refining responses throughout all disaster management stages. The integration of geographic information system (GIS) as well as remote sensing (RS) into disaster management fosters situational awareness, enhanced planning, analysis, and recovery efforts. RS and GIS are widely acknowledged as pivotal tools in disaster management support. The visualization capability, imagery based on satellite, and AI-driven analyses can facilitate swift governmental decision-making in the case of natural disasters.

Keywords Disaster management · Artificial intelligence · Urban disaster · Resilience

S. T. Hassan (✉)
Adamas University, Kolkata, West Bengal, India
e-mail: syed.hassan@adamasuniversity.ac.in

K. Mukherjee et al. (eds.), *Urban Climate*, The Urban Book Series,
https://doi.org/10.1007/978-3-032-17895-4_14

14.1 Introduction

Urban areas around the world are increasingly vulnerable to disasters due to rapid urbanization, climate change, and infrastructural complexities (Ruth and Coelho 2015). Building resilience in urban settings is essential to mitigate the adverse impacts of both natural and man-made disasters. In recent years, artificial intelligence (AI) has emerged as a transformative tool in enhancing urban disaster resilience through predictive analytics, real-time monitoring, and intelligent decision-making systems (Rane et al. 2024; Venkadesh et al. 2024; Srivastava and Maity 2023). AI technologies, such as machine learning, cloud computing, IoTs, computer vision, and natural language processing, enable early warning systems, optimize resource allocation during emergencies, and support rapid post-disaster recovery strategies (Yang et al. 2024; Haldorai et al. 2024; Velev and Zlateva 2023). AI-driven models can analyze satellite imagery to assess damage extent or use social media data to locate distressed populations in real time (Rane et al. 2024; Venkadesh et al. 2024). Integrating AI into disaster risk management frameworks allows cities to shift from reactive to proactive approaches, thereby strengthening urban resilience.

A disaster represents an event capable of causing damage to a community, whether through human life fatalities, environmental destruction, or economic losses. It surpasses the societal ability to promptly respond or manage its impact (Shaluf 2007). Contemporary geospatial technology and artificial intelligence systems have achieved remarkable levels of sophistication, exhibiting substantial potential in crisis scenarios. The blueprint for disaster response planning significantly hinges on various factors such as geographic features, weather patterns, ecological aspects, and the availability of resources and equipment (Alexander 2015; Hamza et al. 2023). To bolster resilience in emergency relief efforts, this is advisable to employ management science and operations research criteria while accounting for the effects of resource allocation on affected populations (Hussain et al. 2023; Mizrak 2024).

Despite numerous studies assessing the effectiveness of artificial intelligence in disaster management, it's important to recognize the considerable divergence in crisis response situations between different countries, including India. Therefore, there exists a critical necessity to identify and prioritize essential data crucial for managing natural disaster crises effectively. Strategies aimed at minimizing the impact of catastrophes encompass preventive measures, vulnerability reduction, preparedness, and resilience in disaster management (Paton and Johnston 2001). Many scholars rely on geographic information system and artificial intelligence to map flood hazards and vulnerability to flooding in spatial terms (Tamiru et al. 2021; Chan et al. 2022; Nguyen 2023). Geographic information systems play a pivotal role in streamlining the management, integration, and dissemination of spatial data, enabling strategic planning and real-time decision-making in hazard preparedness and flood crisis management (Chtrabhuj et al. 2024). These systems prove highly comprehensive and capable in addressing flood crisis management concerns. Adopting a structured approach in the planning process becomes essential to enable swift and precise decisions. This paper presents a review of the practical benefits of artificial intelligence and its practical applications in disaster management, aiming to reduce the impact of disasters effectively.

14.2 Methodology

The objective of this study was to perform a comprehensive review of literature to explore the role of AI in enhancing disaster management and generating practical insights to fulfill the study's aim. A search string comprised of relevant keywords was employed to scrutinize studies published between 2016 and 2022, specifically focusing on the intersection of artificial intelligence and various facets of disaster management. Moreover, reputable science citation index journals were consulted to select pertinent studies and identify diverse methodologies employed in integrating AI within the realm of disaster management. This research involved a thorough review to investigate the involvement of AI in urban disaster management. The Scopus database served as the primary resource for identifying relevant previous studies within the domain. Alongside Scopus, studies were also sourced from Web of Science (WoS) and Science Direct databases, ensuring access to the most current and credible literature on AI and its practical implications. This study was carried out in May 2023, the research employed the advanced based search option (refer to Fig. 14.3). Initially, 2113 publications were retrieved, with 918 duplicates identified across the three databases and subsequently removed. Through a meticulous assessment of titles and abstracts, 766 studies were excluded. Subsequently, the remaining articles underwent further scrutiny, resulting in the exclusion of 91 studies based on their nature and scope. Ultimately, 134 studies that fully met the established criteria were included in this review. Additionally, this review presents key demographic and statistical findings sourced from articles available in the journal databases and various countries.

A diverse selection of peer-reviewed research articles in credible journals and their respective studies were compiled to encompass a wide range of research on the intersection of artificial intelligence and disaster management (refer to Table 14.1, Fig. 14.1).

Table 14.1 Search strings in journal database

Source	String
Science direct web of science	("Disaster Management", OR "Urban Disaster", OR "Machine Learning", OR "Artificial Intelligence")
Scopus	Title-ABS-key (Disaster Management & Artificial Intelligence) And (LIMIT-to(PUBYEAR, 2022, 2021, 2020, 2019, 2018, 2017, 2016) and "Urban Disaster", "Resilience" Artificial Intelligence","Disaster Risk Reduction","Machine Learning")

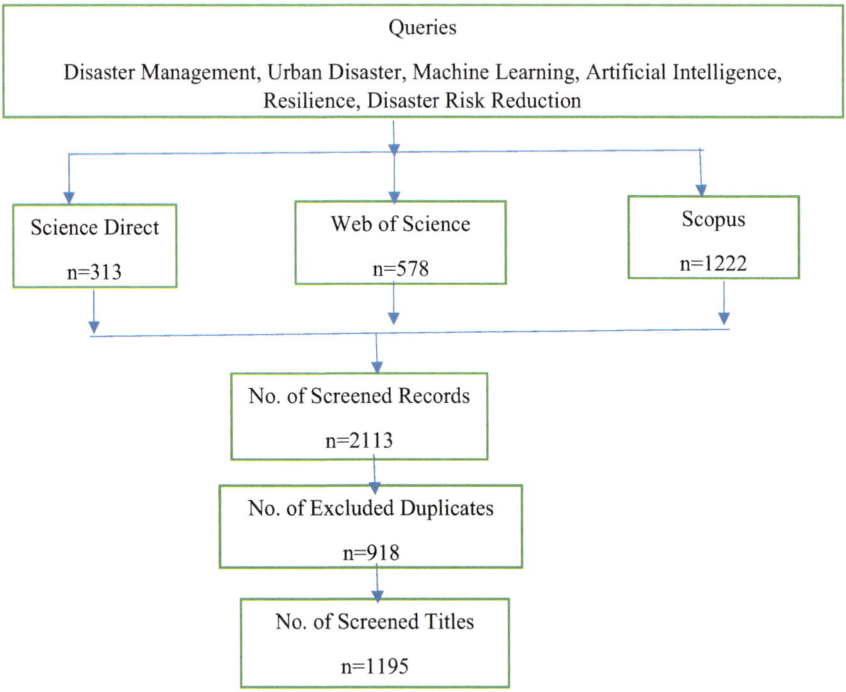

Fig. 14.1 Detailed screening process

The disaster management cycle (DMC) comprises of four different phases aimed at mitigating the effects of disasters (Fig. 14.2). This cycle serves as a framework for planning and strategizing to minimize the impact of disasters. Moreover, it offers guidance to governmental bodies, civil society, agency organizations, and businesses, enabling them to implement measures and foster adaptability within the society and community.

The proposed process delineates four different stages within disaster management, comprising interconnected activities which are further categorized into three segments: Pre-disaster, during or ongoing, and aftermath or post-disaster activities. These activities within the disaster management cycle are orchestrated to diminish the risks associated with both human and physical losses.

Pre-disaster activities encompass mitigation efforts, involving the development of preventive laws and the establishment of standards to combat potential catastrophes or emergencies. This phase emphasizes thorough preparation aligned with available resources, recognizing preparedness as a pivotal stage within the disaster management cycle (DMC) to minimize the impact of disasters. During any disaster, responses as well as rescue operations take precedence. The primary goal is to administer initial help and aid, provide humanitarian efforts and relief, and evaluate the initial extent of damage incurred. In the post-disaster scenario, the recovery activities commence, encompassing comprehensive community and societal development programs aimed

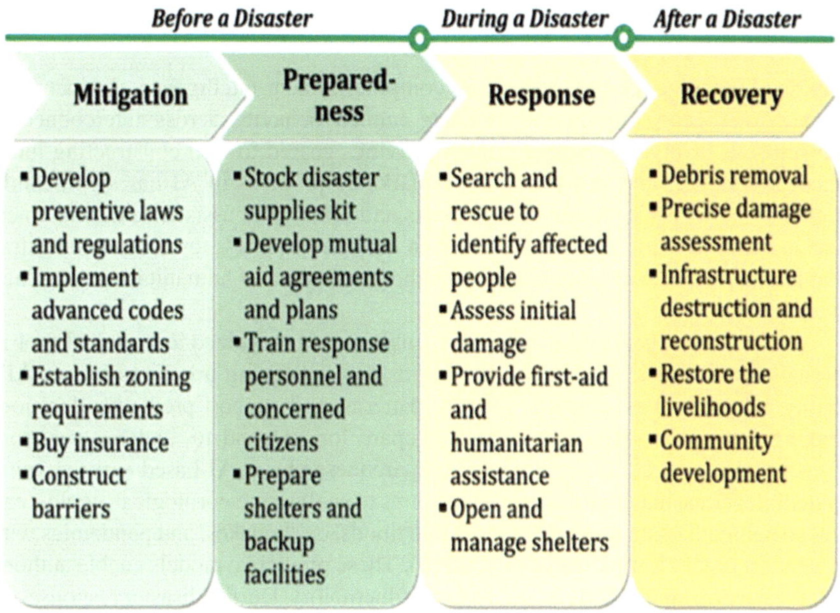

Fig. 14.2 Phases of urban disaster management

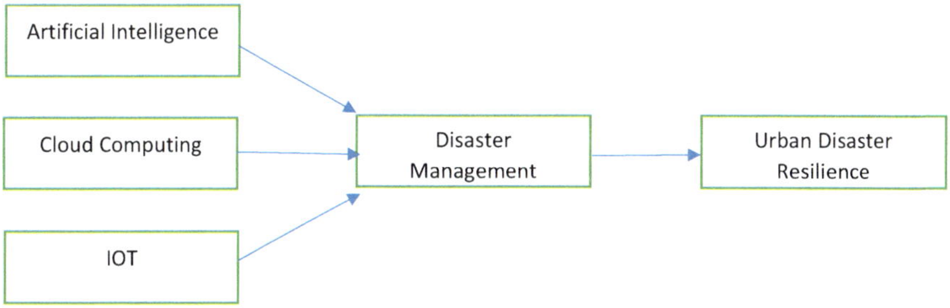

Fig. 14.3 Conceptual model

at restoring livelihoods. Furthermore, meticulous damage assessments are conducted to ascertain precise levels of impact.

14.3 Literature Review

Artificial intelligence (AI) represents computer system intelligence, distinct from human intelligence, aiming to simulate human behavior across interconnected machines. It involves computer-related activities geared toward constructing intelligent systems. In the last ten years, remarkable progress in AI has significantly improved our capacity to forecast disasters and offers vital assistance during emergencies. The evolution of AI is evident in disaster readiness, crowdsourced information platforms, rescue operations, and the distribution of humanitarian aid (Chen et al. 2020).

Artificial intelligence (AI) has increasingly been recognized as a pivotal tool in strengthening disaster management systems and enhancing urban resilience. AI's ability to process vast datasets, identify patterns, and support predictive decision-making has revolutionized how cities prepare for, respond to, and recover from disasters (Vinuesa et al. 2020). In the preparedness phase, AI-based early warning systems leverage machine learning algorithms to analyze meteorological, geological, and social media data to predict events like floods, earthquakes, and pandemics with greater accuracy (Kankanamge et al. 2020). These predictive models enable authorities to take preemptive actions and reduce vulnerability. During disaster response, AI facilitates real-time situational awareness through satellite imagery analysis, drone surveillance, and natural language processing of emergency calls or social media feeds (Imran et al. 2015). AI tools have been used to map post-disaster damage and identify critical zones for immediate relief operations. Chatbots and AI-powered communication systems also assist in disseminating timely information to affected populations.

In the recovery and resilience-building stages, AI supports urban planners in reconstructing infrastructure by simulating scenarios and optimizing resource allocation. Smart city technologies, integrated with AI, continuously monitor urban systems to detect stress signals and automate response mechanisms, thereby reducing long-term risk exposure (Zhou et al. 2020).

This report delves into the utilization of AI technologies such as machine learning, robotics, drones, deep learning, algorithms, and sensors, particularly focusing on their application in prediction of disaster and expediting relief and rescue operations. Robotics, though existing for decades, has recently evolved into truly automated and artificially intelligent machines, owing to advancements in sensor and computing technologies. Contrasting robotics, machine learning is a relatively modern facet of AI. Algorithms based on machine learning can execute specific tasks without explicit instructions; instead, they learn from patterns and data inputs, classifying as a form of AI. A model using convolutional neural network (CNN), networks can detect changes in satellite images, pinpointing disaster-affected areas to streamline relief-based efforts (Hartawan et al. 2019). Airborne drone robots venture into disaster or accident zones for thorough inspection and assistance (Kas and Johnson 2020). Machine learning, a sophisticated software category, interprets patterns from diverse data types and utilizes that information to predict outcomes in novel situations. The application of AI aims to enhance and optimize the efficiency of disaster management processes (Kurwakumire et al. 2019). AI tools like auto-sensors help in emergencies by facilitating information exchange through knowledge frameworks known as ontologies, providing essential data to disaster areas, and providing multi-agent platforms for live assistance and simulated situations. Effective emergency and disaster management relies on acquiring, integrating, and leveraging information from disparate sources to make informed decisions.

AI serves as a significant force multiplier in safeguarding civilian populations during disasters and catastrophe and is undoubtedly the future solution of disaster management. However, successful adoption of AI in disaster management necessitates government-based agencies to establish a plan of development ensuring efficient and productive deployment (David et al. 2023). Utilizing social networking platforms and digital technologies serves as an effective means for the public to disseminate and consume information. During natural or man-made disasters, social media becomes a rapid and widespread means of information sharing. While social media can enhance "situational awareness" during crises, analyzing vast and fast-paced data streams manually poses significant challenges (Lai and Tsai 2019).

In contemporary times, people extensively use social media platforms to disseminate various data types, including textual messages, images, and videos, amid natural or man-made calamities. Social media data have proven invaluable in numerous crisis scenarios such as floods, wildfires, earthquakes, civil unrest, and nuclear disasters. For instance, during the 2011 Japan earthquake aftermath, more than 100 million tweets regarding the tragedy were posted in just one single day. Focusing on publications from year 1991 to year 2018 in the database of WorldCat, outlines the role of AI in disaster management, among the phases of disaster management, the response phase garnered the highest number of publications during this period. AI's widespread application primarily concentrates on the response phase, swiftly analyzing vast data sets to facilitate enhancement in decision-making effectiveness and enhancing overall disaster management. Artificial intelligence, particularly, machine learning (ML), assumes a pivotal role in processing vast volumes of social and online media data during calamities, transforming this influx of information into understandable and credible insights. AI applications are adept at mimicking human-like cognitive processes within substantial datasets, significantly enhancing the efficiency of different tasks, processes, recognition of data pattern across extensive data sets (Jung et al. 2020). While multiple studies have explored disasters through the lens of analyzing data gathered from social media using various ML techniques, few researchers have systematically studied recent advancements, especially focusing on leveraging big data analytics in AI for disaster management. Deep learning stands out as a prominent methodology within prediction systems, notably in emergency management, wherein techniques like time series analysis are employed to mitigate disaster risks. The integration of hierarchical time series with deep learning algorithms represents another approach addressing forecasting challenges.

Kumar and Sud (2020), introduced "DHARA," a flood-supported mobile application leveraging long short-term memory (LSTM), CNN and AI-based applications. This innovative tool aimed to predict flood probabilities, facilitating early warning preparedness and preemptive restoration strategies. Additionally, a research initiative implemented a model for rainfall prediction, forecasting data by examining environmental attributes associated with rainfall. Artificial intelligence employs various methods like regression (nonlinear, linear, and logistic), aimed at forecasting hazards and evaluating potential risks and their impact. Support vector machines (SVM) also contribute to timely prediction and risk assessment. Hierarchical clustering, neural networks, fuzzy clustering, k-means clustering, and principal component analysis have been instrumental in devising and comparing strategies, training systems, and evacuation procedures during disaster (Yao et al. 2020). Moreover, deep learning-based techniques, encompassing deep neural networks and convolutional neural networks (CNN), play a crucial role in disaster mapping. These techniques aid in damage evaluation and assessment, managing disaster information systems, and fostering collaboration among all support during emergencies. Advanced AI methods such as Deep Q-networks and genetic algorithms are increasingly utilized for assessing loss and estimating repair costs.

Another approach involves leveraging multiple or combined deep learning algorithms to achieve faster and more accurate forecasts. However, forecasting future outcomes, especially for natural disasters, often demands substantial amounts of data. High-performance techniques become crucial due to the dense and nonlinear features within these large datasets (Erdelj et al. 2017).

14.4 Cloud Computing

Cloud computing has emerged as a critical enabler in enhancing disaster management and promoting urban resilience. Its core features—scalability, on-demand access, distributed infrastructure, and high data storage capacity—make it well-suited for managing the complexities and uncertainties associated with disaster scenarios (Agrawal et al. 2011). Research indicates that cloud computing supports all phases of disaster management: preparedness, response, recovery, and mitigation, by enabling rapid data processing, fostering inter-agency coordination, and ensuring the availability of critical services during emergencies (Celik and Corbacioglu 2010).

In the preparedness phase, cloud-based platforms facilitate the collection, storage, and analysis of massive datasets from sensors, satellites, and historical disaster records. These platforms support simulation models and early warning systems that forecast potential risks and assess urban vulnerability. For instance, geographic information systems (GIS) and remote sensing tools integrated with cloud infrastructure allow real-time mapping and risk visualization, helping urban planners design more resilient infrastructure (Sengul and Albayrak 2020).

During the response phase, cloud computing plays a vital role in maintaining communication and data access even when local infrastructure is compromised.

Emergency response teams can access real-time data, situational dashboards, and logistics tools via cloud platforms to coordinate rescue and relief efforts more effectively. Mobile apps backed by cloud services also enable crowdsourced data collection, helping authorities identify, and prioritize areas in need (Yousefi et al. 2018).

In the recovery and mitigation phases, cloud solutions support post-disaster assessments, documentation, and decision-making. Cloud-based storage ensures that critical records—such as infrastructure blueprints, health records, and property details—remain safe and accessible for rebuilding efforts. Moreover, smart city initiatives that rely on Internet of things (IoT) devices often use cloud platforms for continuous monitoring and predictive maintenance, thereby reducing the risk of future disasters (Zhou et al. 2021).

The elasticity of cloud computing allows government agencies and urban managers to scale resources as needed, avoiding large upfront investments in hardware while maintaining flexibility during crises. Additionally, its collaborative tools enable multi-agency coordination by offering shared platforms for data and document management, which is crucial during cross-regional or international disasters.

Despite its advantages, the literature also identifies challenges in adopting cloud computing for disaster resilience. Data security, privacy, and regulatory compliance are major concerns, especially when handling sensitive personal or geographic information. Dependence on Internet connectivity also poses risks in regions where digital infrastructure is weak or likely to be disrupted during disasters. Nonetheless, advances in hybrid and edge computing offer potential solutions by decentralizing critical operations closer to the point of need.

Cloud computing is a computational paradigm that delivers a range of IT services through interconnected, cost-effective computing units over IP networks. This enables users, regardless of their location, to access these services instantly over the internet, eliminating the requirement for local hardware or software installations. Leveraging virtually limitless computing power and storage, cloud computing technologies are particularly advantageous for geodata-and Web-GIS applications. In the implementation of smart cities, cloud computing forms the backbone for the digital technologies that enhance citizen well-being, increase operational efficiency, and promote sustainability (Wan et al. 2014), thereby contributing to the disaster resilience of these urban centers. In times of disasters, cloud computing facilitates the secure storage and swift relocation of data and computational resources, including software and algorithms, to alternative physical locations. This includes the backup of entire systems (Alazawi et al. 2014). The plethora of disaster-related data from various sources like sensors, satellite imagery, and social media can be stored and analyzed efficiently. This assists cities in proactively identifying hazards and disaster risks, facilitating the implementation of effective preparedness measures (Ujjwal et al. 2014). Cloud computing plays a vital role as the underlying infrastructure, offering readily available computing resources to meet the changing needs of real-time disaster and hazard analysis, emergency response, and disaster coordination. Additionally, cloud computing offers a refined solution for disaster simulation and modeling, contributing to the identification of existing vulnerabilities for the improvement of resilient smart city infrastructure.

However, in actual disaster scenarios characterized by the breakdown of communication infrastructure and electricity, the use of cloud services becomes challenging. To address these challenges, it is recommended to incorporate emerging disruptive technologies such as fog and edge computing, as well as network topology.

14.5 Internet of Things

The concept Internet of things (IoT) refers to the interconnection of physical objects, including mobile devices, sensors, buildings, and vehicles, to the Internet. This connectivity enables these objects to gather and exchange data among themselves (Tsai et al. 2014). Various IoT applications encompass smart infrastructure such as traffic control, parking facilities, radiation sensing, roads monitoring, smartphone detection, avalanche and landslide prevention, and snow level monitoring (Goel et al. 2021). The evolution of smart city solutions integrating IoT technologies has provided comprehensive insights into urban infrastructures with high temporal and spatial resolutions (Harrison et al. 2016). The widespread deployment and relatively low energy consumption of IoT devices facilitate the creation of extensive, self-organizing networks for emergency communication over extended durations (Li et al. 2019). In the realm of disaster management, IoT-enabled systems, coupled with evolving artificial intelligence tools and data analytics, serve as early warning mechanisms. They play a crucial role in locating victims and facilitating potential rescue operations (Ray et al. 2017).

The Internet of things (IoT) has emerged as a transformative technology in improving urban resilience and disaster management. With its ability to collect and transmit real-time data through interconnected sensors and devices, IoT enhances situational awareness and supports proactive decision-making in urban governance. The integration of IoT into city infrastructure allows for the continuous monitoring of environmental and structural parameters, which contributes to the overall resilience of urban systems (Al-Turjman et al. 2019). IoT facilitates the development of smart urban environments by connecting critical systems such as water, energy, transportation, healthcare, and housing. In the context of disasters, this interconnectedness ensures that essential services remain functional or can be restored quickly. Real-time data from IoT networks enable authorities to understand urban stress points, assess vulnerabilities, and prioritize resource allocation efficiently. For instance, smart energy grids and utility systems equipped with IoT sensors can optimize energy distribution during periods of crisis, minimizing disruptions and aiding recovery efforts (Zanella et al. 2014). IoT empowers citizens by providing them access to live information about safety, infrastructure, and public services. This participatory approach strengthens the relationship between citizens and authorities, fostering a more resilient social framework. Literature also highlights several concerns, including data privacy, high implementation costs, and challenges related to data standardization and interoperability. Despite these limitations, IoT continues to evolve as a foundational component of urban resilience frameworks. By enabling adaptive responses and data-driven urban planning, IoT not only improves disaster management but also promotes sustainability and liveability in cities facing increasing environmental and social challenges.

14.6 Conceptual Model

The model (Fig. 14.3) emphasizes that the integration of these technologies leads to more robust, adaptive, and data-driven disaster management systems. In turn, such systems are fundamental to achieving urban resilience—defined as a city's ability to withstand, respond to, and recover from adverse events. By linking technological innovation to urban resilience through the mediating role of disaster management, the model provides a conceptual pathway for future empirical research and policy development. The conceptual model developed in this study illustrates the pivotal role of advanced technologies—specifically artificial intelligence (AI), cloud computing, and the Internet of Things (IoT)—in enhancing disaster management and fostering urban disaster resilience. Based on extensive review of contemporary literature, the model outlines a layered framework where these technologies act as enabling tools that converge on effective disaster management, which in turn contributes to building resilient urban ecosystems.

Artificial intelligence contributes by enabling predictive analytics, risk assessments, and decision-support systems that improve the anticipation and handling of disasters. Machine learning algorithms, for instance, can analyze historical and real-time data to identify patterns and generate early warnings, allowing timely interventions. Additionally, AI facilitates resource optimization, damage assessment, and citizen engagement during and after disaster events.

Cloud computing enhances disaster management by providing scalable, reliable, and secure data infrastructure. It ensures that critical information is accessible anytime, anywhere—especially during crises when physical infrastructure may be compromised. Cloud platforms support inter-agency collaboration, storage of high-volume data from diverse sources, and real-time analytics, which are essential for informed disaster response and long-term resilience planning.

The Internet of things, with its network of sensors and connected devices, complements AI and cloud systems by capturing real-time data from urban infrastructure and the environment. From monitoring air quality and structural integrity to tracking utility networks and population movement, IoT adds a layer of responsiveness and adaptability to disaster management frameworks.

14.7 Conclusion

This study delves into the pivotal role of artificial intelligence system (AI) in disaster management and how its applications can significantly enhance disaster response. Specifically, this research comprehensively examined various AI applications to delineate the methods and approaches adopted across different phases of disaster management. The findings underscore the value of AI models in fortifying disaster management strategies, particularly focusing on response and mitigation efforts. The expanding scope of geospatial technology continually provides timely solutions, offering a robust agenda for geographic information science (GI Science). Therefore, it becomes essential to deliberate on the geographic implications when considering disaster management. Presently, remote sensing (RS) and geographic information systems (GIS) stand as powerful tools, enabling a deeper comprehension of emergency scenarios.

Looking ahead, we anticipate that AI technology will leverage increasingly intricate multispectral datasets, significantly aiding in mitigating the effect of disasters. The amalgamation of remote sensed data and spatial analysis via GIS has propelled the emergence of a groundbreaking informatics science. Consequently, artificial intelligence holds immense potential in addressing both natural and human-induced disasters. Yet, the success of any AI platform remains intricately linked to proprietary data management and analytical capabilities embedded within the underlying algorithms. Additionally, the effectiveness of these systems relies on other crucial technological components that augment the success of efficient management teams. In conclusion, the findings underscore that a synergistic application of AI, cloud computing, and IoT is not just desirable but necessary for modern cities to address increasing environmental uncertainties. The model presented serves as a foundation for further exploration into the operationalization of these technologies, highlighting the urgent need for investment, regulatory frameworks, and capacity building to support technology-driven urban resilience strategies.

References

Abid M, Isleem HF, Shahzada K, Khan AU, Kamal Shah M, Saeed S, Aslam F (2021) Seismic hazard assessment of shigo kas hydro-power project (Khyber Pakhtunkhwa, Pakistan). Buildings 11:349

Agrawal D, El Abbadi A, Das S, Elmore AJ (2011) Database scalability, elasticity, and autonomy in the cloud. In: International conference on database systems for advanced applications. Springer Berlin Heidelberg, Berlin, Heidelberg, pp 2–15

Alazawi Z, Alani O, Abdljabar MB, Altowaijri S, Mehmood R (2014) A smart disaster management system for future cities. In: Proceedings of the 2014 ACM international workshop on wireless and mobile technologies for smart cities, pp 1–10

Alexander DE (2015) Disaster and emergency planning for preparedness, response, and recovery. Oxford University Press

Al-Turjman F, Salama R, Altrjman C (2023) Overview of IoT solutions for sustainable transportation systems. NEU J Artif Intell Internet Things 2(3)

Celik S, Corbacioglu S (2010) Role of cloud computing in disaster management. Int J Inf Syst Cris Response Manag 2(1):25–34
Chan SW, Abid SK, Sulaiman N, Nazir U, Azam K (2022) A systematic review of the flood vulnerability using geographic information system. Heliyon 8(3)
Chatrabhuj, Meshram K, Mishra U, Omar PJ (2024) Integration of remote sensing data and GIS technologies in river management system. Discov Geosci 2(1):67
Chen J, Li Q, Wang H, Deng M (2020) A machine learning ensemble approach based on random forest and radial basis function neural network for risk evaluation of regional flood disaster: a case study of the Yangtze river delta, China. Int J Environ Res Public Health 17:49
Ci T, Liu Z, Wang Y (2019) Assessment of the degree of building damage caused by disaster using convolutional neural networks in combination with ordinal regression. Remote Sens 11:2858
David A, Yigitcanlar T, Li RYM, Corchado JM, Cheong PH, Mossberger K, Mehmood R (2023) Understanding local government digital technology adoption strategies: a PRISMA review. Sustainability 15(12):9645
der Sarkissian R, Zaninetti JM, Abdallah C (2019) The use of geospatial information as support for disaster risk reduction; contextualization to Baalbek-Hermel Governorate/Lebanon. Appl Geogr 111:102075
Erdelj M, Natalizio E, Chowdhury KR, Akyildiz IF (2017) Help from the sky: leveraging UAVs for disaster management. In: IEEE pervasive computing, vol 16. Institute of Electrical and Electronics Engineers Inc., Piscataway, NJ, USA, pp 24–32
Fernandez P, Mourato S, Moreira M (2016) Social vulnerability assessment of flood risk using GIS-based multicriteria decision analysis. A case study of Vila Nova de Gaia. Geomat Nat Hazards Risk 7:1367–1389
Goel SS, Goel A, Kumar M, Moltó G (2021) A review of Internet of Things: qualifying technologies and boundless horizon. J Reliab Intell Environ 7(1):23–33
Haldorai A, Babitha Lincy R, Suriya M, Balakrishnan M (2024) Climate resilience via smart technologies over natural disaster. In: Artificial intelligence for sustainable development. Springer Nature Switzerland, Cham, pp 383–406
Hamza A, Elemam W, Abuosekken M (2023) The role of strategic planning in environmental crisis management article review and case study. Int J Environ Stud Res 2(4):85–104
Harrison K (2016) Who is the assumed user in the smart city?. In: Designing, developing, and facilitating smart cities: urban design to IOT solutions. Springer International Publishing, Cham, pp 17–32
Hartawan DR, Purboyo TW, Setianingsih C (2019) Disaster victims detection system using convolutional neural network (CNN) method. In: 2019 IEEE international conference on industry 4.0, artificial intelligence, and communications technology (IAICT). IEEE, pp 105–111
Huang Q, Cervone G, Zhang G (2017) A cloud-enabled automatic disaster analysis system of multisourced data streams: an example synthesizing social media, remote sensing and Wikipedia data. Comput Environ Urban Syst 66:23–37
Hussain A, Masood T, Munir H, Habib MS, Farooq MU (2023) Developing resilience in disaster relief operations management through lean transformation. Prod Plan Control 34(15):1475–1496
Imran M, Castillo C, Diaz F, Vieweg S (2015) Processing social media messages in mass emergency: a survey. ACM Comput Surv 47(4):1–38
Jena R, Pradhan B, Beydoun G, Alamri AM, Ardiansyah, Nizamuddin, Sofyan H (2020) Earthquake hazard and risk assessment using machine learning approaches at Palu, Indonesia. Sci Total Environ 749:141582
Jung D, Tuan VT, Tran DQ, Park M, Park S (2020) Conceptual framework of an intelligent decision support system for smart city disaster management. Appl Sci 10:666
Kankanamge N, Yigitcanlar T, Goonetilleke A, Kamruzzaman M (2020) Can AI and big data analytics improve disaster response and recovery? Int J Disaster Resil Built Environ 11(2):123–138
Kas KA, Johnson GK (2020) Using unmanned aerial vehicles and robotics in hazardous locations safely. Process Saf Prog 39(1):e12066

Kumar TVV, Sud K (2020) AI and robotics in disaster studies. Palgrave Macmillan, Singapore
Kurwakumire E, Kuzhazha S, Muchechetere P (2021) Informal cadastres as enabling tools for disaster risk management. Eur J Geogr 12(2)
Lai J-S, Tsai F (2019) Improving GIS-based landslide susceptibility assessments with multi-temporal remote sensing and machine learning. Sensors 19:3717
Lee M-FR, Chien T-W (2020) Artificial intelligence and internet of things for robotic disaster response. In: Proceedings of the 2020 international conference on advanced robotics and intelligent systems (ARIS), Taipei, Taiwan, 19–21 August 2020, pp 1–6
Li X, Wang Q, Lan X, Chen X, Zhang N, Chen D (2019) Enhancing cloud-based IoT security through trustworthy cloud service: an integration of security and reputation approach. IEEE Access 7:9368–9383
Mizrak KC (2024) Crisis management and risk mitigation: strategies for effective response and resilience. Trends Chall Pract Contemp Strat Manag 254–278
Nguyen HD (2023) Spatial modeling of flood hazard using machine learning and GIS in Ha Tinh province Vietnam. J Water Clim Chang 14(1):200–222
Oakley M, Himmelweit SM, Leinster P, Casado MR (1848) Protection motivation theory: a proposed theoretical extension and moving beyond rationality—the case of flooding. Water 2020:12
Park S, Oh Y, Hong D (2017) Disaster response and recovery from the perspective of robotics. Int J Precis Eng Manuf 18:1475–1482
Paton D, Johnston D (2001) Disasters and communities: vulnerability, resilience and preparedness. Disaster Prev Manag Int J 10(4):270–277
Puttinaovarat S, Horkaew P (2020) Internetworking flood disaster mitigation system based on remote sensing and mobile GIS. Geomat Nat Hazards Risk 11:1886–1911
Rane N, Choudhary S, Rane J (2024) Artificial intelligence for enhancing resilience. J Appl Artif Intell 5(2):1–33
Ray PP, Mukherjee M, Shu L (2017) Internet of things for disaster management: State-of-the-art and prospects. IEEE Access 5:18818–18835
Ruth M, Coelho D (2015) Understanding and managing the complexity of urban systems under climate change. In: Integrating climate change actions into local development. Routledge, pp 317–336
Sengul H, Albayrak S (2020) Cloud GIS for disaster management: a spatial decision support system. J Geogr Syst 22(3):393–413
Shaluf IM (2007) An overview on disasters. Disaster Prev Manag Int J 16(5):687–703
Srivastava A, Maity R (2023) Assessing the potential of AI–ML in urban climate change adaptation and sustainable development. Sustainability 15(23):16461
Sulaiman N, Abid SK, Chan SW, Nazir U, Mahmud NPN, Latib SKK, Hafidz HF, Roslan AF, Rahim NA (2020) Need for resilience healthcare facilities management (RHFM) in malaysia's public hospitals. A critical literature review. Proc Int Conf Ind Eng Oper Manag 59:2336–2348
Syifa M, Kadavi PR, Lee C-W (2019) An artificial intelligence application for post-earthquake damage mapping in Palu, central Sulawesi Indonesia. Sensors 19:542
Tamiru H, Dinka MO (2021) Artificial intelligence in geospatial analysis for flood vulnerability assessment: a case of Dire Dawa watershed, Awash Basin Ethiopia. Sci World J 2021(1):6128609
Tsai CW, Lai CF, Vasilakos AV (2014) Future internet of things: open issues and challenges. Wirel Netw 20(8):2201–2217
Ujjwal KC, Garg S, Hilton J (2020) An efficient framework for ensemble of natural disaster simulations as a service. Geosci Front 11(5):1859–1873
Velev D, Zlateva P (2023) Challenges of artificial intelligence application for disaster risk management. Int Arch Photogramm Remote Sens Spat Inf Sci 48:387–394
Venkadesh P, Divya SV, Marymariyal P, Keerthana S (2024) Predicting natural disasters with AI and machine learning. In: Utilizing AI and machine learning for natural disaster management. IGI Global, pp 39–64
Vinuesa R, Azizpour H, Leite I, Balaam M, Dignum V, Domisch S, Nerini FF (2020) The role of artificial intelligence in achieving the sustainable development goals. Nat Commun 11(1):1–10

Wan Z, Hong Y, Khan S, Gourley J, Flamig Z, Kirschbaum D, Tang G (2014) A cloud-based global flood disaster community cyber-infrastructure: development and demonstration. Environ Model Softw 58:86–94

Yang J, Hou H, Hu H (2024) Exploring the intelligent emergency management mode of rural natural disasters in the era of digital technology. Sustainability 16(6):2366

Yao J, Wang P, Wang G, Shrestha S, Xue B, Sun W (2020) Establishing a time series trend structure model to mine potential hydrological information from hydrometeorological time series data. Sci Total Environ 698:134227

Yousefi H, Ghassemian M, Bardsiri VK (2018) Cloud computing for disaster management. Procedia Comput Sci 132:1042–1050

Zanella A, Bui N, Castellani A, Vangelista L, Zorzi M (2014) Internet of things for smart cities. IEEE Internet Things J 1(1):22–32

Zhou W, Wang C, He S (2020) Smart cities and disaster risk reduction: AI-based approaches. Int J Disaster Risk Reduct 45:101460

Zhou J, Huang G, Li H (2021) Smart cities and the role of cloud computing in resilience. Sustain Cities Soc 67:102748

Chapter 15
Environmental Education and its Impact on Promoting a Clean Environment

Parama Kundu and Ugyen Namdel

Abstract This research investigates the capacity of environmental education to increase awareness, shape eco-friendly attitudes, and promote sustainable behavior among people, particularly in institutional and community contexts. With the growing environmental issues such as pollution, global warming, and inappropriate waste disposal, the research emphasizes the potential of formal and informal environmental education programs to bring about change. A quantitative design was applied using a systematic questionnaire administered to 200 samples from urban and semi-urban regions. Descriptive statistics and inferential testing (independent samples t-test and Welch's t-test) were utilized on SPSS to examine effects of education exposure and institutional participation toward environmental awareness and behavioral modification. The findings revealed that the participants exposed to environmental education were significantly more aware ($p < 0.001$; Cohen's d $= 1.17$), and the students in institutions offering environmental programs possessed more environmentally friendly behavior ($p = 0.001$; Cohen's d $= 0.70$). This research differs from the existing literature because in this research, the twofold function of education and institutional action was empirically validated across different demographic populations. The study discovers that the integration of environmental education in curriculums and community events can inculcate long-term environmental stewardship. The study offers valuable lessons for policymakers, teachers, and sustainability professionals who want to maximize environmental awareness through targeted education interventions.

Keywords Environmental education · Clean environment · Eco-friendly behavior · Institutional initiatives · Awareness

P. Kundu (✉)
Adamas University, Kolkata, West Bengal, India
e-mail: parama.geog@gmail.com

U. Namdel
Paro College of Education, Royal University of Bhutan, Paro, Bhutan

K. Mukherjee et al. (eds.), *Urban Climate*, The Urban Book Series,
https://doi.org/10.1007/978-3-032-17895-4_15

15.1 Introduction

Environmental education is a central mechanism for altering attitudes toward the natural world, awareness, understanding, and pro-environmental behavior toward preservation of the environment (Acosta Castellanos 2022). With an expanding number of global environmental concerns including pollution, climate change, deforestation, and waste generation, an empowered, active populace is more desperately needed (Ardoin 2023). Environmental education provides individuals with the knowledge and skills to address environmental problems and encourages active participation in activities that contribute to a cleaner and healthier environment (Dillon 2023). Environmental education bridges the gap between theory and practice and teaches lifetime values of sustainability and ecological stewardship (Farrukh 2022). The impacts of environmental education extend far beyond the classroom to policymaking and business (Hoffmann 2022). By increasing environmental literacy, individuals are likely to adopt environmentally sustainable practices like appropriate disposal of waste, recycling, energy saving, and water conservation (Huang 2023). Environmental education also enhances the sense of responsibility and stewardship, motivating individuals to participate in environmental conservation activities and influence others through good practices (Nation & Feldman, 2021).

15.1.1 *Background and Motivation*

The accelerated process of urbanization, industrialization, and population growth has caused environmental degradation, pollution, global warming, and resource depletion, which endanger ecosystems, human health, and socio-economic development (Onopriienko 2021). Environmental education is a sure way to reduce these issues by evoking consciousness, instilling proper behavior, and instigating sustainable actions (Reddy 2021). It aims to provide individuals, particularly young people, with information, attitudes, and skills to enable them to make environmentally conscious choices and take effective action for environmental protection (Rooney 2022).

Environmental education encourages the culture of environment stewardship within formal and non-formal processes of learning (Seikkula-Leino 2021). Environmental education was found to create environmentally educated citizens who are prone to adopting eco-friendly lifestyles as well as act proactively on conservation activities (Tan 2023). Educational institutions, local communities, and governments throughout the world have integrated environmental education into school curriculum in order to incorporate sustainability as well as environment ethics from early days (Uralovich 2023). As the globe battles with intricacies of environmental problems, an increasing urgency to assess and improve the potency of environmental education initiatives in fostering cleanliness, reducing waste, and sustainable resource management arises (Yadav 2022).

15.1.2 Objectives of the Research

The research objectives of the study are as follows:

- To evaluate the effectiveness of environmental education programs in increasing awareness and knowledge about environmental cleanliness among students and communities.
- To analyze the role of educational institutions in promoting eco-friendly behaviors and practices that contribute to a cleaner environment.
- To assess the long-term impact of environmental education on individuals' attitudes, values, and behaviors related to environmental conservation and cleanliness.

15.2 Review of Literature

Environmental education significantly reinforces clean-living behavior by way of awareness, accountable actions, and healthy habits. It reinforces environmental literacy, disaster preparedness, and sensitivity, especially among communities. Young people and adolescents are assisted by participatory tools like internet-based animations. Environmental education results in improved environmental quality due to sustainable behaviors and reduced carbon dioxide emission. Experiential and scientific approaches bring about long-term behavioral change, resulting in a cleaner, healthier, and safer future.

15.2.1 Environmental Education and Clean-Living Behavior

Environmental education is at the core of shaping individuals' attitudes and behaviors toward having a sustainable and clean environment. It provides individuals with basic knowledge as well as feelings of responsibility that instigate pro-environmental behavior in social and ecological settings. Ma et al. (2023) established that environmental education has a positive impact on pro-environmental behavior, with the observation that people who had educational exposure were likely to participate in activities that minimize pollution and keep their environment clean. Yusuf et al. (2022), in a correlational study on post-tsunami Aceh, Indonesia, reported that environmental education enhanced the environmental literacy, awareness of disasters, and sensitivity of members of the community. These factors together encouraged cleanliness-of-living, with a focus on bringing education into recovery and sustainability initiatives. Combined, these studies show that environmental education not only increases awareness but also develops concrete, cleanliness-of-living behaviors in various populations and contexts.

15.2.2 Youth Engagement Through Environmental Education

Engaging youth in environmental education is essential to create long-term ecological responsibility and sustainable practices. Environmental ideas at a young age do much to form children's attitudes, values, and habits toward the environment, in order that sustainability turns into an automatic element of their perspective. Van De Wetering et al. (2022) in a detailed meta-analysis identified that environmental education profoundly improved environmental knowledge, attitudes, and behaviors in children and teenagers, affirming its robust effect during the critical years. This was supported by Desy et al. (2021), who identified the effectiveness of computer and visual learning media, proving that web animation videos employed at primary school level raised students' interest, motivation, and knowledge of environmental topics. Combined, these researches stress that early and engaging educational interventions have the capacity to equip younger generations with the knowledge and determination to pursue ecologically aware lives.

15.2.3 Environmental Education for Sustainability and Quality

Environmental education has a transformative function in promoting wider sustainability goals and overall environmental quality. It not merely raises public awareness but also generates measurable results such as lower emissions and sustained behavioral change. Wu et al. (2023) demonstrated that higher stages of environmental education significantly increased public consciousness, reduced carbon emissions, and cultivated sustainable behaviors on individual and community levels, particularly in a low-carbon setting. Fang et al. (2023) also presented the "living environmental education" idea that combines scientific data with experiential learning processes. Their study showed that such interactive, experiential learning approaches facilitate stronger behavioral changes and longer-lasting environmental stewardship. Overall, these studies highlight the importance of integrating theoretical understanding with practical experience to facilitate behavioral transformation and attain long-term environmental sustainability.

15.2.4 Research Gap

Whereas previous work highlights the importance of environmental education in developing awareness and green behavior, the overwhelming majority of studies target narrowly defined age brackets or interventions. Few systematic studies have a wide, empirical analysis of how combined formal and informal environmental education influences actual real-world green behaviors in community and institutional settings across urban and semi-urban zones. This study closes the gap by analyzing data collected from 200 participants using SPSS to gauge how various types of education approaches and institutional initiatives impact environmental knowledge and behavior and thus offer effective recommendations for increasing sustainable education policy.

15.3 Research Methodology

This section describes the methodological framework used to study the effect of environmental education on ensuring a clean environment. It discusses the research design, target population, sampling techniques, data collection tools, statistical methods, and ethical concerns. The methodology has been planned to deliver reliable, valid, and transparent research findings through descriptive and inferential approaches, as shown in Fig. 15.1 research framework.

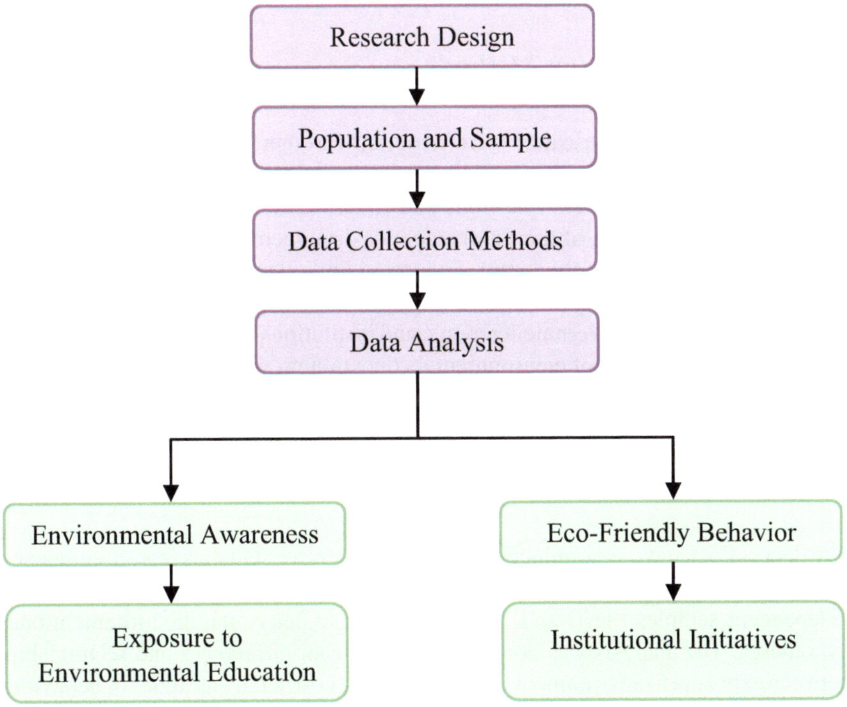

Fig. 15.1 Research framework

15.3.1 Research Design

The present study uses a descriptive and analytical research design to explore the impact of environmental education on the encouragement of a clean environment. The design is suitable for determining the extent of awareness, attitude, and behavior toward environmental cleanliness as it is and examining the role of the educational efforts in encouraging green behavior.

15.3.2 Population and Sample

The population of the sample of the study consists of students and residents from Kolkata, West Bengal, who have been exposed to environmental education initiatives. The 200 respondents were randomly sampled in such a way that representation is not one-sided among any specific demographic category. The sample consists of respondents from educational institutions (like schools and colleges) as well as community groups that are engaged with environmental education campaigns.

15.3.3 Data Collection Methods

Data collected in this research were acquired through a combination of structured questionnaires and corresponding published literature. The questionnaires contained both closed-ended questions and Likert-scale items, where respondents could indicate the degree of their agreement or disagreement on a scale (e.g., from strongly disagree to strongly agree), thus providing quantifiable measures of their knowledge, attitudes, and practices toward environmental cleanliness. Furthermore, published documents, government reports, and institutional reports were used to offer contextual interpretation of environmental education programs and their effects.

15.3.4 Tools and Techniques for Data Analysis

The data collected were examined using both descriptive statistics (e.g., frequencies, percentages, means, and standard deviations) and inferential statistics, including independent samples t-tests and chi-square tests, to determine the aforementioned hypotheses. The analysis was done to find significant differences and relationships between exposure to environmental education and eco-friendly attitudes or behaviors.

15.3.5 Variables and Measurement

The main variables of the study were conceptualized and measured as follows:

- **Environmental Awareness and Knowledge**: Measured using a composite score calculated from answers to Likert-scale questions regarding grasping the value of a clean environment and human activity impacts.
- **Eco-friendly Behaviors**: Measured using Likert-scale items for activities like segregation of waste, recycling, taking part in cleanliness campaigns, and conservation of energy/water.
- **Exposure to Environmental Education**: Categorical variable showing respondents' exposure to formal (school/college), informal (workshops/campaigns), both, or none of environmental education.
- **Institutional Environmental Initiatives**: A dichotomous variable measuring whether the respondent's school or community organization actively encourages environmental education programs.

15.3.6 Data Coding and Scoring Procedure

Likert-scale question responses were quantitatively coded from 1 (strongly disagree) through 5 (strongly agree). Awareness and green behavior composite scores were determined by taking average responses to respective items. Hypothesis testing involved respondents being classified according to:

- **Exposure** (Yes = formal/informal/both; No = none).
- **Institutional Initiative** (Yes = presence of institutional programs; No = absence).

15.3.7 Statistical Assumptions

Before performing inferential analysis, assumptions pertinent to the analysis were tested.

- Normality of the composite scores was checked by visual inspection and Shapiro-Wilk test.
- Homogeneity of variances was checked with Levene's Test. For Hypothesis 1, equal variances were assumed ($p > 0.05$). For Hypothesis 2, the assumption was violated ($p < 0.05$), and Welch's t-test was applied to check robustness.

15.3.8 Hypotheses of the Study

Hypothesis 1: Impact of environmental education programs on awareness and knowledge of environmental cleanliness.

Hypothesis 2: Influence of educational institutions' initiatives on eco-friendly behaviors.

15.3.9 Ethical Considerations

The research always followed ethical study procedures. Participants gave informed consent. The anonymity and confidentiality of responses were promised to respondents. Personal identifiers were not collected and, further, participation was on a voluntary basis.

15.4 Data Analysis and Interpretation

This section summarizes the findings from the questionnaire data, including respondent demographics, descriptive statistics of environmental awareness, attitudes, and behavior, as well as hypothesis testing. Both descriptive and inferential techniques were employed using SPSS to examine the effect of environmental education and institutional efforts on awareness and environmentally friendly practices.

15.4.1 Demographic Profile of Respondents

Table 15.1 is important because it provides the demographic distribution of the 200 respondents, hence justifying the diversity and balance of the sample with respect to gender, age, educational level, occupation, and location, thus making the sample valid. This wide-scale representation guarantees that the study encompasses a broad spectrum of viewpoints from students and teachers to the general public in urban, semi-urban, and rural settings, thus enhancing the reliability and generalizability of the research results.

In order to be able to make sense of the sample composition and place it into perspective for further analysis, demographic information about the 200 participants in the study was investigated. The data have been collected on a structured questionnaire and are portrayed in Table 15.1. Demographic break-up involves gender, age, educational qualification, occupation, geography, and experience with environmental education programs.

Table 15.1 Demographic profile of respondents (N = 200)

Demographic variable	Category	Frequency (n)	Percentage (%)
Gender	Male	98	49.0
	Female	102	51.0
Age group	15–24 years	84	42.0
	25–34 years	60	30.0
	35–44 years	38	19.0
	45 years and above	18	9.0
Education level	Secondary school	50	25.0
	Higher secondary	60	30.0
	Undergraduate degree	54	27.0
	Postgraduate and above	36	18.0
Occupation	Student	110	55.0
	Teacher/educator	28	14.0
	Community member/volunteer	32	16.0
	Others (self-employed, etc.)	30	15.0
Location	Urban	112	56.0
	Semi-urban	48	24.0
	Rural	40	20.0
Exposure to environmental education	Formal (school/college curriculum)	96	48.0
	Informal (campaigns/workshops)	58	29.0
	Both	30	15.0
	None	16	8.0

Demographic analysis of 200 respondents indicates a gender-balanced distribution, a young age group (mainly 15–34 years), and the majority being students (55.0%). A majority of participants had secondary education and above, with urban dwellers being the largest category (56.0%). For environmental education, 48.0% had formal exposure, 29.0% informal, 15.0% both, and 8.0% had no exposure.

15.4.2 Descriptive Statistics on Environmental Awareness, Attitudes, and Practices

Table 15.2 is important because it gives descriptive statistics mean scores and standard deviations for some important variables like environmental awareness, attitudes, behavior, and the impact of educational institutions. Quantifying the knowledge and green practices of the respondents, the table highlights strengths such as high levels of awareness and pinpoints gaps such as relatively low behavior scores. This transposition of qualitative impressions to quantifiable data adds a higher level of academic intensity to the study and offers a more distinct idea of where environmental education is working and where additional intervention may be needed.

Table 15.2 Descriptive statistics for awareness, attitudes, and practices regarding environmental education and cleanliness (N = 200)

Variable	Item statement	Mean	Standard deviation (SD)	Interpretation
Awareness level	I am aware of the importance of a clean environment	4.45	0.63	High awareness
	I understand how human activities impact the environment	4.32	0.71	High awareness
	I have received environmental education (formal/informal)	4.05	0.88	Moderate-to-high awareness
Attitudes toward environmental protection	I believe it is my responsibility to maintain environmental cleanliness	4.48	0.58	Strongly positive attitude
	Environmental cleanliness should be prioritized in school/ community activities	4.26	0.69	Positive attitude
	I support policies that promote environmental protection	4.11	0.74	Positive attitude
Eco-friendly behaviors	I segregate waste at home/ school/workplace.	3.95	0.91	Moderate behaviors
	I participate in cleanliness drives or awareness campaigns	3.76	1.04	Moderate behaviors
	I regularly reuse or recycle items	3.88	0.95	Moderate behaviors
	I conserve energy and water consciously	4.03	0.84	Moderately high behaviors
Influence of educational institutions	My school/college promotes environmental education activities	4.09	0.77	Positive institutional influence
	Environmental clubs/programs have helped improve my environmental behavior	3.92	0.89	Moderate institutional impact
	I am more likely to adopt eco-friendly habits due to education programs	4.12	0.72	Positive behavioral shift due to education

This section shows the descriptive statistical analysis of the significant variables employed in measuring the respondents' awareness, attitudes, and practices toward environmental education and hygiene. The response was gathered based on a 5-point Likert scale where 1 equates to strongly disagree and 5 to strongly agree. Table 15.2 depicts the mean values and standard deviations for each of the items together with their corresponding interpretations.

The results reveal high levels of environmental awareness, with mean values of 4.45 on cleanliness and 4.32 on awareness of human impact. Awareness of environmental education is modestly high ($M = 4.05$). Attitudes are also highly pro-environmental, indicated by a high personal responsibility measure ($M = 4.48$) and policy and community support (all >4.0). Behavioral habits such as segregation of waste ($M = 3.95$), recycling ($M = 3.88$), and conservation of resources ($M = 4.03$) reflect uniform green behavior. Institutional initiatives also reflected a positive impact on behavior, with a score between 3.92 and 4.12.

15.4.3 Hypothesis Testing

In order to achieve the research aims, two hypotheses were tested by applying independent samples t-tests in SPSS to examine the effects of environmental education on awareness and behavior. The analysis tested whether respondents' exposure to educational programs and institutional activities significantly affected environmental knowledge and practice, with specific results presented in the subsequent sections.

15.4.3.1 Impact of Environmental Education on Awareness and Knowledge

The initial hypothesis of the study sought to explore whether environmental education programs have an impact on students' and community members' awareness and knowledge about environmental cleanliness. The answers were grouped according to exposure to environmental education (Yes/No), and the generated awareness scores were compared using an independent samples t-test in SPSS.

Tables 15.3, 15.4 and 15.5 are important in that they all test Hypothesis 1 statistically by examining the effect of environmental education on awareness levels. Table 15.3 shows the group statistics comparing awareness scores in exposed and non-exposed groups, Table 15.4 performs an independent samples t-test showing a very significant difference ($p < 0.001$), and Table 15.5 estimates a large effect size (Cohen's $d = 1.17$). Collectively, these tables present firm empirical evidence to reject the null hypothesis and thus affirm that environmental education really does improve environmental awareness and further uphold the validity of the study's primary claim.

Hypothesis 1 Statement.

Table 15.3 Group statistics

Exposure to environmental education	N	Mean awareness score	Std. deviation	Std. error mean
Yes	184	4.32	0.51	0.037
No	16	3.71	0.64	0.160

Table 15.4 Independent samples t-test results

Levene's test for equality of variances	t	df	Sig. (2-tailed)	Mean difference	95% CI (lower)	95% CI (upper)
F = 2.944, p = 0.088	3.858	198	0.000 **	0.61	0.29	0.92

A p-value in statistics tells you how compatible the data is with a given assumption, usually the null hypothesis. In this case, the null hypothesis H_{01}, that there is no "Impact of Environmental Education Programs on Awareness and Knowledge of Environmental Cleanliness" on the students and community members. Levene's Test shows variances is equal to 0.088 (p = 0.088 0.05), which indicates that it is not significant. This can be interpreted as there is an Impact of Environmental Education Programs on Awareness and Knowledge of Environmental Cleanliness on students and community members or the alternative hypothesis is true

Table 15.5 Effect size (Cohen's d calculation)

Pooled SD	Formula	Cohen's d
≈ 0.52	d = (4.32–3.71) / 0.52 = **1.17**	**1.17**

H_{01} (Null Hypothesis): Environmental education programs do not significantly increase awareness and knowledge about environmental cleanliness among students and community members.

H_{11} (Alternative Hypothesis): Environmental education programs significantly increase awareness and knowledge about environmental cleanliness among students and community members.

As evident from Table 15.3, those who were exposed to environmental education programs possessed a greater mean awareness score (M = 4.32) than those who were not exposed (M = 3.71). This indicates a potentially strong influence of environmental education on increasing awareness and knowledge.

Levene's Test shows variances are equal ($p = 0.088 > 0.05$) and so equals variances. The t-test is statistically significant (t (198) = 3.858, $p < 0.001$), which establishes a significant difference between awareness scores of the two groups. The difference in means of 0.61 is indicative of strong environmental education influence on enhancing awareness regarding cleanliness and environmental accountability.

The test confirms that environmental education has a considerable impact on raising awareness, with a significant effect size (d = 1.17) and high statistical significance ($p < 0.001$). The null hypothesis H_{01} is rejected, indicating the critical importance of such programs in enhancing knowledge and inducing responsible environmental conduct.

15.4.3.2 Influence of Educational Institutions on Eco-Friendly Behaviors

The second hypothesis of the research sought to examine whether educational institutions that apply environmental education programs have a significant impact on students' and community members' eco-friendly behavior. The analysis was conducted using an independent samples t-test to test behavior scores among respondents from institutions that do and do not apply environmental education programs.

Tables 15.6, 15.7 and 15.8 are essential for hypothesis testing 2, as they explore the effect of environmental education programs within institutions on environmentally friendly behavior. Table 15.6 contrasts behavior scores among respondents from institutes with and without these programs, Table 15.7 employs Welch's t-test to validate a statistically significant difference (p = 0.001), and Table 15.8 indicates a moderate to large effect size (Cohen's d = 0.70). These results show that institutional involvement significantly affects sustainable behavior, supporting the instrumental effectiveness of organized environmental initiatives at developing environmentally responsible behavior.

Hypothesis 2 Statement.

Table 15.6 Group statistics

Institutional environmental initiatives	N	Mean behavior score	Std. deviation	Std. error mean
Yes	168	4.01	0.54	0.042
No	32	3.61	0.71	0.126

Table 15.7 Independent samples t-test results

Levene's Test for equality of variances	t	df	Sig. (2-tailed)	Mean difference	95% CI (lower)	95% CI (upper)
F = 4.228, p = 0.041	3.290	198	0.001 **	0.40	0.16	0.64

A p-value in statistics tells you how compatible the data is with a given assumption, usually the null hypothesis. In this case, the null hypothesis H_02, that educational institutions that implement environmental education initiatives do not significantly influence the promotion of eco-friendly behaviors and practices. Levene's Test showed unequal variances (p = 0.041), which indicates that it is slightly statistically significant. This can be interpreted that institutional factors help to shape eco friendly conduct

Table 15.8 Effect size—Cohen's d

Pooled SD	Formula	Cohen's d
≈ 0.57	(4.01–3.61) / 0.57 = 0.70	0.70

H_{02} (Null Hypothesis): Educational institutions that implement environmental education initiatives do not significantly influence the promotion of eco-friendly behaviors and practices.

H_{12} (Alternative Hypothesis): Educational institutions that implement environmental education initiatives significantly influence the promotion of eco-friendly behaviors and practices.

As revealed in Table 15.6, individuals from institutions with environmental education programs indicated a higher mean score on eco-friendly behavior (M = 4.01) than those from institutions without these programs (M = 3.61). This indicates a significant difference in behavior depending on the environmental activities in institutions.

Levene's Test showed unequal variances ($p = 0.041$), thus the Welch's t-test was utilized. The result of the test was statistically significant at the level 0.001 with the t-value equal to 3.290 and mean difference being 0.40 points. This is clear evidence that institutional factors significantly shape eco-friendly conduct.

The findings show that institutional environmental education has a large to moderate effect size (d = 0.70) and high statistical significance (p = 0.001), as it significantly affects environmentally friendly behavior. The null hypothesis H_{02} is thus rejected since it is confirmed that active institutional action is a strong motivator of sustainable habits and environmental stewardship.

15.5 Results and Discussion

This section presents the study findings and provides an in-depth explanation in terms of the research objectives, literature, and empirical data. It is organized into several subsections for the purpose of clarity such as descriptive findings interpretation, hypothesis testing, comparison with existing studies, and challenges identified.

15.5.1 Interpretation of Descriptive Findings

The descriptive findings of the study show that the respondents have a great awareness of environmental cleanliness and its significance. A mean score of 4.45 on the significance of cleanliness shows that most of the participants appreciate having a clean environment a great deal. Additionally, a mean of 4.32 on environmental impact awareness shows that the respondents have a great deal of knowledge about how their activities impact the environment. Attitudinally, the data are strong in environmental protection. The belief in personal responsibility mean score of 4.48 indicates that participants not only comprehend environmental problems but also feel personally responsible for helping to solve them. This suggests a mature attitude and positive orientation toward environmental stewardship.

Behaviorally, the respondents indicated practicing a variety of environmentally friendly behaviors like recycling, saving energy, and sorting out waste. Although the behaviors were rated as moderate to high, this difference indicates that even though there is awareness and intention, the level of behavior change can be influenced by accessibility, motivation, or institutional support.

The research also emphasized the positive impact of educational institutions, especially schools, toward fostering environmentally responsible behavior. The participants admitted that school-based environmental programs greatly influenced their habits and attitudes, recognizing the efficacies of schooling interventions in spurring sustainability.

15.5.2 Hypothesis Testing Summary

Table 15.9 is important because it concisely summarizes hypothesis test results by displaying p-values, effect sizes, and conclusions for both hypotheses in one viewport. The integrated presentation enables reviewers, researchers, and policymakers to immediately understand the empirical robustness of the results, ensuring that environmental education and institutional interventions do indeed effectively impact awareness and behavior. It solidifies the validity of the study as a whole by definitively showing the effective statistical confirmation of both central hypotheses. Hypothesis test results using SPSS are presented in table with both null and alternative hypotheses, test statistic, p-value, and conclusion.

Table 15.9 indicates that both hypotheses were confirmed. For H_1, environmental education improved awareness and knowledge considerably ($p < 0.001$, large effect size d = 1.17). For H_2, institutional programs considerably improved eco-friendly behavior ($p = 0.001$, moderate to large effect size d = 0.70). Hence, both null hypotheses were rejected.

Table 15.9 Summary of hypothesis testing

Hypothesis	Test used	*p*-value	Effect size (Cohen's d)	Conclusion
H_1	Independent samples t-test	< 0.001	1.17 (large)	H_0 rejected. H_1 accepted: Environmental education significantly enhances awareness and knowledge
H_2	Welch's t-test	0.001	0.70 (moderate to large)	H_0 rejected. H_1 accepted: Institutional initiatives significantly improve eco-friendly behaviors

15.5.3 Comparison of Findings with Existing Studies

Table 15.10 is important since it contrasts the current research findings with the most important earlier studies, hence its uniqueness and contribution to the field. In contrast with the previous studies that mostly addressed single groups or were not statistically tested, this study distinctly combines the double effect of both environmental education and institutional action. The table highlights the quantitative rigor of the study by presenting effect sizes and hypothesis testing, distinguishing it from previous studies that were more qualitative or narrowly focused in their analyses. To assess the contribution of current research, results are compared with a sample literature.

Table 15.10 provides a comparison of studies that emphasizes the originality of the current study against previous studies. Although previous research by Ma et al. (2023), Yusuf et al. (2022), and others examined different aspects of environmental education like change in behavior, response to disaster, and sustainability, these studies focused on either certain groups (like children or affected disaster communities) or were not statistically validated rigorously. By comparison, the current research

Table 15.10 Comparative analysis of current findings and prior research

Author(s)	Year	Method	Objective	Findings	Superiority/ novelty
Ma et al.	2023	Quantitative, moderation analysis	Role of education in cleanliness	Education boosts pro-environmental behavior	Focused behavior; lacks institutional scope
Yusuf et al.	2022	Correlational (post-disaster)	Link education with disaster recovery behavior	Education improved literacy and behavior	Post-disaster focus; no hypothesis testing
Van De Wetering et al.	2022	Meta-analysis	Impact on youth behavior	Increased knowledge and behavior in children	Youth-centered; no broader community analysis
Wu et al.	2023	Empirical (low-carbon economy)	Education's role in sustainability	Reduced emissions, promoted sustainability	Policy-level focus; lacks direct behavior test
Desy et al.	2021	Experimental (web-based)	Use of animation to enhance learning	Interactive tools improved environmental learning	Tech-driven; child focus, not behavioral outcomes
Present study	**2025**	Descriptive + inferential (SPSS)	Education and institutional impact on awareness and behavior	Education ($d = 1.17$); institutional support ($d = 0.70$) enhance impact	Integrates education + institutional angle; strong statistical base

(2025) is distinctive in its ability to merge educational and institutional impacts on awareness and behavior with a robust descriptive and inferential design (SPSS), with large effect sizes ($d = 1.17$ and $d = 0.70$), providing a more thorough and statistically supported contribution.

15.5.4 Contribution to Policy and Practice

The results of this study provide important inputs to both policymaking and actual implementation through the identification of the change-inducing potential of environmental education in bringing about sustainable behaviors. The research conclusively shows that systematic educational exposure, either through schools, colleges, or community platforms, has a quantifiable positive impact on the environmental attitudes and behaviors of individuals. This leads to support for the integration of environmental education into national and regional curricula, not as an additive but as a fundamental part of formal education. Furthermore, the success of these institutional initiatives like environmental clubs, awareness campaigns, and school-based sustainability programs again highlights the importance of active policy support. Policymakers are therefore urged to support, finance, and establish these community-based programs to ensure that awareness about the environment is ingrained from a young age and reinforced by communal action. The programs not only encourage current environmentally friendly behavior but also long-term environmental responsibility, and as such, form an important resource for combating more universal environmental issues.

15.5.5 Identified Challenges and Limitations

The research encountered a number of limitations and practical issues which must be addressed in future work.

- The number of participants was restricted to 200, and this could potentially impact the validity of results being generalized.
- Concentration in cities and semi-urban regions does not necessarily present a true rural scenario.
- Self-reporting information could be compromised by social desirability bias.
- Quantitative tools alone were applied; application of mixed-methods would generate deeper insights.
- Time and logistical restrictions reduced the number of institutions and regions included.

15.6 Conclusion and Recommendations

This research strongly supports the fact that environmental education has an important role in raising the public's awareness regarding environmental issues. Participants who were exposed to educational programs whether at school, college, or through community programs indicated a far better understanding of environmental cleanliness and sustainability than participants with no exposure.

In addition, the research highlights that environmental education affects behavior and attitude favorably, promoting individuals to adapt environmentally friendly behaviors in their day-to-day lives. Such adaptations involve improved waste disposal methods, lower plastic use, and taking an active part in sanitation drives, demonstrating the catalytic role that formal and informal learning spheres play.

Statistical findings from the study, including large effect sizes and significant p-values, validate the impact of institutional involvement. Institutions that integrate environmental values into policy and practice encourage students and community members to become more accountable, showing that organizational support optimizes the effectiveness of educational efforts.

By completing an essential lacuna in present literature, this research highlights the need for collective action by educational institutions and authorities. It proposes that environmental consciousness must be not merely learned but consolidated by concrete, visible initiatives at the institutional level.

The study provides a solid basis for policymaking and practical steps toward consolidating environmental education. It compels policymakers, educators, and stakeholders to develop inclusive programs that impart sustainability and sanitation in the social fabric to ensure long-term ecological accountability in communities.

Based on this study's findings, the following is suggested to enhance the effectiveness of environmental education and promote sustainable behavior:

- Incorporate environmental education as a mandatory part of school and college courses to develop early awareness and accountability.
- Urge institutions to adopt practical eco-friendly activities, like clean-up campaigns and recycling, to complement theoretical knowledge.
- Offer periodic training and workshops for teachers to refresh their skills and knowledge in imparting effective environmental education.
- Develop community-based awareness programs involving local leaders, NGOs, and citizens to promote sustainable practices.
- Create and disseminate accessible learning content through technology (e.g., apps, videos, games) to reach various age groups effectively.
- Urge policymakers to support and fund research and institutional initiatives on environmental awareness and behavior change.

AI Discloser No AI tool has been used for writing the manuscript

References

Acosta Castellanos PM, Queiruga-Dios A (2022) From environmental education to education for sustainable development in higher education: a systematic review. Int J Sustain High Educ 23(3):622–644. https://doi.org/10.1108/IJSHE-04-2021-0152

Ardoin NM, Bowers AW, Gaillard E (2023) A systematic mixed study review of civic engagement outcomes in environmental education. Environ Educ Res 29(1):1–26. https://doi.org/10.1080/13504622.2022.2083920

Desy S, Ika L, Arifin M, Nurzengky I, Arita M, Musril Z, Rossi I (2021) Web-based animation video for student environmental education at elementary schools. Int J Interact Mob Technol (IJIM) 15(12):4–14. https://doi.org/10.3991/ijim.v15i12.23247

Dillon J, Herman B (2023) Environmental education. In: Handbook of research on science education. Routledge, pp 717–748

Fang WT, Hassan AA, LePage BA (2023) The living environmental education: sound science toward a cleaner, safer, and healthier future. Springer Nature, p 279. https://doi.org/10.1007/978-3-030-92919-1

Farrukh M, Ansari N, Raza A, Wu Y, Wang H (2022) Fostering employees' pro-environmental behavior through green transformational leadership, green human resource management and environmental knowledge. Technol Forecast Soc Chang 179:121643. https://doi.org/10.1016/j.techfore.2022.121643

Hoffmann R, Muttarak R, Peisker J, Stanig P (2022) Climate change experiences raise environmental concerns and promote green voting. Nat Clim Chang 12(2):148–155. https://doi.org/10.1038/s41558-021-01284-0

Huang CC, Li SP, Chan YK, Hsieh MY, Lai JCM (2023) Empirical research on the sustainable development of ecotourism with environmental education concepts. Sustainability 15(13):10307. https://doi.org/10.3390/su151310307

Ma L, Shahbaz P, Haq SU, Boz I (2023) Exploring the moderating role of environmental education in promoting a clean environment. Sustainability 15(10):8127. https://doi.org/10.3390/su15108127

Nation MT, Feldman A (2021) Environmental education in the secondary science classroom: how teachers' beliefs influence their instruction of climate change. J Sci Teach Educ 32(5):481–499. https://doi.org/10.1080/1046560X.2021.1911631

Onopriienko K, Onopriienko V, Petrushenko Y, Onopriienko I (2021) Environmental education for youth and adults: a bibliometric analysis of research. In: E3S web of conferences, vol 234. EDP Sciences, p 00002. https://doi.org/10.1051/e3sconf/202123400002

Reddy C (2021) Environmental education, social justice and teacher education: enabling meaningful environmental learning in local contexts. S Afr J High Educ 35(1):161–177. https://doi.org/10.20853/35-1-4112

Rooney T, Blaise M (2022) Rethinking environmental education in a climate change era: weather learning in early childhood. Routledge. https://doi.org/10.4324/9781003139411

Seikkula-Leino J, Jónsdóttir SR, Håkansson-Lindqvist M, Westerberg M, Eriksson-Bergström S (2021) Responding to global challenges through education: entrepreneurial, sustainable, and pro-environmental education in Nordic teacher education curricula. Sustainability 13(22):12808. https://doi.org/10.3390/su132212808

Tan CK, Nurul-Asna H (2023) Serious games for environmental education. Intgr Conserv 2(1):19–42. https://doi.org/10.1177/27504877231158024

Uralovich KS, Toshmamatovich TU, Kubayevich KF, Sapaev IB, Saylaubaevna SS, Beknazarova ZF, Khurramov A (2023) A primary factor in sustainable development and environmental sustainability is environmental education. Casp J Environ Sci 21(4):965–975. https://doi.org/10.22124/cjes.2023.6886

Van De Wetering J, Leijten P, Spitzer J, Thomaes S (2022) Does environmental education benefit environmental outcomes in children and adolescents? A meta-analysis. J Environ Psychol 81:101782. https://doi.org/10.1016/j.jenvp.2022.101782

Wu Y, Wan J, Yu W (2023) Impact of environmental education on environmental quality under the background of low-carbon economy. Front Public Health 11:1128791. https://doi.org/10.3389/fpubh.2023.1128791

Yadav SK, Banerjee A, Jhariya MK, Meena RS, Raj A, Khan N et al (2022) Environmental education for sustainable development. In: Natural resources conservation and advances for sustainability. Elsevier, pp 415–431. https://doi.org/10.1016/B978-0-323-85590-4.00009-3

Yusuf R, Yunus M, Maimun M, Fajri I (2022) Environmental education: a correlational study among environmental literacy, disaster knowledge, environmental sensitivity, and clean-living behavior of post tsunami disaster in Aceh communities, Indonesia. Pol J Environ Stud 31(1):1–9. https://doi.org/10.15244/pjoes/135020

Chapter 16
Rethinking Climate Change, Extreme Events, and Resilience in the South Pacific: A Systematic Review

Mumtaz Alam and Kunal Avishek Gounder

Abstract The South Pacific region is confronted with escalating issues arising from climate change and extreme weather events. This research paper delves into a significant shift in the underlying framework that guides our approach to these matters, explicitly emphasizing augmenting resilience. This research provides novel perspectives on tackling the intricate issues of climate change and its effects on the South Pacific by exploring new techniques, sustainable practices, and community engagement. This study offers a comprehensive perspective on resilience building in the South Pacific, focusing on the interconnectedness of environmental, social, and economic factors. By examining the relationship between these factors, the study aims to contribute valuable insights into the field of climate adaptation and disaster preparedness in the region.

Keywords Climate change · Extreme events · Resilience · South Pacific · Paradigm shift

16.1 Introduction

Given the increasing prevalence of climate-related problems in contemporary society, it is pertinent to reexamine the concept of climate from a scientific perspective, considering its historical development and the fundamental origins of these concerns. The island nations of the South Pacific face serious threats to their social and economic standing, resilience, and environmental sustainability due to the rapid consequences of climate change. Rising sea levels, coastal erosion, flooding, and saltwater intrusion have all harmed the environment and the native land's resources,

M. Alam (✉)
Department of Social Sciences, College of Humanities, Education and Law, Fiji National University, Suva, Fiji
e-mail: mumtaz.alam@fnu.ac.fj

K. A. Gounder
College of Humanities, Education and Law, Fiji National University, Suva, Fiji

K. Mukherjee et al. (eds.), *Urban Climate*, The Urban Book Series,
https://doi.org/10.1007/978-3-032-17895-4_16

vital to South Pacific communities' livelihoods and social structures. Because the Pacific is so vulnerable to flooding and tropical cyclones, homes, public spaces, and food accessibility are more at risk. While earlier research has examined the Pacific's overall susceptibility to climate change, little work has been done to examine local circumstances surrounding climate adaptation and disaster preparedness precisely.

Limited information is available regarding the specific impacts and the adaptation and resilience tactics used by coastal communities in the South Pacific, which are especially vulnerable to the effects of climate change. Pacific island nations are now at the forefront of the effects of climate change due to their extreme vulnerability to flooding, increasing sea levels, diminishing fisheries, and severe weather, particularly tropical cyclones. Pacific nations like Kiribati and Tuvalu mostly rely on their natural resources, including fisheries, for economic survival. However, the threat posed by climate change has created uncertainty. Understanding community reality is crucial to developing adaptation methods that support the sustainability of Pacific communities.

16.2 Climate Change in the South Pacific

The main effects of climate change in the South Pacific are rising sea levels that altered patterns of sediment deposition, heightened danger of inundation and coastal floods, accelerated erosion, and saltwater intrusion in watersheds and underground aquifers. These are all modifications to the coastal zone's physical environment, and depending on the social and geographic contexts, they may have different consequences for ecosystems and human society. Nonetheless, the South Pacific Island Countries share a certain degree of vulnerability. The term "climate change" describes a long-term shift in the planet's overall weather patterns brought on by widespread human activity like deforestation, the generation of carbon dioxide, and fast development. The global surface temperature reached a peak of 1.1 degrees Celsius over the period between 2011 and 2020 compared to 1850–1900 due to the deliberate release of greenhouse gases into the atmosphere (IPCC 2023). Additionally, the Intergovernmental Panel on Climate Change (IPCC) Report notes a 12% increase in global net anthropogenic greenhouse gas emissions in 2019 compared to 2010. This demonstrates that even in the face of the catastrophic effects of climate change, human activity is still increasing, exacerbating the phenomenon's effects. Climate change devastates Earth, causing changes in precipitation patterns, ocean currents, storm intensity, sea level rise, and ocean acidification (Bennett & Dearden 2013).

The IPCC Report (2023) states that in extremely vulnerable areas, such as the Pacific Islands, the death toll from natural catastrophes has grown by a factor of 15. Smaller island nations face significant risks from climate change, including rising sea levels, coastal floods, coastal erosion, and habitat destruction (Prasad & Mkumbachi 2021). The impact is greater in larger landmass countries than in smaller island governments because of landmass limits and elevation. Sustainable development has been hampered by the Pacific's small geographic boundaries, where population growth is yearly despite spatial shrinkage brought on by sea level rise, saltwater intrusion, and salinization (Veitayaki et al. 2007). According to Veitayaki (2002), Pacific Islanders used their traditional wisdom to address environmental problems in the past. However, modernization and development effects have diminished the older generation's traditional knowledge of environmental issues (Veitayaki 2010). The Pacific of today has deteriorated its environment as a result of development. Climate change has been exacerbated by extensive land clearing for infrastructure construction with little regard for sustainable planning. Climate change's various environmental impacts have brought about social and economic limitations. Due to the necessity of food and water security for human survival, people living in Small Island Developing States (SIDS) have been forced to move (Turvey 2007).

16.2.1 Extreme Events in South Pacific

The extreme events caused by climate change in the South Pacific have far more destruction than is described in words. Small Island nations have been facing extreme challenges that have been forcing people to leave their native land and migrate to new land. Tropical cyclones are becoming intensified and more frequent due to climate change, as South Pacific countries are in the World's latest ocean, the Pacific Ocean. A significant expansion of the warm ocean is a factor that causes the formation of tropical cyclones. Extensive research by Sun et al. (2017) has elaborated that an increase in Sea Surface Temperature (SST) makes tropical cyclones stronger and more prominent, resulting in more significant destruction. Figure 16.1 is the diagrammatic representation providing a synopsis of the potential mechanisms underlying the effects of global SST warming on Tropical cyclone size. Projections have implied that global SST will continuously increase in the long term, making tropical cyclones stronger and more extensive. The risk posed by tropical cyclones to South Pacific Island Countries is immense, affecting a large portion of countries GDP and slowing down economic growth. South Pacific has experienced many disastrous tropical cyclones, such as Tropical Cyclone Winston, which are remembered for their catastrophic destructions.

Water security is a concerning global issue, and South Pacific Island Countries are no strangers to it. Underground coastal aquifers are being intruded by saltwater due to rising sea levels, hence affecting the availability of fresh water for consumption. A study by Lal and Datta (2019) elaborates that the excessive withdrawal of freshwater from the aquifers has led to saltwater intrusion, causing an increase in the salinity in freshwater. Since it is essential for survival, the increasing population and water demand have increased the pressure on natural freshwater aquifers. Sharan, Lal and Datta (2021) further added that rapid industrialization is also a factor in Pacific Island Countries that have excessive demand for groundwater, increasing freshwater pumping. As groundwater and seawater interact hydraulically, intrusion typically happens when the water context of the aquifer is lower than average. Figure 16.2 summarizes the process of saltwater intrusion in the coastal aquifer system. Significant groundwater abstraction results in additional saline water intrusion seeping in

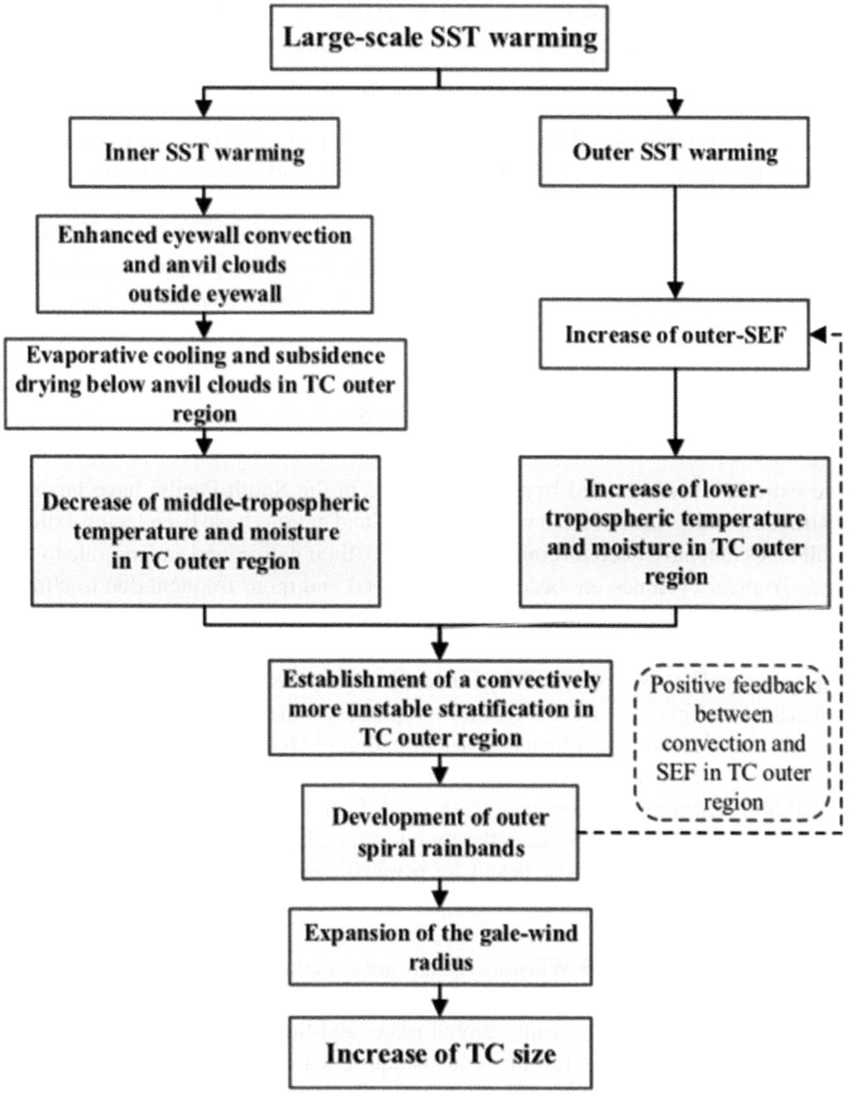

Fig. 16.1 Schematic diagram summarizing the possible mechanisms responsible for the impact of large-scale SST warming on TC size. *Source* Sun et al. (2017)

and the creation of ascendence and depression zones, which further reveal the degree of saline water intrusion. Excessive groundwater withdrawal from limited aquifers causes massive depression cones to form, with a radius of more than 100 km, which in turn causes a significant amount of saltwater intrusion (Lal & Datta 2021).

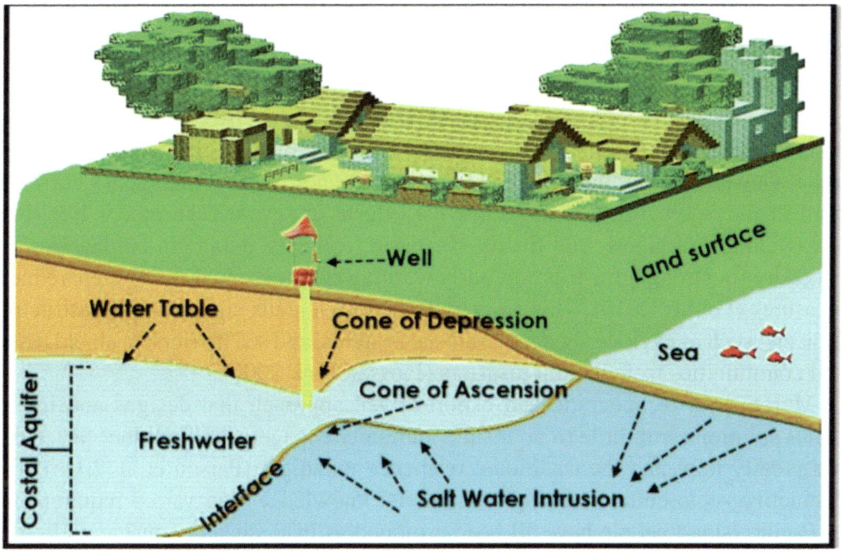

Fig. 16.2 Diagram representation of saltwater intrusion in coastal aquifers. *Source* Sharan, Lal and Datta (2021)

El Nino and La Nino have been factors causing extreme weather events such as flooding, tropical cyclones, and drought, leading to many socio-economic and environmental challenges. El Nino is an extreme event describing a range of phenomena where the Pacific Ocean warms more than average temperatures. Changes in SST have a cascading influence on precipitation, wind directions and speed, ocean currents, and cyclone developments (Kelman 2019). The changes started in the Pacific region and spread to other parts of the world via teleconnections. El Nino occurs every 3–7 years and lasts 12–18 months, ranging in different strengths (Kelman 2019). During the period of El Nino, an increase in temperature caused higher rates of evapotranspiration, resulting in heavy precipitation such as rain, causing floods. Ocean warming also increases the risk of tropical cyclones, as mentioned earlier. Whereby La Nina causes stronger and more intense trade winds which intensify the warm pool in the Western Pacific. This results in SST in the Central and Eastern Pacific dropping, causing more rain in the Western Pacific. Countries such as Solomon Islands, Vanuatu, Fiji, Tonga, Niue, and parts of Papua New Guinea experience higher rainfall, whereas countries like Kiribati, Nauru Tuvalu, and Cook Islands experience lower rainfall during La Nina. Pacific has experienced a triple-dip La Nina from 2020 to 2022 (Malsale 2023).

16.3 The Standard Perspective on Resilience

Resilience has historically been seen from a limited perspective, emphasizing technical solutions and top-down approaches to endure and recover from the impacts of natural disasters or extreme weather events triggered by climate change. This traditional viewpoint has influenced most of the resilience theories and tactics currently used in the South Pacific. To reduce the effects of severe occurrences like tropical cyclones, droughts, and floods, the basic resilience paradigm emphasizes the necessity for strong infrastructure, early warning systems, and disaster preparedness measures (Pelling 2010). This strategy frequently entails significant infrastructure projects, such as cyclone shelters, drainage systems, and sea barriers, to shield assets and communities from natural disasters (Janssen et al. 2006).

Moreover, a technocratic and export-driven approach that designs and implements solutions with little to no local community engagement is another factor that has greatly impacted the traditional resilience paradigm (Parson et al. 2016). The distinctive socio-cultural settings, traditional knowledge, and varied requirements of Pacific Island people have all been critiqued as inadequately considered by this top-down strategy.

Traditional knowledge in the Pacific has solutions to adverse weather events, reducing the impacts. For example, Anthony Liuvaie Freddie, a Master of Architecture candidate at Victoria University of Wellington (2018), concentrates on utilizing Niue's natural characteristics to create a sustainable construction resilient to cyclones. He investigated research methods from other Pacific countries after realizing little material was on Niuean architecture. He discovered that the Ovava (banyan) tree and other islands' natural characteristics may be the foundation of a construction resistant to cyclones. Using the Ovava tree as the focal point of his design, Anthony used cyclone-resistant building practices from other Pacific countries, such as planting coastal trees and bamboo, to fortify the site against cyclone damage. Reliance on imported materials, which contributed significantly to the damage and fatalities during the hurricane, will be decreased.

16.3.1 Materials and Methods

This study used a thorough examination of the literature and secondary data sources to look at climate resilience methods in the South Pacific region. The research method was meant to combine what we already know with new ideas on how to improve the region's ability to adapt. The study used a lot of different types of secondary sources, such as peer-reviewed academic journals, reports from international climate groups, government policy documents, and specialist online resources. Climate scientific publications, regional development studies, and documents from both government and non-governmental groups that work in the South Pacific were some of the most important sources.

The study's main goal was to create better resilience frameworks to deal with the South Pacific's growing climate vulnerability and extreme weather occurrences. The research method focused on finding and studying new ways to adapt to climate change, with a focus on sustainable development practices, community-based adaptation strategies, and creative solutions that rethink how we deal with environmental problems. The strategy used a systems-thinking approach to look at how ecological, social, and economic elements affect regional resilience in a complicated way. This detailed analytical methodology helped the study show how complex the climate adaptation problems and prospects are in the South Pacific's many different island populations.

This study's goal is to find useful information that can help climate adaptation programs and disaster preparedness plans in the South Pacific region by looking at the interconnected factors that affect the region's climate resilience in a systematic way. The technique was made to connect theoretical knowledge with real-world uses for making policies and putting them into action in communities.

16.3.2 Conceptual Framework

The conceptual foundation of this study is the paradigm shifts toward strengthening resilience in the face of increasing climate change concerns in the South Pacific area. Environmental, social, and economic aspects are all considered together because of the intricate relationships that determine susceptibility and ability to adapt. Emphasizing the need for holistic methods that transcend traditional sectoral lines, the framework incorporates lessons from the literature on disaster preparedness, community participation, and climate change adaptation. Crucial components include learning what makes people vulnerable, finding ways to adapt, encouraging sustainable habits, and building community resilience via shared decision-making. Efforts to effectively adapt to climate change and reduce catastrophe risk in the South Pacific may be guided by this framework, which views resilience as a multi-dimensional and dynamic concept to provide thorough knowledge of the interplay between human and environmental systems.

16.4 Results and Discussion

16.4.1 Observed Impacts

16.4.1.1 Tropical Cyclone

Han et al. (2024) state that throughout the last 35 years, there has been a rise in SST and TC intensity due to global warming and these increases are expected to continue. There is still a lack of a satisfying link between climate change and TC. However, the current understanding is that due to global warming, the SST increases, contributing to the formation of climate change. El Niño and La Niña contribute, changing the wind speed and direction. An extended climatic baseline has been made available to investigate the relationship between TC activity and climatic change through the growing use of geological proxies to ascertain historical and prehistorical patterns of TC variability (Chand et al. 2020). Scientific advancements are being made to improve the ability of TC prediction in the future. Reconstruction of TC activities from the past has been conducted by many scholars (Woodruff et al. 2012; Haig et al. 2014) to show the trend of change in intensity and frequency of TC. However, this has been lacking when reconstructing specifically for the South Pacific. According to Chand et al. (2020), the leading cause of the interannual variability of TC in the Pacific in the current climate is El Nino Southern Oscillation (ENSO); however, it is unknown how much ENSO affected TC in historic and prehistoric eras, raising questions about how susceptible TC was to ENSO before the modern era. Reconstruction of TC from prehistoric and historic in the South Pacific will provide crucial information about the link between natural climate variability and TC. Knutson et al. (2010) stated that neither the overall number of TC nor the number of severe TC showed any discernible trend between 1981 and 2007.

The graph above shows a declining trend in the total number of cyclones and the frequency of severe TC. Figure 16.3 supports Knutson et al. (2010). Confidence in this trend is low due to low data from the South Pacific is a concern. The trend might change as the data improves in the future. Interestingly, the recent research by Chand et al. (2017) integrated the climate model from CMIPS experiments, which projects changes in ENSO-driven TC toward the end of the twenty-first century. Specifically, Fig. 16.4 illustrates that in the Pacific, tropical cyclone frequency increases by about 40% during future-climate El Niño events relative to present-climate El Niño events and decreases during future-climate La Niña events around a collection of small island nations (e.g., Fiji, Vanuatu, and Marshall Islands) (Chand et al. 2020). Their findings bear significant consequences for the vulnerable small island states in the South Pacific with regard to climate change and adaptation strategies. According to the study, large-scale sources of climatic variability, including ENSO, are critical in defining regional-scale changes in tropical cyclone features, even if the total frequency of tropical cyclones is anticipated to decrease in the future climate.

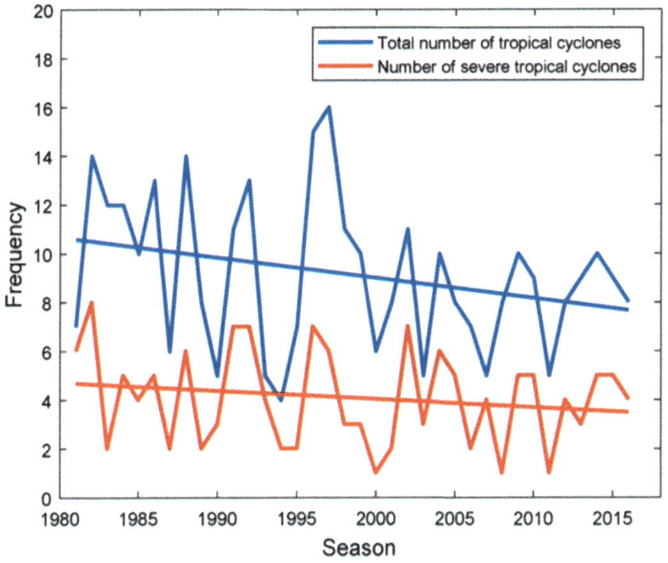

Fig. 16.3 Trend of the total number of TC and severe TC in the South Pacific over the period of 1981/1982–2016/2017. *Source* Chand et al. (2020)

16.4.1.2 Saltwater Intrusion

Saltwater intrusion is an upcoming challenge for South Pacific countries as rising sea level leads to saltwater intrusion into freshwater aquifers. As mentioned earlier, growing population and industrialization is increasing the demand for freshwater, resulting in pumping out of more waterleads to groundwater depletion. Hence, the saltwater enters the aquifers. Figure 16.5 summarizes the water sources for 12 Pacific Island Countries, from which 12% of countries use the desalination process to purify the water for consumption as the saltwater has intruded into the source. The Cook Islands is one of the PICs where saltwater intrusion is a significant concern, especially in the coastal aquifers. Kiribati and Nauru face similar issues. Desalination plants are installed to purify the water and meet the demand of the increasing population and for other usages such as agriculture and industrialization.

16.4.1.3 Climate Resilience: A Paradigm Shift

The South Pacific Islands are a natural paradise surrounded by a vast ocean and are in grave danger due to the global threat of climate change. Frequent droughts, intensified cyclones, and rising sea levels are just the tip of the iceberg. However, the effects of climate change are beyond the environment, as it affects the complex web of Pacific life, upsetting customs, putting pressure on social systems, and escalating economic vulnerabilities. A comprehensive strategy that acknowledges the

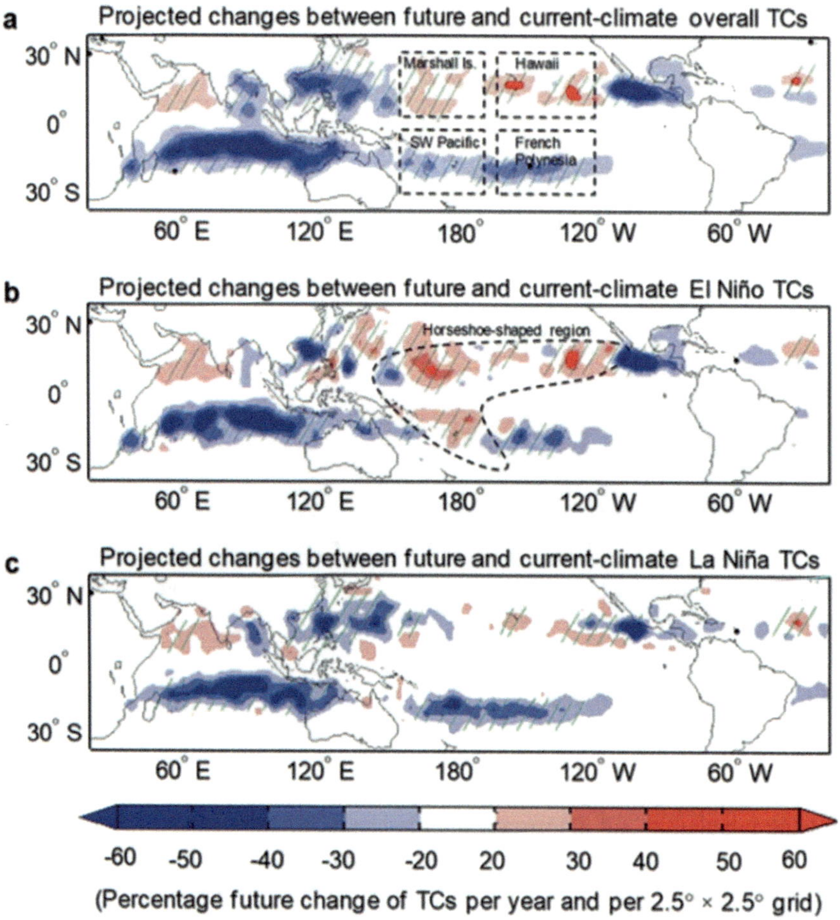

Fig. 16.4 Projected changes in tropical cyclone density between the late twentieth and late twenty-first centuries for **a** overall climatology, **b** El Niño years, and **c** La Niña years. *Source* Chand et al. (2020)

interdependence of these environmental, economic, and social elements is necessary to develop real resilience. Traditional approaches have frequently divided these components into discrete parts, leading to disjointed and, eventually, unproductive outcomes. It is about time for a paradigm change that combines sustainable development, cultural empowerment, and conservation of the environment into a cohesive whole to build resilient populations in the South Pacific. South Pacific communities possess rich traditional ecological knowledge (TEK), which is at the core of this integrated approach. Based on a thorough grasp of the regional ecosystems, South Pacific communities have created complex systems for preserving their land and resources efficiently for decades. These ancient customs, which range from

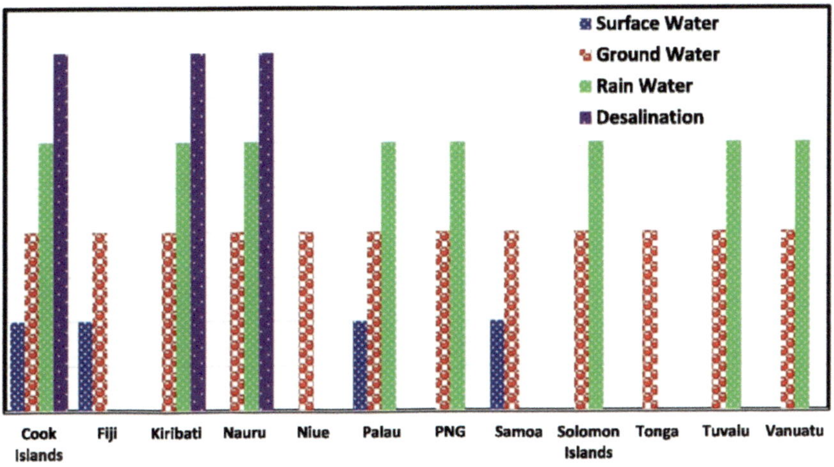

Fig. 16.5 Water sources for Pacific Island Countries. *Source* Sharan, Lal and Datta (2021)

agroforestry methods to customary marine protection, provide a food security and environment conservation model in an era of climate change.

Nonetheless, TEK is deeply embedded in the Pacific Nation's social and cultural fabric, making it more than a set of methods. Indigenous knowledge systems manifest the ideals, worldviews, and lifestyles that have made South Pacific communities successful in their settings. Efforts to promote resilience that do not know and incorporate this socio-cultural component face the danger of weakening the pillars supporting real resilience. Innovative projects are already showcasing the effectiveness of this integrated strategy throughout the region. In Fiji, Vusama village is using traditional knowledge of salt making to increase their revenue as the community faces a decline in crabs, which used to be a significant source of income (Kakal 2021). In Ba, Navala is a traditional village that has preserved its traditional house called Bure, made from local bamboo, reeds, ferns, pandanus leaves, and hardwood. The traditional approach of post-disaster recovery is more sustainable and speedier since it uses locally found materials and encourages communal buildings. This is because the materials and labor required for reconstruction are readily accessible (Aquino et al. 2021). Current et al. (2019) further add that traditional buyers could control the surrounding temperature to fit Fiji's tropical environment, maintaining homes cool in the summer and warm in the winter. These elements are absent from modern design, and some construction materials, such as roofing iron, present a risk during TC (Fig. 16.6).

Aquino et al. (2021) also criticized Fiji's government for rebuilding using contemporary housing structures and materials and did not recognize indigenous construction design in Fiji's building codes and standards. In 2015, Vanuatu recorded a low death toll after TC Pam as they used traditional cyclone houses "gamali" as community shelter during the cyclone (Handmer & Iveson 2017). Nunn et al. (2024) also

Fig. 16.6 Traditional Bure in Bua, Fiji. *Source* Nunn and Kumar (2019)

contributed that a traditional house in the Federated States of Micronesia (FSM), "faluw" (meeting house, similar to bure, Fig. 16.7), can survive cyclones' force gust have specific roof designs, flexible structural elements and walls that allow wind to flow through relatively unobstructed; Samoa "fale" on the other hand does not have walls but has adjustable mats to be used as walls when it is needed.

Extensive research by Nunn et al. (2024) proves that traditional knowledge could be used to predict the weather. In the Pacific Islands, some bird behaviors like the manu-ni-cagi in Fiji, the mansiroboe in northern Vanuatu, or the lofa talamatangi soaring over the plains in Tonga are frequently seen as indicators of impending weather changes. Wet weather forecasting also uses observed insect behavior, especially that of ants. The roles of cicadas, fruit flies, and cockroaches are comparable on certain islands. Other animal habits, such as hermit crabs in Samoa excavating their tunnels deeper than usual or turtles depositing their eggs further inland in Vanuatu, are used to forecast rainy or cyclonic weather. One typical kind of tarot (tarot) for

Fig. 16.7 Faluw in FSM, with the underlying stone platform to protect against sea level rise. *Source* Nunn et al. (2024)

longer-range forecasts of unusually rainy seasons is the finding that hymenopterans construct their hives or nests near the ground in thick vegetation. This activity can start three to four months before a tropical storm, alerting locals to take necessary precautions, such as stockpiling food. Certain pigeon species also utilize their nest height to forecast the activity of cyclones (Nunn et al. 2024).

For resilience on water security, residents of Oneisomw Island in FSM have restored their traditional water wells by cleaning, placing concrete coverings over the wells to keep debris and pollutants away, and planting vegetative buffer strips around streams and wells to stabilize eroded banks and lesson sedimentation (Mcleod et al. 2019). Due to globalization, most of the population depends on the water services provided by the authorities, putting pressure on the water supply. The resilient method to cope with water security is digging wells inland to access water. Nunn et al. (2024) also added that there is evidence that shows Pacific people have skills to deal with drought and knowledge to reduce evapotranspiration, such as in Marshall Island, use of coconut trees' hollowed-out bases, or "Emma," to retain rainwater dropping from the foliage above.

Food security is another critical challenge since it is a basic need, but its vulnerability is very high in the South Pacific. Climate-Smart Agriculture (CSA) is the way forward, and traditional knowledge is the pillar of its success. Pacific community works with NGOs to increase food security by discovering salt and drought-resilient crops as salinity increases and drought becomes prone. Some traditional methods to protect the plant from drought are shading crops with palm leaves, agroforestry to maintain shade, and composting seaweed (Mcleod et al. 2019). Marine resources are also depleting due to ocean acidification and increased SST. Low-cost aquaculture initiatives are also being carried out in Ahus, such as adapting Palau clam farming methods to local conditions to ensure food security and reseeding local reefs with clam larvae to restore the local wild population. Locals in Tamil, Yap constructed a plant nursery to revegetate erosion-prone coastal regions, such as Nipa palms, and food crops using conventional composting methods. The nursery improves community health, diversifies food supplies, lessons the impact of coastal erosion, and lessons reliance on declining coastal fisheries (Mcleod et al. 2019).

16.5 Recommendations

16.5.1 Strengthen Policy Framework

Creating national and local policies that appreciate traditional ecological knowledge (TEK) and make incorporating it into contemporary scientific methods more accessible is critical. These regulations should set up procedures that allow local populations to engage in catastrophe risk reduction and climate adaptation actively. Involving TEK will motivate ancestors to keep their knowledge alive and pass it on to the younger generations, as it could be used in the future for resilience. Furthermore, it is crucial to establish favorable conditions for ecosystem-based adaptation and nature-based solutions, such as conservation and sustainable resource management. Usually, in the South Pacific, more priority is given to development over protecting the environment. Hence, it is important that this ideology needs to change in the Pacific. Moreover, it motivates for green development that is more sustainable for the environment.

16.5.2 *Increase in Funding*

Expanding the financial and technical support options available to community-led resilience projects, especially those incorporating TEK and indigenous practices, is critical. Funding capacity-building initiatives that enable local communities to create and carry out context-specific adaptation plans is equally crucial. The impact can be enhanced by facilitating knowledge-sharing networks and platforms that spread successful models and best practices throughout the South Pacific area. Many organizations, such as UNDP and FAO, are interested in these projects. Hence, the government must express their interest to these organizations in projects like involving TEK. Pacific has many unemployed youths who can benefit from these types of projects.

16.5.3 *Prioritize Research and Study*

Funding for long-term monitoring and assessment is required to determine the efficacy, sustainability, and potential for replication of community-driven resilience efforts. It is equally necessary to research to learn more about the socio-cultural elements driving adaption processes and integrating TEK with contemporary methodologies. To ensure that efforts to build resilience are inclusive and equitable, it is important to investigate the connections between human rights, sustainable development, and climate resilience. Conducting research will allow us to understand the issue and plan strategies better accordingly. The outcome of research needs to be integrated into the making of policies and plans for an effective outcome.

16.6 Conclusions

Due to the increasingly severe effects of climate change, including tropical cyclones, droughts, and sea level rise, the South Pacific region is facing previously unheard-of difficulties. Pacific Island communities face risks that traditional, top-down engineering solutions cannot solve, especially when those solutions lack community input. To acknowledge and capitalize on these communities' priceless traditional ecological knowledge (TEK), the research promotes a paradigm change toward community-driven resilience. We can create comprehensive, culturally rooted solutions that address resilience's interrelated environmental, social, and economic aspects by fusing TEK with contemporary scientific knowledge and encouraging participatory decision-making processes.

The potential of this integrated approach is exemplified by regional examples, where communities are reviving indigenous practices to improve climate adaptation and disaster preparedness. Examples include customary resource management techniques, agroforestry systems, and traditional housing designs. These programs support sustainable livelihoods, preserve cultural heritage, and increase environmental resilience. All parties involved in community-driven resilience must work together to fully realize its potential, including international organizations, governments, businesses, and civil society. Scaling up successful models and empowering local communities require enabling institutional and legislative frameworks, specific financing methods, capacity-building support, and cross-sectoral engagement.

The South Pacific provides a ray of hope and inspiration as the globe struggles with the grave threat posed by climate change. We can create a path toward true resilience firmly anchored in these island communities' circumstances and assets by embracing Indigenous knowledge systems' wisdom, encouraging community leadership, and fusing environmental stewardship, cultural preservation, and sustainable development. This paradigm shift honors the diverse cultural fabric that has supported life in the South Pacific for many centuries while offering a practical strategy for improving climate adaptation. We can negotiate the rising waves of climate change while maintaining the priceless legacy that makes this area so singular and precious by acknowledging and elevating the voices of individuals who have protected these islands through centuries of environmental challenges.

References

Aquino DHM, Wilkinson S, Raftery GM, Mannakkara S (2021) Inclusive resilience: incorporating the indigenous into the picture of resilient reconstruction. In: Djalante R, Bisri MBF, Shaw R (eds) Integrated research on disaster risks. Disaster risk reduction. Springer, Cham. https://doi.org/10.1007/978-3-030-55563-4_16

Bennett NJ, Dearden P (2013) A picture of change: using photovoice to explore social and environmental change in coastal communities on the Andaman coast of Thailand. Local Environ 18(9):983–1001

Chand SS, Tory KJ, Ye H, Walsh KJE (2017) Projected increase in El Niño-driven tropical cyclone frequency in the Pacific. Nat Clim Chang 7:123–127

Chand SS, Dowdy A, Bell S, Tory K (2020) A review of South Pacific tropical cyclones: impacts of natural climate variability and climate change. Clim Chang Impacts Pac:251–273

Currenti R, Pearce T, Salabogi T, Vuli L, Salabogi K, Doran B et al (2019) Adaptation to climate change in an interior pacific Island village: a case study of Nawairuku, Ra, Fiji. Hum Ecol 47:65–80

Haig J, Nott J, Reichart G (2014) Australian tropical cyclone activity lower than at any time over the past 550–1,500 years. Nature 505:667–671

Han C, Bowen M, Sutton P (2024) The response of the upper ocean to tropical cyclones in the South Pacific. J Geophys Res Oceans 129(4):e2023JC020627

Handmer J, Iveson H (2017) Cyclone pam in Vanuatu: learning from the low death toll. Aust J Emerg Manag 32(2):60–65

IPCC (2023) Summary for policymakers. In: Climate change 2023: synthesis report. Contribution of working groups I, II and III to the sixth assessment report of the intergovernmental panel on climate change [Core Writing Team, H. Lee and J. Romero (eds.)]. IPCC, Geneva, Switzerland, pp 1–34. https://doi.org/10.59327/IPCC/AR6-9789291691647.001

Janssen MA, Schoon ML, Ke W, Börner K (2006) Scholarly networks on resilience, vulnerability and adaptation within the human dimensions of global environmental change. Glob Environ Chang 16(3):240–252

Kakal Z (2021) Traditional knowledge as a powerful lever for climate resilience. UNDP. Retrieved May 27, 2024, from https://www.undp.org/pacific/blog/traditional-knowledge-powerful-lever-climate-resilience

Kelman I (2019) Pacific Island regional preparedness for El Niño. Environ Dev Sustain 21(1):405–428

Knutson TR et al (2010) Tropical cyclones and climate change. Nat Geosci 3:157–163

Lal A, Datta B (2019) Optimal groundwater-use strategy for saltwater intrusion management in a Pacific Island country. J Water Resour Plan Manag 145(9):04019032 https://doi.org/10.1061/(ASCE)WR.1943-5452.0001090

Malsale P (2023) La Niña impacts felt throughout the Pacific, La Niña impacts felt throughout the Pacific | Pacific Environment. Available at: https://www.sprep.org/news/la-nina-impacts-felt-throughout-the-pacific. Accessed 23 May 2024

Mcleod E, Bruton-Adams M, Förster J, Franco C, Gaines G, Gorong B, James R, Posing-Kulwaum G, Tara M, Terk E (2019) Lessons from the Pacific Islands–adapting to climate change by supporting social and ecological resilience. Frontiers in Marine Science, 6:289

Nunn PD, Kumar R, Barrowman HM, Chambers L, Fifita L, Gegeo D et al (2024) Traditional knowledge for climate resilience in the Pacific Islands. Climate Change, Wiley Interdisciplinary Reviews, p e882

Nunn P, Kumar R (2019) Measuring peripherality as a proxy for autonomous community coping capacity: a case study from Bua Province, Fiji Islands, for improving climate change adaptation. Soc Sci 8(8):225

Parsons M, Glavac S, Hastings P, Marshall G, McGregor J, McNeill J et al (2016) Top-down assessment of disaster resilience: a conceptual framework using coping and adaptive capacities. Int J Disaster Risk Reduct 19:1–11

Pelling M (2010) Adaptation to climate change: from resilience to transformation. Routledge

Prasad RR, Mkumbachi RL (2021) University students' perceptions of climate change: the case study of the University of the South Pacific-Fiji Islands. Int J Clim Chang Strat Manag 13(4/5):416–434

Sharan A, Lal A, Datta B (2021) A review of groundwater sustainability crisis in the Pacific Island countries: challenges and solutions. J Hydrol 603:127165

Sun Y, Zhong Z, Li T, Yi L, Hu Y, Wan H et al (2017) Impact of ocean warming on tropical cyclone size and its destructiveness. Sci Rep 7(1):8154. https://doi.org/10.1038/s41598-017-08533-6

Turvey R (2007) Vulnerability assessment of developing countries: the case of small Island developing states. Dev Policy Rev 25(2):243–264

Veitayaki J (2002) Taking advantage of indigenous knowledge: the Fiji case. Int Soc Sci J 54(173):395–402

Veitayaki J (2010) Using traditional knowledge to address climate change: the Fiji scenario. Conservation International

Veitayaki J, Tawake A, Bogiva A, Meo S, Ravula N, Vave R et al (2007) Addressing human factors in fisheries development and regulatory processes in Fiji: the Mositi Vanuaso experience. Ocean Yearbook Online 21(1):289–306

Victoria University of Wellington (2018.) A new way for niue: building cyclone-resistant sustainable architecture for the pacific: Available at: https://www.wgtn.ac.nz/architecture/about/news/news-archives/2018-news/a-new-way-for-niue.Accessed: 25 May 2024

West FJ, Sophie F (2024) Pacific Islands. Encyclopedia Britannica. https://www.britannica.com/place/Pacific-Islands

Woodruff J, Sriver RL, Lunf DC (2012) Tropical cyclone activity and western North Atlantic stratification over the last millennium: a comparative review with viable connections. J Quat Sci 27:337–343